ECO-EFFICIENCY IN INDUSTRY AND SCIENCE

VOLUME 32

More information about this series at http://www.springer.com/series/5887

Harry Lehmann
Editor

Factor X

Challenges, Implementation Strategies and Examples for a Sustainable Use of Natural Resources

Managing editors
Mandy Hinzmann, Nick Evans, Terri Kafyeke,
Stephen Bell, Martin Hirschnitz-Garbers (Ecologic Institute)
Martina Eick (German Environment Agency)

Springer

Editor
Harry Lehmann
Factor X/10 Club
German Environment Agency
Dessau-Roßlau, Germany

ISSN 1389-6970
Eco-Efficiency in Industry and Science
ISBN 978-3-319-50078-2 ISBN 978-3-319-50079-9 (eBook)
https://doi.org/10.1007/978-3-319-50079-9

Library of Congress Control Number: 2017954331

Printed on acid-free paper

This Springer imprint is published by Springer Nature
The registered company is Springer International Publishing AG
The registered company address is: Gewerbestrasse 11, 6330 Cham, Switzerland

Foreword

Dear reader,

The high standard of living that we enjoy depends entirely on the quality and availability of natural resources. Driven by global population growth and increasing economic performance, mineral raw materials and fossil fuels are being extracted in ever greater quantities. The extraction of these resources has profound environmental impacts, such as the destruction of ecosystems and habitats as well as air, water and soil pollution.

For this reason, we pursue various measures for more efficient resource use and management, with the aim of keeping the negative effects of resource use within reasonable bounds. The overarching objective is a transition towards sustainability—a resource-efficient, low-carbon economy, in which natural capital is protected and enhanced and the health and well-being of citizens is safeguarded.

Therefore, it is of utmost importance that all countries urgently adapt their economies by increasing resource efficiency, reducing resource consumption in absolute terms and abandoning resource-intensive consumption patterns in favour of resource-efficient lifestyles. The economical use of raw materials not only reduces pressures on the environment but also creates economic opportunities for individual companies and strengthens the economy as a whole.

The development of resource policy requires a skilful combination and bundling of different measures and instruments since there is no uniform policy approach that meets the different requirements. Rather, the respective objectives, action requirements, target groups and policy levels must be addressed using specific policy approaches.

Some aspects of sustainable resource use are, even in political circles, not well understood. To shed light on this critical topic and inform the ongoing political process, we have invited a wide range of relevant stakeholders from the fields of science, politics, business and technology to share their experiences and views on how to achieve the sustainable use of natural resources.

Because of the thematic diversity, it is not surprising that the contributions in this book describe a range of developments in resource efficiency, from incremental improvements to profound change and transformation. At the same time, all political levels are addressed: the authors consider global megatrends and comprehensive resource policies as well as regional and national efforts, such as the European Union's Circular Economy Package and Germany's Resource Efficiency Programme. In addition, the numerous practical cases detail best practice examples of resource use in urban and rural areas, manufacturing companies and private households. Of particular interest are unusual and innovative ways of thinking, such as contributions on the path to degrowth for a sustainable society or the husbandry of the finance system and natural resources.

This book is intended for anyone interested in the sustainable use of natural resources. It provides insights into awareness raising and policymaking and contains references to practical developments that will accompany us on the path of transition towards a more sustainable society.

Thus, it is my hope that this book will attract a great deal of attention.

Regarding the production of this third *Factor X* book, I like to thank the teams responsible for the coordination and supervision, namely Mandy Hinzmann, Martin Hirschnitz-Garbers, Terri Kafyeke, Nick Evans and Stephen Bell from Ecologic Institute as managing editors; Krithika Radhakrishnan, Catalina Sava and Fritz Schmuhl from Springer; and Martina Eick who managed the process on the part of the German Environmental Agency. And of course I specially and warmly thank all the authors of the book.

President of the German Environment Agency Maria Krautzberger
Dessau/Berlin, Germany

Contents

Part I
Challenges

Chapter 1
Factor X – 25 Years – "Factor X Concept" Is Essential for Achieving Sustainable Development

Harry Lehmann, Friedrich Schmidt-Bleek, and Christopher Manstein

Abstract A dematerialisation of industrialised countries by a Factor of 10 (minus 90%) was first suggested 25 years ago in order to achieve sustainable economic development worldwide by 2050. The Factor 10 postulate was a response to two realities: first, anthropogenic material flows have increased dramatically since the first Industrial Revolution, and second, the richest countries consume significantly more natural resources per capita than the world's poorest countries. Twenty-five years later these facts have not changed in principle, and a global per capita consumption of three to eight tonnes of primary raw material must be reached in this century. Today the term "Factor X" is often used instead of "Factor 10", because the necessary dematerialisation is different from country to country. Industrialised countries have higher targets. The article describes the beginning of the Factor X postulate in the early 1990s as well as developments thereafter and discusses today's options and challenges for tomorrow.

Keywords Earth system policy • Anthropogenic material flows • Dematerialisation • Factor 10 / X • Resource efficiency policy • Protection and efficient use of natural resource • Material-cycle societies • New wealth model • Happiness

---

H. Lehmann (✉)
Factor X/10 Club, German Environment Agency, Dessau-Roßlau, Germany
e-mail: harry.lehmann@uba.de

F. Schmidt-Bleek
Factor X/10 Club, International Factor 10 Institute, Berlin, Germany
e-mail: bio@schmidt-bleek.com

C. Manstein
German Environment Agency, Dessau-Roßlau, Germany
e-mail: christopher.manstein@uba.de

H. Lehmann (ed.), *Factor X*, Eco-Efficiency in Industry and Science 32,
https://doi.org/10.1007/978-3-319-50079-9_1

1.1 The Beginning – A Systemic Approach to "Earth Systems Policy"

The Rio Declaration on Environment and Development was signed by more than 170 countries in 1992. "Recognizing the integral and interdependent nature of the Earth, our home", the Declaration proclaims 27 principles of future sustainable development. In Principle 15 it is written: "In order to protect the environment, the **precautionary** approach shall be widely applied by States according to their capabilities. Where there are threats of serious or irreversible damage, lack of full scientific certainty shall not be used as a reason for postponing cost-effective measures to prevent environmental degradation." In light of this it was clear that only a systematic approach and analysis of the "System Earth" could derive the adequate policies needed solve global ecological problems.

We live in an almost closed system called Earth. This system interacts with the rest of the proximate universe almost exclusively through the exchange of energy in the form of radiation. Thus, the Earth and its inhabitants are linked in a system of mutual dependence. From a human point of view, the Earth's "survival" system consists of two domains—the anthroposphere and the biosphere. In the anthroposphere, humans consume minerals, ores, water and air, burn energy carriers, harvest biomass, hunt and fish, thus creating "wealth" and waste at all levels. The biosphere is the domain in which flora and fauna seek to survive according to the rules of evolution and according to the given anthropogenic circumstances. The survival of human beings is dependent on both sub-systems.

Science today still knows very little about these sub-systems, and has hardly even begun to investigate some parts of them (e.g. the deep sea). However, one of the most researched aspects is the Earth's climatic system. Massive expenditure on personnel and technology has been required to establish reliable prognoses of the future behaviour of the climatic system. Science has learned from these analyses that the biosphere and anthroposphere are extremely complex systems permanently undergoing reorganisation. The basic laws governing the behaviour of the subsystems are nonlinear. Because of the predominantly nonlinear dynamics, these systems can behave chaotically and are subject to massive changes in a short time period as a result of seemingly minor causes. Nevertheless, science provides only limited information on the effects of human action, either regarding the intensity of the biosphere's reaction or the time scales involved.

Today, the changing climatic system and global warming are the most discussed environmental concerns, but there are many other known problems, including radioactive contamination, water pollution and use, air pollution, acid rain, land use, land soil destruction, degradation of land, chemical contamination, deforestation, waste deposition, the ozone hole, monocultures and destruction of biodiversity. Moreover, there is a high probability of numerous additional problems that we have still not realized even exist.

Even if in a "thought experiment" we could have a full understanding ("God's view") of both systems—of the flows in the bio- and anthroposphere, of all interactions, of the actual status of all parameters and last but not least a computer-like ability to run such a complex model—this would still not be sufficient to make predictions about all the effects of human actions. The dominating nonlinear dynamics and the ability of reorganisation will alter the system beyond what is predictable—our "God's view" will become obsolete and the predictions of the future behaviour of the system derived from our all-knowing model will become inaccurate.

At this stage in history, and perhaps for all time to come, our actions must be guided by the recognition of how little we know about our planet's "survival" system and its susceptibility. As a precautionary measure, following Principle 15 in the Rio Declaration, we should therefore attempt to minimise anthropogenic effects on this system. We should strive to prevent as far as possible any negative consequences, assuming that an undisturbed biosphere will continue to exist in a way humanity can survive and live in an agreeable manner. This precautionary principle must constitute the main guideline for all human activity if sustainable development is to be our primary aim.

To argue that somehow, we can "repair" the biosphere later is both, arrogant and irresponsible. First, this implies the assumption that we are capable of repairing a system which science has thus far failed to fully comprehend, and, secondly, it ignores the fact that such global effects as climate change are frankly phenomena, which are beyond "repair".

Returning to the simple picture of the "System Earth", the goal is to minimise the impact of humankind on the biosphere. Minimising anthropogenic effects on the biosphere entails a measurement of the consequences of human actions.

There are two types of interactions within this Earth's survival system. First, one can gauge the effects of anthroposphere on the biosphere by investigating its "**outputs**". This is done today with a lot of effort in climate change research, by looking at the effects of different types of agriculture on biodiversity or through an examination of the "riskless" limit of emissions of various types of chemicals. It is an important and established mainstream method of ecological policymaking in the last decades.

There is an inherent limitation of such a policy approach. It can only take into consideration singular effects, i.e., those which are already known—or in some cases—suspected. The environmental policies of the last decades have been essentially reparative, "re-active" policies. Once recognised as hazardous or toxic, substances were withdrawn from the market at considerable expense, filtered out of waste gases, incinerated as residues or prevented through costly process restructuring. In most cases, the limit values set for such hazardous substances are oriented to human health. However, while the combatting of pollutants will remain an important priority in the future, it is hardly possible to derive stable and generally applicable principles for the ecological reform of industry from such substance-specific procedures.

Second, one can gauge the material flows into the anthroposphere and the use of land area and water by humankind. From an anthropogenic point of view, these are "**inputs**". There are different indicators used in today's discourse but the most common is a measure “resource productivity”. As long as indicators are linear and easy to apply they can be used to develop a proper policy.

There is not—and probably will never be—a model that adequately links link anthropogenic inputs and outputs. This is not least due to the fact that the anthroposphere is highly complicated and the different sub-sectors of the anthroposphere are highly interactive. A highly theoretical approach is to measure the changes of entropy through the anthroposphere. Yet, aside from the fact that the majority of people will have trouble understanding such an indicator, there are principal theoretical problems with calculating real values.

“**Size**”: The biosphere needs undisturbed areas to live, to survive, to readjust, to move and evolve. In former times, the anthroposphere was small compared to the biosphere. The biosphere is finite—the anthroposphere has grown dramatically. Population growth and the industrial revolution have eaten up resources and led to exponential growth in the size of the anthroposphere leaving less and less room for the biosphere.

We must also measure the **velocity** of the impact. How fast are these changes in “input” and “output”? How fast do we change the composition of the sub-system (e.g. increasing the greenhouse gases in the atmosphere)? The biosphere can react, rearrange and try to find a new system state. In all these categories, humankind has intervened in System Earth, resulting in changes in operations, previous functions and services of the biosphere. Such changes can last for long periods and some are irreversible.

Based on the previous analysis and on the precautionary principle the “**Factor X**” concept was formulated in the beginning of the 1990s. The concept revolves around fundamental principles:

- respecting that all humans—within the current generation, across all continents and over the generations—have the same right to the fulfilment of basic needs and requirements;
- recognizing that humankind needs a functioning biosphere—with enough “space”.

To achieve this we must:

- significantly decrease the land use and the input into the anthroposphere;
- define limits of the anthroposphere; and
- develop a “new wealth” model (happiness, having access to services, “well-being” rather than “well-having”, i.e., owning artefacts).

1.2 How Big Is the X? – “An Eco Safety Factor”

During the early 1990s ecologically visible problems seemed to demand a reduction of environmental pressures by at least 50%. In a growing world, population and growing desires for prosperity—particularly in the developing countries—demands

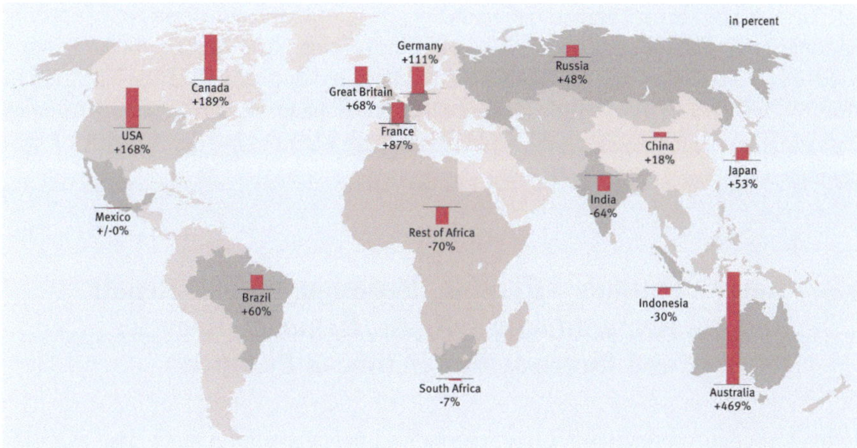

Fig. 1.1 International comparison between per capita raw material consumption (RMC) and the global average, 2011 (Source: Lutter et al. 2016)

a fivefold increase of economic output. Together these lead to the first X—the "**Factor 10**" as postulated by Friedrich Schmidt-Bleek (1992). Several authors analysed, discussed and worked out the idea of a Factor of 10, publicised in 1993 in the "Fresenius Environmental Bulletin" (Schmidt-Bleek et al. 1993).

The Factor 10 Club (1994) was met with considerable international recognition by representatives of business, policy and industry. The idea was that resource efficiency goals of countries should be high enough to bring about a sufficient decoupling of resource use from human development. The goals must influence the rules of the economy and change lifestyles—the way we produce and consume goods and services (Schmidt-Bleek 1992, 1993; Lehmann and Schmidt-Bleek 1993).

In 1995, the authors of the book "Factor Four" (Weizsäcker et al. 1997), took a somewhat different approach. They looked at what was at hand in terms of technology or at least easily conceivable across the board, including energy. Assessing the range of opportunities, they thought that an increase in resource/energy efficiency by a factor of two and a doubling of the generated wealth would be feasible in the foreseeable future.

Calculation with a modified "World3-91 Model" showed the necessity of higher factors to bring the system into dynamic equilibrium. This started a discussion as to whether a "Great Transformation" (absolute reduction of resource use levels) or a gradual approach (relative improvements in the resource productivity) was the right strategy (Fig. 1.1).

If we desire a sustainable level of raw material consumption and use, we must apply the precautionary principle. This means reducing raw material consumption wherever possible, making use of all productivity (efficiency) potentials and avoiding rebound effects (boomerang effects). We arrive at the currently discussed corridor of global per capita consumption between three and eight tonnes of primary raw material use (e.g. Bringezu 2015; UNEP 2016).

This worldwide ecological sustainability goal allows the development of specific national, regional and sectorial reduction factors. Today the term "Factor X" is often used instead of "Factor 10", because the necessary dematerialization is different from country to country. Industrialised countries have higher targets. Estimates indicate that such goals could—and must—be reached by the middle of this century.

1.3 Today's Options – Rethink, Redesign, Refuse, Repair, Reduce, Remanufacture, Reuse, Remodel, Recycle, Recover and Increase the Lifetime of Products

In the last years, policymakers have recognised the necessity to react. One example is the recent declaration from the leaders of the G7 countries:

> The protection and efficient use of natural resources is vital for sustainable development. We strive to improve resource efficiency, which we consider crucial for the competitiveness of industries, for economic growth and employment, and for the protection of the environment, climate and planet. Building on the 'Kobe 3R Action Plan', and on other existing initiatives, we will continue to take ambitious action to improve resource efficiency as part of broader strategies to promote sustainable materials management and material-cycle societies. (Leaders' Declaration G7 Summit, 7–8 June 2015, Schloß Elmau, Page 17 ff.)

Awareness is growing about the challenges as well as the necessary political, societal and business strategies. The number of good practice examples is growing all over the world. Since 2007 the "International Resource Panel" of UNEP has aimed to help nations to use natural resources sustainably without compromising economic growth and human needs. Moreover, further policies exist on the international, European, national (e.g. Germany and the German Resource Efficiency Program, ProgRess and ProgRess II (BMUB 2012, 2016)) regional and city level.

Science and engineers are increasingly working on the field of resource productivity and numerous standards and norms have been developed and formulated. NGOs like the "Factor X Club" conferences like the "World Resources Forum" or the "European Resources Forum" serve as neutral, international platforms for debate on global resource consumption issues, advocating innovation for resource productivity. Members of the business community are beginning to redesign their models and new business strategies are creating revenues from the quality of services rather than by selling material products.

Detailed analysis of the actual status of Germany's use of natural resources indicate that we are not really decoupling and lowering the use of resources (Lutter et al. 2016) (Fig. 1.2).

There are additional challenges such as the necessary transformation of the energy supply in Germany and the rest of the world into a fully renewable and sustainable system. In this process, we must take into account the resources needed for the transformation and to run and maintain such a system (see the ongoing study by the German Environment Agency "Greenhouse Gas Neutral and Resource Efficient Germany"). Another challenge is the increasing need for housing and

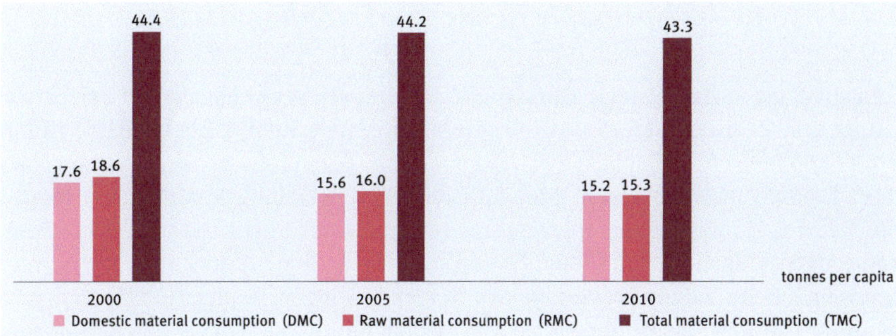

Fig. 1.2 Development of per-capita raw material use in Germany using different indicators, 2000–2010 (Source: Lutter et al. 2016)

infrastructure in some parts of the world. This will also require the use of resources like metals, cement, energy and many critical materials.

1.4 Tomorrow – Urgent Policy Mix for “System Earth”

Bearing all of this in mind, a comprehensive policy mix needs to be designed and implemented urgently. Future-oriented system-policies can no longer focus preferentially—on curing individual symptoms stemming from systemic problems. System policies are as essential for measures designed to protect the environment as they are needed when attempting to seek improvements in pursuing social, economic, financial/fiscal and institutional and health improvements.

Based on today’s current level of knowledge the following important and urgent actions are required:

- Minimize mobilisation and use of natural resources—maximize their productivity;
- Synthesise materials that can replace increasingly scarce natural materials;
- Change to materials that fit into natural material cycles after use;
- Minimise the use and release of toxic substances and radio-nuclides;
- Switch to an energy system based 100% on sustainable renewable resources;
- Stop the emissions of greenhouse gases.

A group of “first mover states” should define national, regional and sectorial goals, including time-lines, for decoupling and lowering the use of resources. These goals must then be periodically reviewed. There should also be an independent panel, tasked with proposing adapted policies.

Such a group of like-minded first movers can be the Group of 7, Group of 20 or a contingent of European countries. However, in the long run it must be the global community that commits to an international agreement or treaty on worldwide per-capita targets for natural resource extraction and consumption—somewhere between three and eight tonnes of primary raw material use.

1.4.1 Some Indispensable Elements of a Policy Mix Are

Changing the fiscal system: markets do not operate perfectly, market prices are wrong due to discounted externalities, relevant information is not available to the actors and innovation barriers exist. Adjusting the fiscal framework is therefore the most fundamental and urgent pre-requisite for approaching a sustainable future. Subsidies that increase the consumption of natural resources must be eliminated, and economic instruments should be deployed to facilitate a shift away from overheads on labour and towards taxing raw materials—which induces job creation and facilitate income redistributing to developing countries where many of the resources come from—and create new markets with tradable permits. Instead of applying value added taxation to final goods it may be more effective to tax natural resources at the point at which they are removed from nature or where they enter the industrial metabolism, i.e., the so called "material added tax".

Minimising the use of resources should be prescribed by law. Resource conservation legislation is still in its infancy owing to the extreme complexity and diversity of resource use and the traditionally medium oriented structure of environmental control legislation. To remedy this situation, a basic resource law needs to encompass the entire supply chain, just as resource protection itself does. The interlocking nature of resource conservation also means that laws in this domain need to address not only environmental regulations, but also a whole host of other legal domains (see the work of the German Environmental Agency on a "Resource Protection Law"; Alsleben et al. 2013).

We must significantly decrease the land take for settlements and transport infrastructures as well as for mining sites. "Land take" means the transformation of agricultural land, forests or still quite natural areas into artificial surfaces where soil is degraded by compression, sealing or complete destruction (this was already discussed in the nineties – e.g. Lehmann et al. 1995).

In the German National Sustainability Strategy, Federal Government set a goal to reduce land take for settlements and transport infrastructures to 30 hectares a day by the year 2020. Actually in the year 2015, daily land take in Germany still amounts to 61 hectares. Despite of a considerable reduction of land take, due to demographic change and many minor steps of different policies, we are still far from reaching the 30 hectares goal. As the enlargement of infrastructures and settlements swallows high amounts of material and energy during construction and also causes high inputs of materials and energy for heating, cooling and illumination as well as for maintenance during their whole lifetime. Reducing land take is crucial for future resource efficiency. The German Climate Protection Plan even states that by 2050 land take should be replaced completely by the recycling of already degraded land.

Also, the Sustainable Development Goals (SDGs) of the United Nations declare in Target 15.3 that we should set course, by 2030, to a degradation neutral world ("By 2030 …. Strive to achieve a degradation neutral world"). Spatial planning and mining law offer considerable potential for greater integration of resource conservation goals and in their implementation these laws could help ensure that resources and raw materials are used more efficiently (see the UBA position paper

"Environmentally sustainable use of the subsurface and resource conservation", Penn-Bressel et al. 2014).

Other instruments and measures should be considered as well, such as information and coordination instruments and command and control mechanisms (e.g. adoption of standards). Legal frameworks should introduce and enforce a common material productivity indicator (label) for all goods and services and initiate a systematic R&D program for gaining information on all ecological, social and economic issues related to natural resources. Moreover, a public information and data centre on all issues related to natural resources should be established.

1.4.2 Happiness, a New Leitmotif

The consumption of goods has become equated with prosperity, and therefore represents a potent leitmotif for prosperity—for well-"having"—for many members of our society. But is this really "prosperity"? It certainly is, in the sense of having or possessing and in the sense of being able to demonstrate one's social rank and status.

But consumption does not necessarily mean prosperity in the sense of well-"being"—which includes intangible elements such as having time for oneself and experiencing the true joys of life. Prosperity in its widest sense means occupation, education, health, security (social and political), the absence of violence, information, liberalness, communication, free time, equal rights for all, the rule of law and environmental quality. Non-material thing, like the enjoyment of intact natural surroundings or the development of one's own personality, will once again rank higher than the possession and consumption of material, tangible goods like, e.g., a car.

For this reason, the path towards a viable future society involves leasing and renting goods, as opposed to owning them; recycling and the efficient use of resources to produce material goods and energy, as opposed to unbridled growth in consumption. These will hopefully become more and more established as a new leitmotif for the majority of society. To this end, modern and traditional knowledge, ethical values, wisdom and spirituality should inspire us to answer the question: "How much is enough?"

Disclaimer This paper does not necessarily reflect the opinion or the policies of the German Federal Environment Agency.

References

Alsleben C et al (2013) Ressourcenschutzrecht. Positionspapier Umweltbundesamt, Dezember 2013

BMUB – Federal Ministry for the environment, Nature Conservation, Building and Nuclear Safety (2012) German resource efficiency program (ProgRess) – program for the sustainable use and conservation of natural resources. Berlin

BMUB – Federal Ministry for the environment, Nature Conservation, Building and Nuclear Safety (2016) German resource efficiency program II (ProgRess II) – program for the sustainable use and conservation of natural resources. Berlin

Bringezu S (2015) Possible target corridor for sustainable use of global material resources. Resources 4:25–54

Factor 10 Club (1994) Declaration of the international resolution

G7 (2015) Leaders' declaration G7 summit, 7–8 June 2015, Schloß Elmau. https://www.g7germany.de/Content/EN/Artikel/2016/01_en/2016-01-20-g7-abschlussbericht_en.html.https://www.g7germany.de/Content/DE/_Anlagen/G7_G20/2015–06-08-g7

Lehmann H, Pareyke R, Pfluger A, Reetz T (1995) Land use in Europe – actual status and a possible sustainable scenario, Wuppertal Texte, Wuppertal Institute

Lehmann H, Schmidt-Bleek F (1993) Material flows from a systematical point of view. Fresen Environ Bull 2:413–418

Lutter S, Giljum S, Lieber M, Manstein C (2016) The use of natural resources. Report for Germany 2016. German Environment Agency. www.umweltbundesamt.de/en/resourcesreport2016

Penn-Bressel G et al (2014) Umweltverträgliche Nutzung des Untergrundes und Ressourcenschonung Anforderungen an eine Raumordnung unter Tage und ein modernes Bergrecht. Positionspapier Umweltbundeamt, November 2014

Schmidt-BleekF(1992)Materialintensität–EinökologischesMaßfürdenVergleichvonMaßnahmen, Produkten und Dienstleistungen. Magazin des WissenschaftszentrumWissenschaftszentrums von NRW, Düsseldorf, 1992

Schmidt-Bleek F (1993) Wieviel Umwelt braucht der Mensch. MIPS – Faktor 10 – das Maß für ökologisches Wirtschaften. Birkhäuser, 1993. München:dtv

Schmidt-Bleek F et al (1993) Fresenius Environ Bull 2(8)

UNEP (2016) Resource efficiency: potential and economic implications. A report of the international resource panel. Summary for policy-makers. Ekins P Hughes N et al

Weizsäcker E Lovins A Lovins H (1997) Factor four. Doubling wealth, halving resource use. Earthscan, London. (The book was first published in 1995 in its German translation ", Faktor Vier")

Chapter 2
Necessities for a Resource Efficient Europe

Leida Rijnhout, Magda Stoczkiewicz, and Meadhbh Bolger

Anyone who believes that exponential growth can go on forever in a finite world is either a madman or an economist.

Attributed to Kenneth Boulding (United States Congress 1973)

Abstract In this article, the authors list some of the main challenges that need to be overcome in order to make the transition to a Europe founded on resource justice, arguing that it is important to move beyond focusing solely on resource efficiency to a focus on reducing absolute resource use. Despite increased awareness about the importance of protecting the environment, mainstream economic theory and practice, as well as mainstream politics and governance, still fails to consider environmental costs; Europe's absolute resource use remains one of the highest globally and it continues to use more than its fair share of resources. Europe is highly dependent on imported resources causing significant negative impacts in third countries, including the Global South. To address this, the authors argue that it is essential to measure and monitor all the resources embodied in a product throughout its full life-cycle, from extraction to consumption taking a consumption-based, or footprint, approach. The authors show how the EU's Resource Efficiency Roadmap, 7th Environmental Action Plan and Circular Economy Package do not sufficiently reflect the justice aspects of resource use or ensure coherence with other policies. They highlight levels which need to be addressed for resource justice agenda including governance, financial tools and structures, social innovation and behaviour change, alternative business models as well as legal and regulatory frameworks. The authors argue that the 2030 Sustainable Development Agenda should be taken as overall framework to support more coherent policymaking. Finally, they argue that we should move beyond 'resource efficiency' to 'resource sufficiency', and fit our economies into "one-planet-lifestyles".

L. Rijnhout (✉) M. Stoczkiewicz • M. Bolger
Friends of the Earth Europe, Mundo-b building, Rue d'Edimbourg 26, 1050, Brussels, Belgium
e-mail: leida.rijnhout@foeeurope.org; madzik2@gmail.com; meadhbh.bolger@foeeurope.org

H. Lehmann (ed.), *Factor X*, Eco-Efficiency in Industry and Science 32,
https://doi.org/10.1007/978-3-319-50079-9_2

Keywords Sustainable lifestyles • Resource sufficiency • Environmental and social justice • Friends of the Earth Europe • Policy coherence • Degrowth

2.1 Introduction

Climate change, wars over water, premature deaths caused by air pollution; it is now impossible to ignore the environmental and social costs of our production and consumption patterns.

During the 1970s, a general awareness of the necessity to protect our environment and to manage and maintain our natural resources for future generations began to arise. Publications such as the Limits to Growth (Meadows et al. 1972), Factor 4 (Lovins et al. 1998) and Small is Beautiful (Schumacher 1974) became essential reading for concerned students and journalists. Nevertheless, this never broke through to mainstream economic theory and practice; especially not in the ivory towers of the economics departments at Universities, governments and business schools.

It is remarkable that mainstream economic theory and practice still fails to integrate environmental costs. It is a scientific fact that both natural resources and the absorption capacity of environmental damage are limited, and to not consider this in economic modelling is undoubtedly irresponsible. The global economic system remains organised in a way that ignores the "polluter pays" principle, with a pricing system that does not reflect the true costs of products on the market.

On a global scale, there are striking figures relating to the exponential growth in consumption of natural resources in the twentieth century (Fig. 2.1), including increases in fossil fuel extraction by a factor of 12, ores and minerals by a factor of 27 and construction materials by a factor of 34 – whilst the population only grew by a factor of 3.7 (Krausmann et al. 2009). The world's richest countries are consuming, on average, 10 times as many materials as the poorest countries (UNEP 2016a).

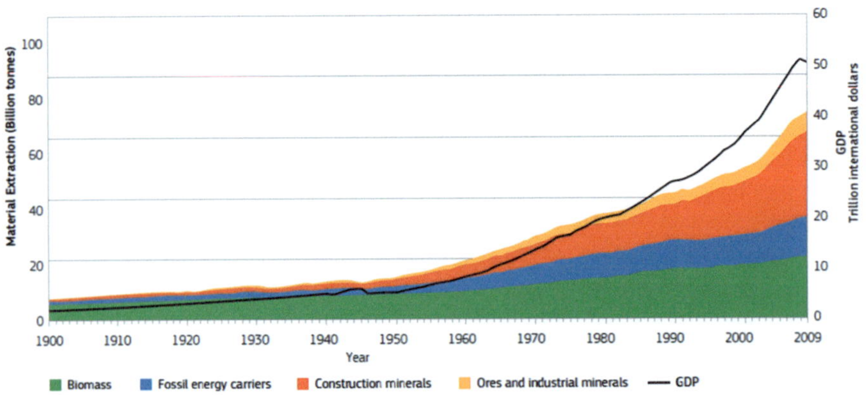

Fig. 2.1 Global material extraction by resource type and GDP (1990–2009) (Source: European Commission 2016a)

It is clear the world is consuming beyond what the planet can sustain – but what is Europe's share in this?

Europe has historically been dependent on resources from the Global South, and the pattern continues today: the flow of natural resources is much greater from South to North than vice-versa. When talking about natural resources, we are not only speaking of minerals and fossil fuels, but also about water, land and forests. We live in an extractive and globalised economy, where most countries considered well developed in economic terms, including the majority of European countries, rely heavily on resources from third countries, including the Global South. This is far from sustainable, but also unjust, as we are blocking development opportunities of communities in the Global South who need resources for their own development needs.

Despite using resources more efficiently, over the past few decades Europe's consumption of raw materials has increased in absolute terms (EEA 2012) (this trend has only recently been interrupted by the economic downturn, but is likely to resume unless action is taken). Europe is still one of the highest consuming continents on the globe and we continue to consume more than our fair share (EEA 2015). In 2010, we had an annual per capita material footprint of 20 tonnes, second only to the United States. By comparison, Africa's footprint was below 3 tonnes per capita (UNEP 2016a). Our absolute rise in resource use, despite increases in productivity and efficiency, is also evidence of the so called "rebound effect" in action and should serve as a warning that focusing on resource efficiency and technological innovation alone might be insufficient.

2.2 Europe's Share

Europe is highly dependent on imported natural resources for our economic activities (EEA 2015). In 2014, the EU imported over 50% of its energy, with four Member States importing over 80% (Eurostat 2016a). According to the most recent data, 40% of the EU land footprint for agricultural products comes from regions outside the EU (Fig. 2.2), increasing to 65% for non-food cropland (FoEE 2016). The picture is the same for raw materials: the EU is highly dependent on imports of many metal ores and natural rubber, and imports nearly 100% of numerous materials including cobalt, tin, iron, bauxite (aluminium) and rare earth elements (European Commission 2016a).

This makes the European economy vulnerable to price fluctuations and increases, – around 40% of European manufacturers' costs are for resources (Greenovate! Europe 2012) – but it is also simply a matter of injustice. We are exporting negative impacts of products consumed in Europe. This includes social impacts such as land grabbing in the Global South in the rush to grow large plantations to supply palm oil to the European market, and environmental impacts such as water stresses in many villages in – for instance – Peru because of the production of asparagus for European consumption.

Fig. 2.2 The EU land footprint (Source: FoEE 2016; Lindsay Noble design)

A lot of attention in the EU is currently focused on improving the recycling and re-use of materials. This is indeed vital, given that over 50% of municipal waste continues to be landfilled and incinerated in Europe (Eurostat 2016b). However, at the same time, it is not the answer to our overconsumption crisis. Demand for raw materials outweighs the volume of recycled or reused materials available on the market (European Commission 2016a). Much of the problem indeed relates to technical limitations in recycling and reuse due to the current design of products and the types and combinations of materials used. However, studies also show that even if we could recycle 100% of our waste, our high consumption rates mean that the demand for virgin resources would remain high and primary extraction would remain necessary. A good example is aluminium – despite high rates of recycling (62% to 95%), our demand is so great that it cannot be met by recycling alone – recycled aluminium supplied only 35% of consumption in Europe in 2008, creating a continuous demand for the virgin resource (Chapman et al. 2013).

All this points to the need to prioritise an absolute reduction in our consumption of resources, and the first step is to measure the resources we consume. In order to account for all resources embodied throughout the full life-cycle of products from extraction to consumption (including of imported products), measurement must take a consumption-based, or footprint, approach. A comprehensive set of resource footprints have developed by researchers in the past decade (SERI 2009), and have gained the support of the EU (European Union 2011, EREP 2014, European Commission 2013, European Parliament 2015) and UNEP (UNEP 2016b), although they have not yet been made into policy. This set of four indicators covers the core resource input categories of materials, water and land area and the output category of carbon emissions. In a recent analysis of these four footprints (Table 2.1), Tukker et al. (2016) state that Europe is the only region in the world where the carbon, water, land and material footprints are all higher than the territorial emissions and resource extraction.

2.2.1 Measuring Our Consumption: Key for Decision Making

Land, carbon, water and material footprints should be a central part of all policy impact assessments and part of an overarching EU framework on sustainable resource management. Currently, impact assessments, which are carried out by the European Commission on legislative proposals, often allow short-term economic considerations to trump longer-term environmental and social issues. There is a worrying trend, pushed by big business lobbies, towards entrenching this bias further. Rather than allowing impact assessments to become a tool for narrow, short-termist economic considerations, they should inform decision-makers about Europe's resource consumption, how this is likely to be affected in future, and consequently, how dependent the EU is becoming on the availability and affordability of resources. Incorporating the four footprints into impact assessments will help to do this, and ensure that unintended negative consequences with regards to other resources are avoided. For example, the process that began in 2003 with the first directive setting EU biofuels targets solely on a carbon basis meant that associated increasing land use was not considered, creating unintended impacts that undermined the targets' efficacy. To avoid similar pitfalls, all policies related to the use of resources must include an assessment of the impacts on the EU's material, land, water and carbon footprints.

Member States need to start measuring their resource efficiency and levels of resource consumption, including through the four footprints, to be able to introduce policies that improve the current situation. Monitoring progress via the European

Table 2.1 Contributions in % of regions to the global Territorial (Terr.) emissions and Footprints (Fp) of consumption, and fraction of the Footprint that is covered (% Cov.) by Territorial emissions and extractions in 2007

Indicator	Carbon (% of global total)			Water (% of global total)			Land (% of global total)			Materials (% of global total)		
Region	Terr.	Fp.	%Cov.	Terr.	Fp.	%Cov.	Terr.	Fp.	%Cov.	Terr.	Fp.	%Cov.
Europe (EU)	16,1	20,2	**80**	7	12,7	**55**	5	12,6	**40**	12,9	18,7	69
United States of America (USA)	16,9	19,8	**85**	13,1	11,5	114	7,8	10,9	**72**	11	13,9	**79**
Asia and Pacific	24,2	22,1	110	42,8	42	104	24,7	25,1	**98**	23,4	22,3	105
China (CN)	24,1	19,2	126	16	14,2	113	8,5	8,9	**96**	22,9	22,6	101
Canada (CAN)	1,8	1,9	**95**	0,4	0,8	**50**	4,3	2,7	159	1,6	1,3	123
Latin America (LAM)	5,5	5,9	**93**	7,4	6,3	117	18,9	15	126	12,3	9,4	131
Australia (AUS)	1,3	1,5	**87**	1	0,8	125	6,6	4,8	138	2,5	1,5	167
Middle East (ME)	6,8	6,1	111	3,5	4,6	**76**	3,3	4	**83**	6,6	5	132
Africa (AFR)	3,2	3,2	100	8,8	8,1	109	20,9	16	131	6,8	5,2	131
Global total (%)[a]	100	100		100	100		100	100		100	100	
Global total (absolute)	38 Gt CO^2-eq.			1660 km^3			88 Mio km^{2b}			66 Gt		

Source: Tukker et al. (2016).

[a]May not add up due to rounding of differences.

[b]Forest (38 Mio km^2) and agricultural land (35 Mio km^2 pastures, and 15 km^2 arable land) only. Settlements and infrastructure, accounting for 3.6 Mio km^2, not included. Values in bold: regions whose footprints

Semester – which is currently overwhelmingly oriented on classic macro-economic considerations – would begin to provide a better overview of Europe's progress on resource use. The European Semester is the yearly cycle of coordination of economic and budgetary policies at EU level, and a tool to implement the broader Europe 2020 Strategy for "smart, sustainable and inclusive growth". It begins with the Commission's Annual Growth Survey, which together with the Commission's country specific recommendations, generally incorporates environmental goals in a very limited way, and mainly where they are seen to facilitate growth, economic or labour market recovery, rather than wider goals set out by Europe 2020, such as resource efficiency. The environmental issues covered mainly relate to climate and energy, while issues like biodiversity, water and waste management are side-lined or absent. The narrow focus of the European Semester must be broadened and adapted to fully align with the 2030 Agenda for Sustainable Development.

2.2.2 *What Is the European Commission Doing?*

In 2011, the European Commission's Roadmap to a Resource Efficient Europe (European Union 2011) put forward a vision of a European economy, which by 2050 "has grown in a way that respects resource constraints and planetary boundaries, […] is competitive, inclusive and provides a high standard of living with much lower environmental impacts."

While the Roadmap set up a clear vision, there are numerous problems with it and its implementation – it does not recognise the justice aspect nor the fact the EU has for a long time been using more resources than its fair share and at the expense of countries in the Global South, and despite being much needed, there is no strategy for the EU to develop a single robust resource use policy and ensure coherence with other policies across the board. Where current European policies and initiatives on resource use and efficiency are present, they are fragmented and split across different departments, lacking shared goals, visions and actions. This is a real concern in a world of limited resources, with rising resource based conflicts, increasing waste production and environmental impacts linked to our production and consumption system.

On its current trajectory, can the EU deliver the necessary transformational change in time? Taking a look at the policy proposals currently on the table and Commission priorities for 2017 (European Union 2016), serious doubts are legitimate. More drastic changes are needed to ensure better measurement and management of the resources we use and to remain within a safe planetary operating space. Even relatively progressive potential actions for the coming years, including improvements to waste legislation, expanding the scope of the Eco-design Directive and investigating the sourcing of more raw materials from within Europe, fall far short of delivering the transformational change needed.

The Roadmap is not perfect, but it was a start. In this document the Commission launched a process of developing the four footprint indicators, with an additional provisional lead indicator for resource productivity (the effectiveness of which is debatable). Among others, there were important points on the need to address markets and prices, taxes and subsidies that do not reflect the real costs of resource use and lock the economy into an unsustainable path, and on land – that by 2020, EU policies take into account their direct and indirect impact on land use in the EU and globally.

However, since its launch there has been little progress on the initiatives. In vital areas such as on indicators to measure the resources we use, there has been no progress – the original Roadmap promised to continue to develop these indicators to become fully consumption-based, yet we have seen none fully developed. For example, material consumption is still being measured by domestic material consumption (DMC), which gives a distorted view as it only takes into account the final weight of physical products imported, not the total embodied weight that goes into all the materials used to produce them; and land consumption is measured by land take within the EU, ignoring the fact that we rely on large amounts of land outside the EU to satisfy our consumption.

Besides the Roadmap, the 7th Environment Action Programme (7th EAP), 'Living well, within the limits of our planet', entered into force in 2014 as a guide for European environment policy until 2020 (European Commission 2013). On the surface, it has an impressive long-term vision, including that nothing is wasted, natural resources are managed sustainably, and biodiversity is protected, valued and restored in ways that enhance society's resilience. In regard to resource use and efficiency, the Programme states that the EU should set a framework for action to improve resource efficiency including targets for reducing the overall lifecycle environmental impact of consumption, in particular in the food, housing and mobility sectors, and indicators and targets for land, water, material and carbon footprints with methodologies to measure these to be developed by 2015. However, none of these actions have been followed through in a meaningful way.

The 7th EAP also contains objectives on the better integration of environmental concerns into other policy areas, to ensure coherence when creating new policy and to maximise the benefits of the EU's environment legislation by improving implementation. However, between empty promises on initiatives, drops, delays and fragmented policies, these objectives need significant work to deliver change.

Currently, the main focus within EU policy related to resource use is the Circular Economy Package (European Commission 2015a), a set of legislative and non-legislative measures on cutting waste and resource use, which was taken forward from the Roadmap. It was first launched in mid-2014, but later that year the incoming Commission scrapped the package with a promise to update and retable it in 2015 – despite the original package having support in the European Parliament and European Council. The re-tabled Package was published in December 2015. It includes proposals on changes to four EU Directives on waste as well a new action plan of 54 initiatives (European Commission 2015b).

One of the biggest downfalls of the current package is its absence of a resource efficiency target and other overarching legislative tools and policies. Many initiatives are focused on social and technological innovation instead of addressing the root problems in our system of production and consumption. Circularity does not in itself deal with the fact that we are, collectively, living beyond our planetary boundaries. Furthermore, many of the actions are being delayed and weakened. For example, the action plan committed to use the Eco-design Directive to make products more readily recyclable, repairable and reusable, yet it is likely that popular household items such as toasters and hair dryers will be excluded, and that there will be delays in addressing mobile phones and washing machines (European Commission 2016b).

2.2.3 A New Framework for Europe: The 2030 Sustainable Development Agenda

The adoption in 2015 of the 2030 Agenda of Sustainable Development (UNGA 2015) can be a game changer in overall policy making. This new global agenda is comprehensive, touching on most relevant topics to achieve a better world for everyone including the protection of the planet. It has the potential to provoke a paradigm shift in the concept of development as such, and stresses more than any other agenda the necessity to achieve coherence in policymaking. It promotes a new approach for policy work, with less focus on end of pipe solutions, charity, incremental change and damage control. Instead, it gives much more space to systemic change, fighting the root causes of poverty, social exclusion, environmental degradation and pushing for a fair distribution of wealth and use of natural resources.

For the EU, achieving coherence in internal and external policies is essential. The cross-border impacts of European internal policies should therefore be taken into account. A lot needs to be done here, as it is clear that European consumption and production patterns are harmful to the Global South. As explained before, the majority of the natural resources we use come from countries outside Europe. These "externalities" of European lifestyle are far from being compatible with poverty eradication and food security in the Global South, and therefore not in line with the 2030 Sustainable Development Agenda. For measuring progress on our responsibility to decrease the use of natural resources, the UN should start to set land, water, and carbon footprint reduction targets, per regions and country, combined with the implementation of measurements and monitoring tools. The International Resources Panel (IRP) of the UNEP should therefore be upgraded and receive the resource and the mandate to coordinate this research and political process.

2.2.4 Time to Go Beyond Resource Efficiency

Despite tentative steps forward from the European Commission, Europe continues to consume more than its fair share of global natural resources, even with improvements in resource efficiency and better waste management, at a high price for ecosystems and people around the world. The high European demand for resources is thus not only environmentally unsustainable, but raises issues of social justice – resource conflicts are growing in number and tension, and three environmental defenders were killed every week in 2015 (Global Witness 2016), merely for protesting against the destruction of the ecosystems they depend upon for their livelihoods.

The economic and wellbeing impacts of our overconsumption are also evident here in Europe. Our lifestyles put pressure on our wellbeing, with high rates of diseases and death from air pollution, stress, obesity and traffic accidents as well as a lack of access to nature.

With all this in mind, we cannot say that we are going into the right direction – and so it is time to go beyond resource efficiency, and look at how to cut resource use in absolute terms.

What is clear across EU policies is that the main focus is on technological innovation, putting the emphasis on eco-design and/or product standards and better ways of managing waste. While these are important steps, they will not give the EU what it needs: an absolute reduction in resource use to sustainable and just levels. As long as the EU's narrative is based on the assumption that increased consumption is fundamental to economic growth, we will never achieve an absolute reduction of resource use – we will merely buy ourselves time before ecological collapse.

Good politics means that we should all be able to live a decent life, within the environmental space of one planet. The main ambition is thus to fit our economies into "one-planet-lifestyles". According to the Global Footprint Network, in 2016 the world population overshot the level of resource use that we can use within 1 year on the 8th of August (Global Footprints Network 2016). For the remainder of that year, we used resources that could not be replaced by nature. Getting into debt with the planet will lead us to ecological bankruptcy.

The huge challenge is to decrease our absolute resource use, moving beyond resource efficiency. A new term which is gaining importance and addresses this challenge is 'sufficiency'. It comprises the questions of how much is enough, and how many resources we have. Based on this, the economy and lifestyles that answer to the most basic needs and achieve wellbeing for all can be developed. This implies a shifting away from material needs towards a new definition of a 'good life'. As such, moving beyond resource efficiency to sufficiency is crucial. We not only have to dematerialise our economy, but our entire concept of wellbeing!

Wolfgang Sachs introduced the concept of sufficiency into the sustainability debate in Germany at the beginning of the 1990s. He came up with the Four E's, from the German Terms *Entschleunigung, Entflechtung, Entrümpelung* and *Entkommerzialisierung.* Translated into English it would be something like the "four lessens": lessen our speed, lessen our distance, lessen our clutter, and lessen the commercialisation of our lives.

The big problem with solely focussing on resource efficiency is that it does not take into account the rebound effect, which undermines the net gains of resource use reduction. Total consumption is increasing much faster that the resource savings from resource efficiency, as is happening in Europe's case as explained above. Therefore, it is impossible to rely only on technological innovation, as social innovation with a focus on changing behaviour and lifestyles is crucial. The politics of sufficiency should create a framework where both technological and social innovation are possible, mainstreamed and upscaled.

2.3 Tackling the Root Causes of Overconsumption

Even if we keep repeating that 'business as usual' (and therefore 'policy-making as usual') is not an option, making real change is extremely difficult. The great transition cannot be achieved without a better understanding of how we are locked into our economic system, what the drivers of this model are, and how we can escape from it. Merely implementing resource efficiency policies will not bring us the desired system change. Therefore, we should go deeper into our thinking and also focus on the following policy levels.

2.3.1 Governance

Governance, combined with policy-making, is an important enabler for sustainable lifestyles, as it designs the structure and the rules of the game for our societies. It has unique powers to establish top down approaches (regulations, taxes, spatial planning etc.), but can also support and sometimes even initiate bottom up approaches (such as upscaling local initiatives and active participation facilities).

Until now, policy has acted as a driver of unsustainable lifestyle patterns, and elaborated a system of unsustainable consumption and production. This includes unsustainably high levels of production stimulated by subsidies, investment facilities, tax regulations, marketing support, educational programs and subsidies for associated research and development. Even if this was not intentional, governments have financially supported the development of consumption infrastructure such as road construction, communication system protocols and shopping malls. Some say that without government support we would not have this consumerist society (Schreurs 2010).

On the other hand, governments have also been integral in implementing policies to support households in times of scarcity - such as during World War II and during the oil crisis in the 1970s, which shows it is possible to have strong governance in favor of collective needs. Various public health campaigns have played a key role in moderating the consumption of alcohol and cigarettes.

Policy and governance mechanisms can be important catalysts for sustainable or unsustainable lifestyles. Governments are the elected defenders and managers of a collective vision and interests of a body of people, and therefore must act responsibly for the long term wellbeing of all citizens. Responsible leadership is critical, and governments have to be held accountable for their actions and decisions. Assessment tools for sustainability are helpful, before and after decisions are made. If governments put sustainable development and wellbeing at the core of their policy and long-term decisions, this will be a very important enabler for sustainable lifestyles.

If governments set priorities correctly, which can be done now by using the 2030 Sustainable Development Agenda as an overall framework, together with the related indicators, we will have a much more balanced dashboard of markers on where to go. This will support more coherent policymaking, shifting away from the focus of the 'jobs and growth' agenda.

2.3.2 Financial Tools

2.3.2.1 Stop Harmful Subsidies and Tax the Bads, Not the Goods

At the moment, Member State governments and the EU are subsidising activities that do not contribute to sustainability. For example, the fossil fuel sector receives up to USD 2 billion a year in exploration subsidies from Germany, Italy, France and the UK (Makhijani 2014). But also, EU structural and cohesion funds in Central and Eastern European countries finance new incineration plants rather than invest in waste prevention, reuse and recycling (CEE Bankwatch Network 2013). Urgent EU action is needed to reverse environmentally harmful subsidies and any action should prioritise the conservation of resources, the prevention of waste, and the re-use of products and materials.

Another key step is a tax shift – Member States can create a significant impact in driving more efficient resource use by increasing taxes on raw materials and products extracted from the environment instead of taxing labour. Shifting the tax burden from labour to resources would help promote more labour-intensive re-use and repair activities, instead of supporting disposable or non-repairable products. It would create much more jobs and less pressure on natural resources.

The EU made a commitment in the Roadmap to a Resource Efficient Europe both to phase out environmentally harmful subsidies by 2020 and to reform environmental tax by calling for a major shift from taxation of labour towards environmental taxation. However, the Roadmap's non-binding nature, and the prioritisation of environmental harmful short-term economic recovery, means that progress has been dismally slow. Nonetheless, the EU already has the commitment and the tools to make good on its intentions. It is time to turn them into a reality.

2.3.2.2 Public Money Versus Private Money

Public spending is very important to support and facilitate societies to achieve wellbeing – public spending includes public goods such as well-functioning public transport, adequate health systems, good education, and culture. As an increasing amount of money goes to private hands and is diverted away from public finances, inequality is growing tremendously. This is not only bad for social justice reasons, but also for the environment and because it lowers investments for collective goods and services. An International Resource Panel (IRP) report (UNEP 2016a) shows it is overconsuming richer groups who consume the majority of the world's natural resources. For that reason, growing inequality is one of the drivers of increasing use of natural resources. There are plenty of reasons to bring back money to the public with a fair tax system, by scrapping tax havens (USD7.6 trillion is offshore (Hardoon et al. 2016) – around the total GDP of the UK and France), and banning currency speculation.

2.3.3 Money Makes the World Go Down

Debt is draining resources worldwide. Our monetary system is highly debt-based, which is therefore one of the main driving forces of economic growth - which in turn leads to increasing consumption of natural resources. New products and substantive increases in global public debt have fueled an unprecedented increase in global financial assets from 1:1 ratio to global GDP in 1980, to a ratio of 4:1 in 2015 (300 trillion versus 75 trillion (Falkenberg 2016)).

A long time ago, only state banks could create money based on the public reserves in national banks. Now all banks can create money, a process that is often based on dubious financial products without any real value underpinning them. At this moment, 95% of all money is virtual and issued by private banks. Debts must be repaid to the private banks, which funnels 'real money' into the virtual economy. In order to obtain more real money, governments are obliged to increase trade and consumption, and the cycle of overconsumption continues.

This competition for money has negative consequences (Tekelova 2011):

- The ludicrous production of cheap goods of poor durability or with inbuilt obsolescence, so that manufacturing jobs are 'protected', and manufacturers can maintain or increase their profit. This has led to rapidly increasing consumption of raw materials as well as increasing levels of pollution and waste production.
- The huge advertising 'industry' building the demand for new products and the latest fashions in order to keep people buying, resulting in increasing levels of borrowing and debt.
- The ridiculous export drives by which every country simultaneously attacks the economies of every other nation, under the pretense that such global free trade improves general wellbeing.

- The externalization of embodied environmental and social impacts allowing, for example, burgeoning transport practices, with similar goods or even their parts (e.g. various parts of a car before assembly), crisscrossing the globe, without taking into account the environmental impact of this in final costs.

In addition to this there is the huge underlying key trend of concentration of wealth based on the increasing return to capital instead of labour. So, beyond these physical or 'real' investments, much of the investment activity in our monetary system takes the form of speculation in property, commodities or asset prices. The financial crisis bore witness to a kind of casino capitalism, gambling on the future, at the expense of financial and social stability. It became apparent through the crisis that sustainability – indeed, basic economic security – depends on a healthy financial system. Prosperity itself depends on a properly functioning money system. Transforming the financial system is a clear priority. Though it is beyond the scope of this paper to expand on that task in detail, it is worth highlighting three particularly important social innovations which are supported strongly by the analysis here.

2.3.4 Social Innovation and Behaviour Change

Due to the rebound effect, resource justice cannot be only achieved through technological innovations. More emphasis should be given to social innovation as the necessary complement to technological innovation in order to achieve systemic, long-lasting social and economic changes. A great deal of research has shown that local and grassroots initiatives have been successful in testing innovative ideas. Social innovations can be initiated by individuals or groups, but also by entrepreneurs. Social entrepreneurs and designers are promising actors, finding new solutions to existing social needs – or market failures. This occurs through the so-called 'acupuncture principle': small scale and local initiatives with potential for systemic change can be identified, and should be supported for up-scaling.

It is a pity that the leverage of social innovation is often so heavily undermined. The EU does not really invest in it directly or indirectly through research funding. This is why it often stays in a niche, where upgrading those initiatives could have much more results than technological efficiency. Social change will also change the narrative of basic needs and wellbeing, and vice versa.

Social innovation faces two main problems. First, social innovation competes with technological innovation for the focus of attention in political and business initiatives for sustainability as well as funding schemes, instead of being regarded as a complement. Second, due to the historically local nature of social innovations they are perceived as having niche impact only, as opposed to being scalable and replicable solutions. A countless number of such initiatives are very promising solutions, however, their potential has yet to be sufficiently explored. Beyond the question of how to scale promising approaches, it is also necessary to establish political, and institutional structures (such as education) to foster them.

2.3.5 Other Business Models

Alternative business models (Fig. 2.3), such as leasing, can transform consumers into users (FoEE 2014). Unlike the traditional model of purchasing, where the buyer is responsible for the products disposal, leasing puts the responsibility on the designer for the disposal of the product, thus incentivising the design of products that are more durable, reusable, easily repairable and upgradable during their use phase, and which are recyclable or naturally decomposable at the end of their lifespan. Manufacturers retain ownership and responsibility for a product during its life, and at the end of it they regain access to components that can be re-used or recycled. Under this model, companies need to consider not only the sale of a product, but also its repair, refurbishment and return, and therefore they will benefit economically from investing in ways to extend their products' lifespans, rather than profiting by launching and marketing new, ever-so slightly 'better' products every year. Providing economic incentives for leasing or other similar business models can prompt manufacturers to design repairable and this more sustainable products (Fig. 2.4).

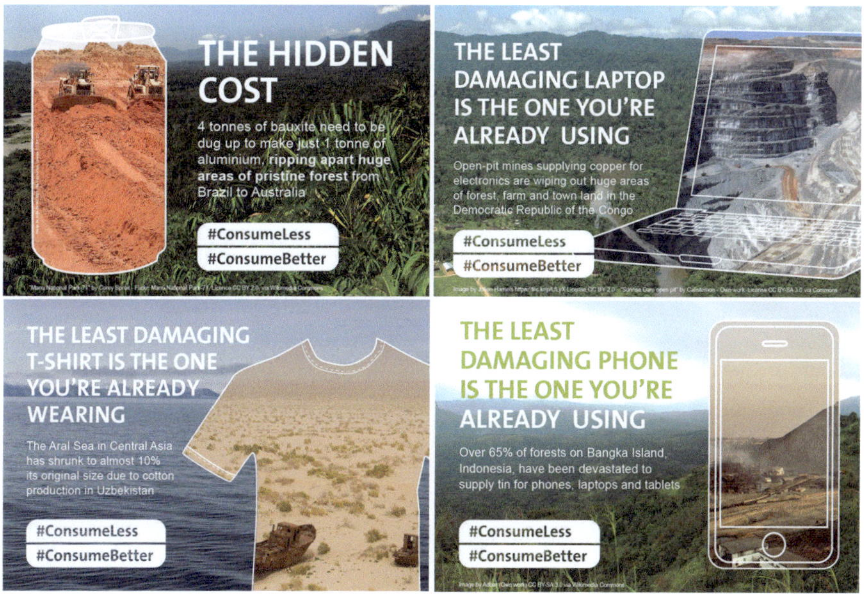

Fig. 2.3 The global impacts of our consumption (Source: Friends of the Earth Europe 2015)

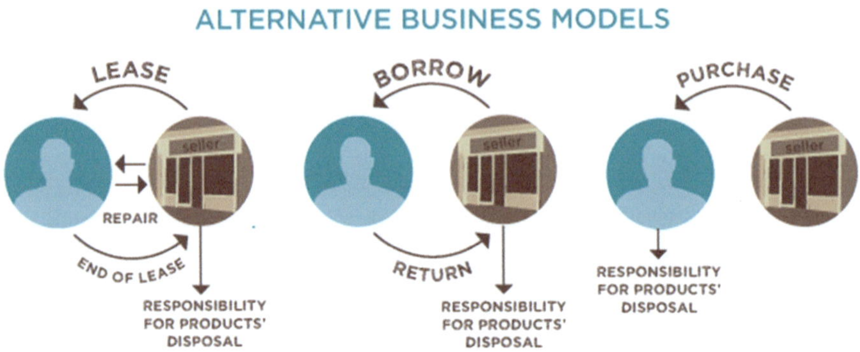

Fig. 2.4 Comparison of alternative business models with traditional model of purchasing (Source: Friends of the Earth Europe 2014)

2.3.6 *Legal and Regulatory Frameworks*

Voluntary improvements on corporate responsibility and financial incentives to push for resource-efficient behaviour and production patterns are of course positive. But what if the unwilling groups of producers and consumers are not moving anywhere? Legally binding agreements and regulations must also be put in place. Several good initiatives have been taken, at the EU global levels, but nevertheless in general environmental law is still quite weak compared to civil law. Many civil society organisations are still demanding a binding UN treaty on corporate accountability. Recognising 'ecocide' as a crime against humanity is another civil society demand which is gathering momentum.

In the outcome of Rio + 20, "The world we want", a reference was made to Principle 10 (Access to information, participation and justice) of Agenda 21 (agreed in 1992) as leverage for environmental democracy. In Europe the Aarhus Convention exists, but is still relatively unknown by most of the environmental and civil society groups. It is an effective tool, that could be better put into practice.

On another level, the EU could be stricter and more innovative on product norms based on environmental performance. Too many energy or natural resources intensive products are still on the market. Products norms and bans are mostly based on issues related to human health. Taking care of the health of Mother Earth should be argument enough for taking some products out of the market, as is done with old-fashioned light bulbs. A whole range of policy opportunities is still open related to product norms and bans.

Last but not least, we would also urge that discussions around the design of resource capping schemes should begin. In order to share existing natural resources in a fairer way and use the scarce resources for basic needs, instead of for the greed of a minority, there seems no other way than to regulate resource use. For that the International Resource Panel of UNEP should have the mandate and the trust to study the stocks of natural resources and design schemes for their management and destiny.

2.4 Conclusion

With this article we have tried to list the main challenges to make the transition towards a resource-efficient Europe. We are deeply convinced that this cannot be done by technological innovation only, but that much more attention to social innovation (voluntary behaviour change) is needed and the drivers of (over)consumption must be tackled.

We do acknowledge that this will be difficult, as vested interests in the current economic system are strong and are an obstacle to this transition, but we hope that future generations of CEOs, politicians and financial managers will put the collective interest higher on their agendas than private gains. As there are no jobs on a dead planet, we should put the planet and people first, above the survival of our current economic system. Resource efficiency is a step forward, but for fair 'One-Planet Living', we also need resource sufficiency.

Acknowledgments The authors would like to thank Sophie Colsell and Paul Hallows for their invaluable contributions to the article.

References

CEE Bankwatch Network (2013) No time to waste: cohesion funds programming for a resource-efficient Europe. CEE bankwatch network

Chapman A, Tercero Espinoza L, et al (2013) Study on critical raw materials at EU level. Oakdene Hollins and Fraunhofer ISI

EEA (2012) Material resources and waste – 2012 update. The European environment state and outlook 2010. European Environment Agency, Copenhagen

EEA (2015) The European environment – state and outlook 2015. European Environment Agency, Copenhagen

EREP (2014) European resource efficiency platform: Manifesto & policy recommendations. European Commission. Accessed 25 Nov 2016. http://ec.europa.eu/environment/resource_efficiency/documents/erep_manifesto_and_policy_recommendations_31-03-2014.pdf

European Commission (2013) Living well, within the limits of our planet. Accessed 25 Nov 2016. http://ec.europa.eu/environment/pubs/pdf/factsheets/7eap/en.pdf

European Commission (2015a) Circular economy strategy. http://ec.europa.eu/environment/circular-economy/index_en.htm

European Commission (2015b) COM (2015) 614 final: closing the loop - an EU action plan for the circular economy. http://eur-lex.europa.eu/legal-content/EN/TXT/?uri=CELEX:52015DC0614

European Commission (2016a) Raw materials scoreboard – European innovation partnership on raw materials. Publications Office of the European Union, Luxembourg

European Commission (2016b) Commission to set out new approach on ecodesign. Accessed 25 Nov 2016. https://ec.europa.eu/energy/en/news/commission-set-out-new-approach-ecodesign

European Parliament (2015) Resolution of 9 July 2015 on resource efficiency: moving towards a circular economy, (2014/2208(INI))

European Union (2011) Communication: roadmap to a resource efficient Europe, 20 September 2011, COM(2011) 571 final

European Union (2016) Communication: commission work programme 2017. Delivering a Europe that protects, empowers and defends. 25 October 2016, COM(2016) 710 final

Eurostat (2016a) Energy production and imports. Accessed 25 Nov 2016. http://ec.europa.eu/eurostat/statistics-explained/index.php/Energy_production_and_imports
Eurostat (2016b) Municipal waste statistics. Accessed 25 Nov 2016. http://ec.europa.eu/eurostat/statistics-explained/index.php/Municipal_waste_statistics
Falkenberg K (2016) Sustainability now! A European vision for sustainability. European Political Strategy Centre
FoEE (2014) Preventing waste: recycling isn't enough for a circular economy. Friends of the Earth Europe, Brussels
FoEE (2016) The true cost of consumption: the EU's land footprint. Friends of the Earth Europe, Brussels
Global Footprints Network (2016) Earth overshoot day. Accessed 25 Nov 2016. http://www.overshootday.org/
Global Witness (2016) On dangerous ground. Accessed 25 Nov 2016. https://www.globalwitness.org/en/reports/dangerous-ground/
Greenovate! Europe E.E.I.G (2012) Guide to resource efficiency in manufacturing. Greenovate! Europe E.E.I.G, Brussels
Hardoon D et al (2016) An economy for the 1%: how privilege and power in the economy drive extreme inequality and how this can be stopped. Oxfam International, Oxford
Krausmann F et al (2009) Growth in global materials use, GDP and population during the 20th century. Ecol Econ 68(10):2696–2705
Lovins HL et al (1998) Factor four: doubling wealth - halving resource use: the new report to the Club of Rome. Earthscan, London
Makhijani S (2014) Subsidizing unburnable carbon: taxpayer support for fossil fuel exploration in G7 nations. Oil Change International, Washington DC
Meadows DH et al (1972) The limits to growth. Earth Island Limited, London
Schreurs J (2010) Living with less: prospects for sustainability. Uitgeverij Genoeg
Schumacher EF (1974) Small is beautiful. Sphere/Abacus, London
SERI (2009) How to measure Europe's resource use: an analysis for Friends of the Earth Europe. Sustainable Europe Research Institute, Vienna
Tekelova M (2011) In-debt to a destructive economy. Article by Green Economy Coalition
Tukker et al (2016) Environmental and resource footprints in a global context: Europe's structural deficit in resource endowments. Glob Environ Chang 40:171–181. Elsevier
UNEP (2016a) Global material flows and resource productivity. An assessment study of the UNEP international resource panel. Schandl H et al Paris, United Nations Environment Programme
UNEP (2016b) Resource efficiency: potential and economic implications. A report of the international resource panel. Ekins P, Hughes N, et al. Paris, United Nations Environment Programme
UNGA (2015) United Nations General Assembly resolution 70/1, Transforming our world: the 2030 Agenda for Sustainable Development, A/RES/70/1 (25 September 2015), available from. undocs.org/A/RES/70/1
United States Congress (1973) Energy reorganization act of 1973: Hearings, Ninety-third Congress, first session, on H.R. 11510. United State Congress, Washington, DC, p 248

Chapter 3
Global Megatrends and Resource Use – A Systemic Reflection

Ullrich Lorenz, Harald Ulrik Sverdrup, and Kristin Vala Ragnarsdottir

Abstract The exponentially developing Global Megatrends of human society can be explained and modelled from a systemic perspective. Results show simultaneous exponential growth for population, energy consumption, raw materials extraction, GDP, pollution and climate change. The authors participated in the development of the WORLD6 system dynamics model to explain these trends and to explore what the future may hold. This was done based on the systemic approach in order to include feedback loops and changes in trends that may occur at later points in time. The modelling results allow to reproduce developments that are referred to as "Global Megatrends" in literature and that are used to proof the concept of the Anthropocene. While in the Global Megatrends literature and the Anthropocene literature exponential growth is based on empirical data we suggest taking a longer and systemic perspective on these global trends. Whether the trends are to level off, decline or crash depends to a large degree on future policies. We can confirm that the "rise and fall scenarios" are principally right in their analysis as well as the root causes. It is evident from our WORLD6 simulations that a business-as-usual scenario will lead to a decline of disruptive nature. It is also evident that the worst case scenario can be avoided by a careful design of policies, using the systems perspective assisted by dynamic model simulations. The German policies for an Energiewende (energy transition) must be linked to a future policy of Ressourcewende (resource transition) and a Nachhaltigkeitswende (sustainability transition). Such policies appear strategically appropriate and well founded in science.

Keywords Global Megatrends • Anthropocene • World model • Limits to growth • System dynamics • Sustainability transition

---

U. Lorenz (✉)
German Environment Agency (UBA), Wörlitzer Platz 1, DE-06844 Dessau-Roßlau, Germany
e-mail: ullrich.lorenz@uba.de

H.U. Sverdrup
Industrial Engineering, University of Iceland, Hjardarhagi 2-6, IS-107 Reykjavik, Iceland
e-mail: hus@hi.is

K.V. Ragnarsdottir
Faculty of Earth Sciences, University of Iceland, Sturlugata 7, IS-101 Reykjavik, Iceland
e-mail: vala@hi.is

H. Lehmann (ed.), *Factor X*, Eco-Efficiency in Industry and Science 32,
https://doi.org/10.1007/978-3-319-50079-9_3

3.1 Introduction

Evidence shows that since the Club of Rome published the report "The Limits to Growth" in 1972, unchecked development of global population and consumption of natural resources has continued. As given in their "standard run" (business as usual scenario), this will lead to serious environmental and thus also economic and social problems, if not to a collapse. While a collapse is not perceived as very likely in the scientific and policy community, the use of natural resources increased significantly since the industrial revolution and in particular from the mid-twentieth century onwards, almost quadrupling to 80 billion tons between 1970 and 2015 (UNEP 2016). While the central aim of the simulations of The Club of Rome (Meadows et al. 1972, 1992, 2004) was to present various scenarios, the aim of their work has never been to predict exact timings and quantities but rather show patterns of developments. Recent studies suggest that pathways in resource consumption, population growth and pollution are fully following the trajectory of the business as usual scenario, which includes tipping points towards a crash (Turner 2008, 2012). Assuming that humanity is able to take measures to either stop these developments or at least ease the severity of a possible downward turning in a timely way, the "eye opening" message from a systems science point of view is that the world system is essentially correctly described. Acknowledging that complex systems follow certain interconnected "rules" this shows that it is urgent to apply systems thinking not only in science but also for policy design. System science offers tools and perspectives to overcome linear thinking, which is the primary foundation of policy across the world. The highly interconnected socio-technical-economical-global ecological systems are classified by feedback structures, delays or ambiguities which require causal link approaches to handle this kind of complexity with seemingly chaotic or counter intuitive behaviours.

Recent work by Sverdrup et al. (e.g. Sverdrup and Ragnarsdottir 2014, 2016; Sverdrup et al. 2014a, b, 2015) includes the upgrade of the original WORLD model towards version 6.[1] WORLD6 was derived from the World3 model built in 1970–1972 by Meadows et al. (1972) from the initial world system dynamics model of Forrester (1971). In these previous version of the WORLD model, population dynamics, food supply, food production, materials resource extraction, energy resource extraction, renewable energy, social feedbacks, economy and investments were linked in the way they are thought to link causally in the real world. In the WORLD3 model, all resources were lumped together as energy and natural resources (i.e. fossil fuels, metals, materials, water), adding them up into one resource "index R". Pollution was also generalized into persistent pollution, adding up all kinds of pollution into one "Pollution index". In WORLD6 this has changed. The present version is a result of the dismantling of the WORLD3 model and an

[1] The update from WORLD5 to WORLD6 model was funded by SimRess project ("Models, potential and long-term scenarios for resource efficiency") by German Federal Ministry for Environment and the German Environmental Protection Agency (FKZ 3712 93 102).

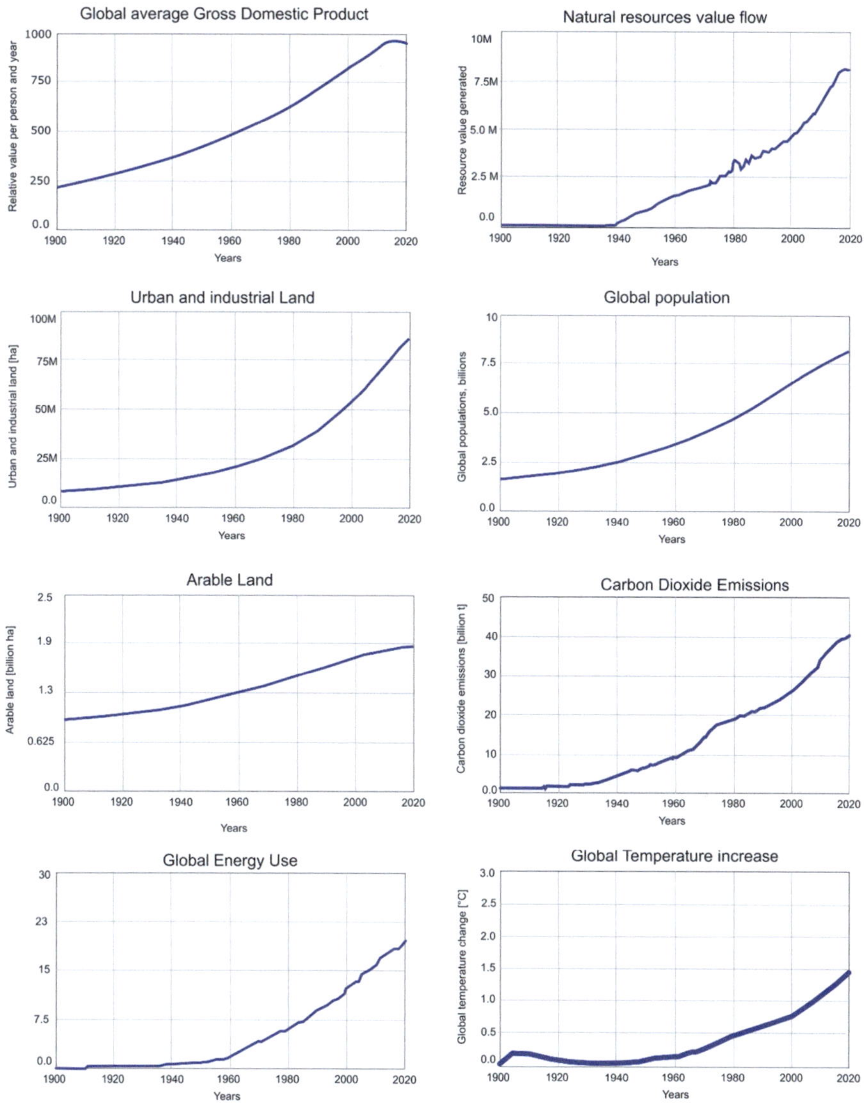

Fig. 3.1 Megatrends created with the WORLD6 model from 1900 to the present (2015) comparable to the trends presented by Steffen et al. (2015) showing typical exponential growth shape

extension and substitution of its resource module (Sverdrup 2016; Sverdrup and Ragnarsdottir 2016). In WORLD6 energy and natural resources are separated into individual categories and a series of metals and materials as well as energy carriers are modelled individually. Figure 3.1 show some central simulation output from 1900 till 2015, indicating strong (exponential) increase of the presented indicators.

The graphs presented in Fig. 3.1 show nearly the same patterns as the indicators compiled in their article "The great acceleration" by Steffen et al. (2015) or in the work by (Monastersky 2015). Both authors show a series of socio- economic trends driving earth system trends as well. Most of these trends (with some exceptions, e.g. marine fish capture that seems to have exceeded its peak and is now in decline) show the same specific exponential growth pattern. This rapid and unprecedented global change driven by human activities is consequently referred to as the Anthropocene, which has recently been suggested as having started either at the beginning of the industrial revolution or in the middle of the twentieth century.

Most of these global changes are also referred to as Global Megatrends (e.g. EEA 2011, 2015; Sadler 1996; Retief et al. 2016). The term 'Global Megatrend' is used to describe large-scale, extensive and often closely interrelated change processes at the social, technological, economic, ecological and political levels that threaten the resilience of social and ecological systems (EEA 2015: 5, EEA 2015: 33 ff). This concept of "(…) Global Megatrends and challenges, related to population dynamics, urbanisation, disease and pandemics, accelerating technological change and unsustainable economic growth add to the complexity of tackling environmental challenges and achieving long-term sustainable development" (EEA 2015, p. 13).

Global Megatrends can hence be described as the attempt to reduce the complexity of global developments to make them more manageable for political decision making. Nevertheless, Global Megatrends could not be seen as unconnected developments in the world. Global Megatrends can potentially raise the awareness of political decision makers, and they can be used as a communication tool both for politicians and the public. But this reduction of complexity can in turn make the audience forget about the interconnected systems in the background. Although, in the moment the consequences of the global trends for e.g. a policy issue are assessed, it is indispensable to understand feedback mechanisms and respective ambivalences of the underlying system (see e.g. Lorenz and Haraldsson 2014, examining the potential impact of Global Megatrends on national scale).

Beginning from the work of Meadows et al., the defining of the Anthropocene by Steffen et al., using the concept of Global Megatrends by EEA or more recent results from the work with WORLD6 among other approaches: all these are attempts to handle the complexity of the highly interlinked global systems and finally design adequate policy approaches to prevent unfavourable developments for the health of nature and the human population.

3.2 The Root Cause of Observed Megatrends

The graphs in Fig. 3.1 show that human activities as observed from 1900 to 2015 mostly follow an exponential growth curve; there is a steady increase in nearly all indicators with a steep increase from around 1950. While the empirical data ends in 2013 (as presented in Steffen et al. 2015), the simulations in WORLD6 continue until 2100 and beyond. These graphs are following a certain pattern: after the

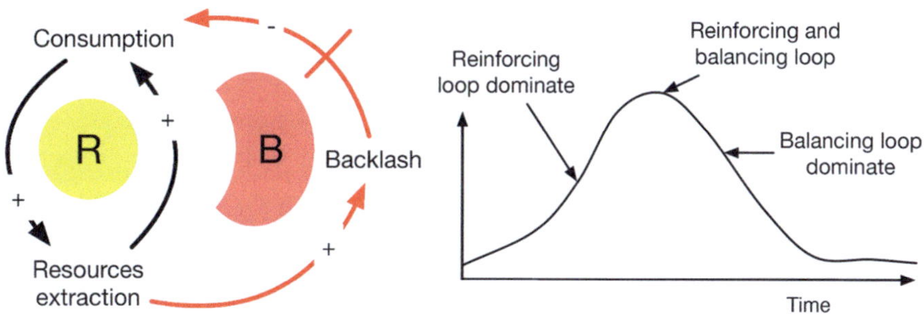

Fig. 3.3 The rise is caused by the consumption driven by a resource that often lead to exponential growth. If the resource is a finite one, the backlash of resource exhaustion will occur, leading to a decline

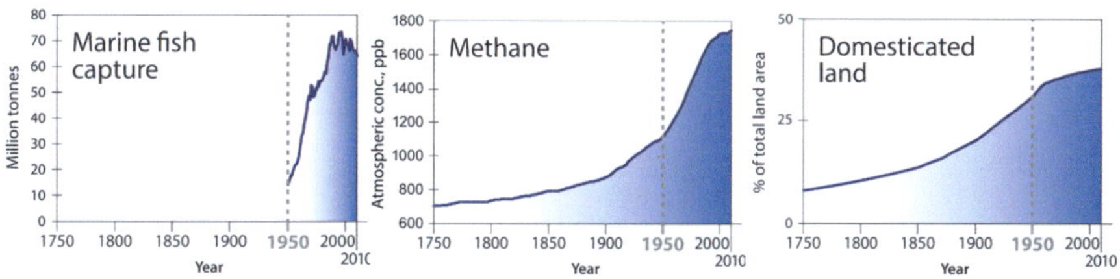

Fig. 3.4 Selected earth system indicators adapted from Steffen et al. 2015, demonstrating that certain indicators already show a broken exponential growth pattern. These include marine fish capture, methane concentration in the atmosphere and domesticated land

single megatrends – seemingly missing the fact that the megatrends belong to the same interconnected system – we provide in this chapter a series of connected megatrends that on one hand drive the use of natural resources (mainly materials) and on the other hand megatrends from the global Earth system that feed back into to socio-economic sphere. The causal loop diagram in Fig. 3.5 together with the basic archetype (Fig. 3.3) is capable to explain and deduce the "rise and fall" of Global Megatrends that are highly relevant for the material use in society. Here we take into account that the megatrends are connected and their inherent systemic characteristics, including resistance to external changes (delays until they actively change – shown with a short line perpendicular to causal lines). The diagram connects some spheres:

1. Civilization: population, demography, infrastructure, wealth, debts;
2. Physics: natural mineral resources, finite resources, solar and wind resources;
3. Natural ecosystems and their ecosystem services.

The basic principles shown in Fig. 3.3 have been further elaborated in Fig. 3.5 with respect to human society, integrating the industrial metabolism and the economy. The yellow cores are different reinforcing loops (R) that create exponential growth in the human system.

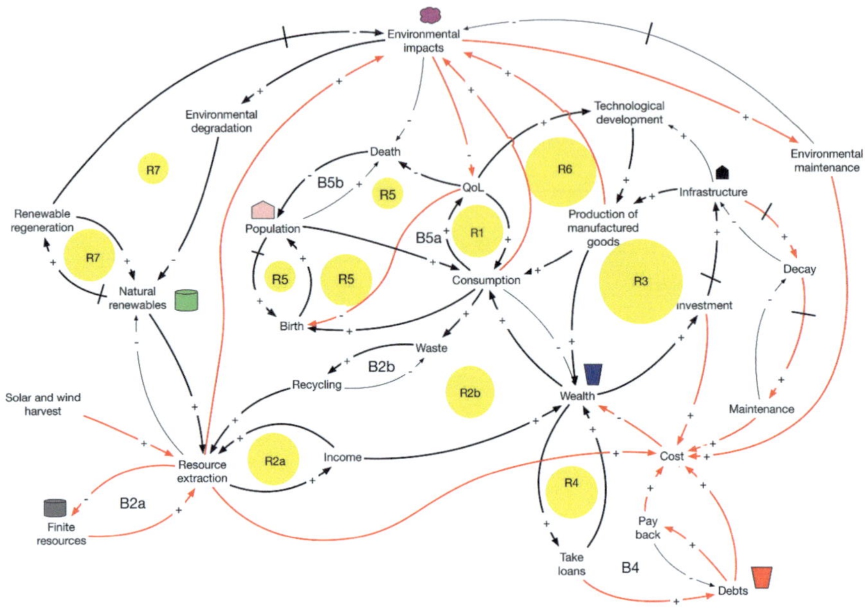

Fig. 3.5 The megatrend "machine" represented in a causal loop diagram. The causal loop diagram shows how humans ran the world from 1750 to about 1990. The observed megatrends can be explained by employing known causal relationships. The *yellow circles* marked with Rs represent reinforcing loops. Together, these loops are responsible for the exponential behaviour of the present system (Figs. 3.1 and 3.2). The zones marked with Bs and the *red lines* show the balancing loops that will eventually break off the exponential growth. This diagram explains all observed megatrends

The reinforcing loop R2a and R2b in Fig. 3.5 are responsible for exponential growth of indicators like resource extraction or wealth. We have shown these indicators in Fig. 3.2. The R2-loops are though connected to balancing loops (B2a and B2b) which will induce the decline of the curves. B2a is referring to limited resources in a finite world. This balancing mechanism becomes mostly apparent for indicators like urban land or industrial land, but also for raw materials as shown later in Fig. 3.6. The balancing Loop B2b introduces recycling as a measure. Recycling could be capable to reduce resource extraction and by this change the decline pattern of the respective resource. Also, this will be highlighted more in Fig. 3.6. Next to the explanatory potential of this kind qualitative analysis with a causal loop diagram the limitations also become apparent: while we are able to generally detect and explain the exponential growth pattern followed by a decline, we cannot say, at which level the system stabilises. The potential impact of the balancing loop depends in the case of recycling on the recycling rate, the level of resource extraction and on the level of consumption. If, for example, consumption level is high and recycling rate is low, resource extraction will still be high and the finite resources will still be eaten up. Staying with this example: only if consumption

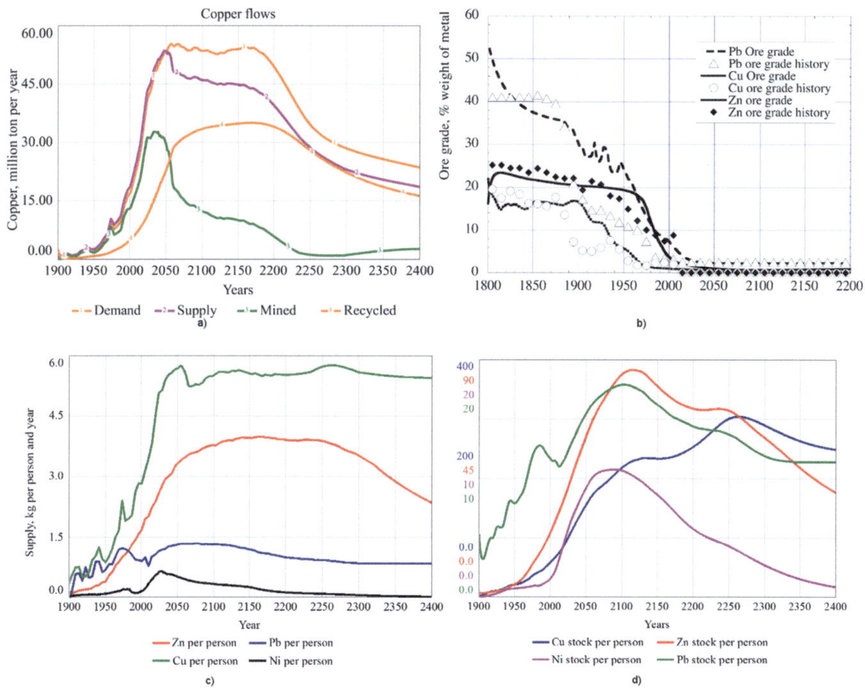

Fig. 3.6 Copper is a typical base metal and is used in very many technologies developed by man; (**a**) shows the rate of mining, supply to society and recycling; (**b**) shows ore grade declining both in the empirical and modelled data; (**c**) and (**d**) show some key metals for human technologies, plotted as amount supplied kg per person and year (**c**) and as stock-in-use (**d**)

would be lower, together with high recycling rates and together with natural renewables (R7) the whole system might stabilise without a full decay of finite resources.

This CLD is a generalised one. You could easily replace e.g. the finite resource with a metal like copper, iron, coal or with any other kind of resources and follow the loops without losing explanatory power.

As an example, one of the typical and crucial base materials for society is copper, which is used in numerous technologies developed by mankind and used for provision of service. Without copper, human society would face a significant challenge. The Fig. 3.6 shows supply to mining, supply and recycling (a) and the steadily sinking quality of the resource we extract from, for copper. Simulated and observed ore grades for copper and some of the metals associated with it in (b), using the WORLD6 model basis run that corresponds to the graphs shown in Fig. 3.2 Copper – as one example of in the long term the finite materials – is one of the limiting factors that shapes the backlash, signalling the initiation of "fall" in the archetype behaviour. Take note to observe that even if the total supply of copper has a peak behaviour (a), it does look a bit different when drawn up as supplied per person or as stock-in use per person. It can be seen in (c) and (d) that the supply will level off and stay constant for copper and lead, but reach a peak and decline for lead

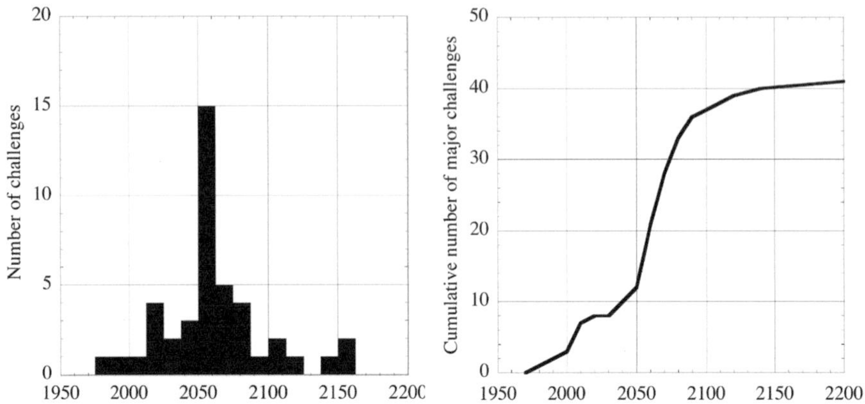

Fig. 3.7 Occurrence of the maximum year for supply to society we get that shows the density of maxima (*left*) and the cumulative distribution of maximum points with time (*right*)

and nickel. The stock-in-use, which tell us how much we have in use for providing services, show a distinct peak for all the metals. The decline is caused by the supply not being able to keep up with maintenance and replacement, suggesting that we have an overshoot in terms of infrastructure as compared to what is sustainable.

Besides the fact that declining resources availability, expressed as declining amount and declining quality, trigger the backlash in the socio-economic system, it becomes apparent that the balancing loop does not start with the peak of the curves as shown in Fig. 3.6 but clearly earlier. This leads to the conclusion that timing of action is crucial. It is evident that until the present day, the general megatrends were increasing in everything. With traditional linear thinking, this may appear to just continue, and no problems are apparent. However, as Fig. 3.5 shows, the world is in no way linear. What the systems analysis of the world system has shown in Figs. 3.3 and 3.5, suggests that this is not an uninterrupted trend of increase, but rather part of a longer trend that looks different as other parts of the system come into play. The professional term for such an archetype is "rise and fall" or "overshoot and decline". All indications from systems analysis on the world system of today suggest strongly, the present society is in an "overshoot and decline" mode. The output of the WORLD6 model indicates in addition that most of the upward trends will reach a plateau or start declining between 2020 and 2050 (Fig. 3.7). In short, the general patterns of the megatrends that we observe today cannot be expected to continue.

3.4 Conclusions

Taking a systemic global perspective makes it clear that all observed developments, including the Global Megatrends, are tied together. They form a highly interconnected and extremely complex system while at the same time we can find that many reports on Global Megatrends show megatrends as single entities and mostly just the first part of the trends.

The complexity and interconnectedness of Global Megatrends makes it nearly impossible to find direct linear cause-effect relationships in the system. There is simply no mechanism to easily fix these global challenges. Additionally, many trends are seemingly inert to changes, due to buffering subsystems, or simply a slow pace of change, determined by the system. Furthermore, delays must be expected due to lags in political knowledge, acceptance and negotiations. In the case of the 1987 Montreal Protocol,[2] it took roughly 15 years from the detection of the problem, until the treaty came into force. Another 30 years later, the first significant effects in the atmosphere[3] can be seen. Simulations suggest a recovery of the ozone layer by 2050. In total, the time span from discovering the problem to fixing it is close to 70 years. The LRTAP[4] protocol offers a similar parallel case, from first negotiation to (1979) to final protocol in force (2004), it took 25 years.

WORLD6 model runs indicate that efficiency measures and recycling lift the supply curve up and thus that the peak time for certain resources moves further into the twenty-first century (the time of maximum supply to society). If the supply maximum point could be reached after the population reaches its maximum value, then the challenge changes in a profound way. The decline in supply would coincide with a simultaneous decline in population, making it far easier to maintain provision per capita. If this is combined with better consumer resource use efficiency, less irreversible losses from the society and more recycling of key materials, then a solution may be in sight. Then the supply per person may change from a "steady increase, peaking and declining" curve to a "rising and levelling off" type of curve. This also shows that addressing global population size as well as managing population in local regions must be a part of the overall strategic agenda.

There are a number of themes that must be addressed in an integrated policy strategy, aiming to target the aforementioned systemic intervention points. In addition to advocating for efficiency measures and promoting recycling, the delays and timings of resources in society must be considered more systematically. Materials must be kept in service for a longer time period, which requires designing longer-lasting products, reparability, reusability, and recyclability. The efficiency (material and energy) of repairing and recycling processes must be optimised. The irreversible consumption/degradation of raw materials (e.g. the burning of fossil fuels, the micro-dissipation of precious metals) must be avoided. The time that materials are bound to infrastructure or landfills must be minimised in order to get as much service out of the materials as possible. All of these issues call for better education at all levels, a focus on vocational training, and new approaches to design.

---

[2]An international treaty designed to protect the Earth's ozone layer by phasing out the production of numerous substances that are responsible for ozone depletion.

[3]The ozone hole not getting bigger.

[4]United Nations Economic Commission for Europe, Long Range Transboundary Air Pollution Protocol.

This will result in a better informed public and help policy makers lead our society into a circular economy (EASAC 2015, 2016a, b).

Disclaimer This paper does not necessarily reflect the opinion or the policies of the German Federal Environment Agency.

References

Boulding K (1966) The economics of the coming spaceship earth. In: Jarrett H (ed) Environmental quality in a growing economy, essays from the sixth RFF forum on environmental quality. The Johns Hopkins Press, Baltimore

Butler JH, Montzka SA (2016) The NOAA Annual Greenhouse Gas Index (AGGI). https://www.esrl.noaa.gov/gmd/aggi/aggi.html. Accessed 6 Dec 2016

EASAC (2015) Circular economy: a commentary from the perspectives of the natural and social sciences. EASAC

EASAC (2016a) Indicators for a circular economy. EASAC policy report EASAC

EASAC (2016b) Priorities for critical materials for a circular economy. EASAC policy report EASAC

Eddington AS (1928) The nature of the physical world. MacMillan. 1935 replica edition: ISBN 0-8414-3885-4, University of Michigan 1981 edition: ISBN 0-472-06015-5 (1926–27 Gifford lectures)

EEA (2011) The European environment — state and outlook 2010: assessment of global megatrends. European Environment Agency, Copenhagen

EEA (2015) European environment — state and outlook 2015: Assessment of global megatrends. European Environment Agency, Copenhagen

Forrester J (1971) World dynamics. Wright-Allen Press, Cambridge

Lorenz U, Haraldsson H (2014) Impact assessment of global megatrends – Two case studies connecting global megatrends to regional topics. Report 6602. Swedisch Environment Agency

Meadows DH, Meadows DL, Randers J, Behrens W (1972) The limits to growth. Universe Books, New York

Meadows D, Meadows D, Randers J (1992) Beyond the limits. Chelsea Green Publishing Company, White River Junction VT

Meadows D, Meadows D, Randers J (2004) Limits to growth: the 30-year update Chelsea Green Publishing Company, White River Junction VT

Monastersky R (2015) Anthropocene: The human age. Nature 519:144–147

Retief F, Bond A, Pope J, Morisson-Saunders A, King N (2016) Global megatrends and their implications for environmental assessment practice. Environ Impact Assess Rev 61:52–60

Sadler B (1996) International study of the effectiveness of environmental assessment – final report: environmental assessment in a changing world: evaluating practice to improve performance. Minister of Supply and Services, Ottawa

Steffen W, Brodgate WJ, Deutsch L, Gaffney O, Ludwig C (2015) The trajectory of the Anthropocene: the great acceleration. Sage Journals, Australia, Australia/Oceania

Sterman JD (2000) Business dynamics: systems thinking and modeling for a complex. World Mcgraw-Hill Education Ltd; Auflage: International edition

Sverdrup HU (2016) Modelling global extraction, supply, price and depletion of the extractable geological resources with the LITHIUM model. Resour Conserv Recycl 112:112–129

Sverdrup H, Ragnarsdottir KV (2014) Natural resources in a planetary perspective. Geochem Perspect 3:129–341

Sverdrup HU, Ragnarsdottir KV (2016) A system dynamics model for platinum group metal supply, market price, depletion of extractable amounts, ore grade, recycling and stocks-in-use. Resour Conserv Recycl 114(41):130–152

Sverdrup H, Koca D, Ragnarsdottir KV (2014a) Investigating the sustainability of the global silver supply, reserves, stocks in society and market price using different approaches. Resour Conserv Recycl 83:121–140

Sverdrup HU, Ragnarsdottir KV, Koca D (2014b) On modelling the global copper mining rates, market supply, copper price and the end of copper reserves. Resour Conserv Recycl 87:158–174

Sverdrup HU, Ragnarsdottir KV, Koca D (2015) Aluminium for the future: modelling the global production, market supply, demand, price and long term development of the global reserves. Resour Conserv Recycl 103:139–154

Turner GM (2008) A comparison of the limits to growth with 30 years of reality. Glob Environ Chang 18:397–411

Turner GM (2012) On the Cusp of global collapse? updated comparison of the limits to growth with historical data. GAIA: Ecol Perspect Sci Soc 21:116–124

UNEP (2016) Resource efficiency: potential and economic implications. A report of the International Resource Panel.: Paul Ekins, Nick Hughes

Chapter 4
Data, Indicators and Targets for Comprehensive Resource Policies

Stephan Lutter, Stefan Giljum, and Martin Bruckner

Abstract Today's most pressing environmental problems, such as climate change, biodiversity loss, land cover conversion, etc. are caused by the overall growth of production and consumption. For a methodologically sound and comprehensive measurement of societal resource use and its environmental, economic and social impacts as well as for monitoring progress towards defined targets, appropriate indicators are needed. In addition to the territorial indicators currently in use, such indicators take into account resources "embodied" in traded goods and services. These "footprint-type indicators" help understand to what extent a country's economy, the environment, and the resource efficiency performance of goods and services are influenced by global value chains. In this book chapter we discuss the state of the art of data and indicator development focussing on three different types of natural resources – raw materials, land, and water. First an overview of methodological options regarding environmental accounting frameworks as well as the calculation of footprint-type indicators is provided. We show empirical trends of resource use and analyse to what extent drivers of global resource use such as the European Union have managed to decouple their economic development from resource use. Methodological requirements and necessary next developments are identified and the ongoing processes as well as empirical analyses linked to the question of how targets for sustainable resource use can be identified in both regards, to the methodological as well as institutional level.

Keywords Footprint-type indicators • Environmental-economic accounting • Evidence-based policy making • Resource use targets

S. Lutter (✉) • S. Giljum • M. Bruckner
Institute for Ecological Economics/Vienna University of Economics and Business (WU), Welthandelsplatz 1, 1020 Vienna, Austria
e-mail: stephan.lutter@wu.ac.at

H. Lehmann (ed.), *Factor X*, Eco-Efficiency in Industry and Science 32,
https://doi.org/10.1007/978-3-319-50079-9_4

4.1 Introduction

The societal metabolism is using natural resources at an unprecedented level (WWF et al. 2012; UNEP 2011; Giljum et al. 2014). The world economy as a whole, progress in developing and emerging economies and the maintenance of quality of life achieved in the developed world require ever more resource throughput. Thereby, today's most pressing environmental problems, such as climate change, biodiversity loss, land cover conversion, etc. are caused by the overall growth of production and consumption (Lutter 2016). During the last decades, global production, trade and consumption patterns have changed significantly. The complexity of these changes, their international or even global scope, as well as the involvement of multi-dimensional cause-effect-impact relationships and time-lags make them very difficult to address (Giljum et al. 2009).

As described in earlier chapters, due to increased resources requirements, resource use and related environmental impacts have already surpassed several planetary boundaries (Steffen et al. 2015; Rockström et al. 2009). Strategies like, for instance, the Commission's "Roadmap to a Resource Efficient Europe" (European Commission 2011) or UNEP's Green Economy Strategy (UNEP 2011) recognise the need to reduce overall resource use and define resource efficiency as one major means to achieve it (see Chap. 2). The Commission, furthermore, clearly identifies the development of specific thresholds and targets for a sustainable resource use as imperative – on the local, national, regional and global level – as well as strategies and recommendations how to implement and evaluate them (European Commission 2011).

For a methodologically sound and comprehensive measurement of resource use and its environmental, economic and social impacts as well as for monitoring progress towards defined targets, appropriate indicators are needed (Giljum et al. 2009). It has been articulated by a large number of stakeholders, including policy makers, civil society and academia, that, in addition to the indicators currently in use, indicators taking into account indirect flows (i.e. resources "embodied" in traded goods and services) are needed (for instance, EREP 2014). The main point of critique is that territorial indicators, such as Domestic Material Consumption (DMC) for the material case or water abstraction for the water case, do not reflect the complete resource use picture. By dislocating resource-intensive industries to other countries and by substituting domestic extraction by imports, countries can apparently reduce their national resource consumption and improve their resource productivity (Lutter 2016). As supply chains of goods and services are increasingly organised on the global level, "footprint-type indicators" allow for complementing the territorial indicators. They help understand to what extent a country's economy, the environment, and the resource efficiency performance of goods and services are influenced by these global value chains (Lutter et al. 2016a).

In this book chapter we discuss the state of the art of data and indicator development for evidence-based policy making in the context of sustainable resource use, focussing on three different types of natural resources – raw materials, land, and water. We start with an overview of methodological options regarding environmental accounting frameworks as well as the calculation of footprint-type indicators and will show empirical trends of resource use measured by means of territorial as well as footprint-type indicators. We further analyse to what extent drivers of global resource use such as the European Union have managed to decouple their economic development from resource use. In a next step, we identify the methodological requirements and necessary next developments to ensure resource use is accounted for in a comprehensive as well as scientifically sound manner. Finally, we link the ongoing processes as well as empirical analyses to the question of how targets for sustainable resource use can be identified in both regards, to the methodological as well as institutional level.

4.2 Resource Accounting and Modelling Methodologies

4.2.1 Resource Accounting

Meaningful resource policies ideally build on solid official data on the appropriation of natural resources by societies. In general these data are accounted for on the national level. To ensure data quality as well as comparability, in the last decades different standards have been developed for the main resource categories.

In the case of *raw materials* the main methodology applied is material flow accounting and analysis (MFA). On the economy-wide level (EW-MFA) MFA builds on concepts of material and energy balancing, which were introduced over 40 years ago (for example, Kneese et al. 1970). The approach accounts for material flows in mass units, i.e. the turnover of mass during a defined period of time – usually one year – and distinguishes domestic extraction of resources, as well as physical imports and exports. Major efforts have been undertaken to harmonise methodological approaches developed by different research teams (Adriaanse et al. 1997; Matthews et al. 2000; Fischer-Kowalski et al. 2011; Krausmann et al. 2015), which resulted in the publishing of methodological guidebooks by EUROSTAT (2001, 2013) and the OECD (2007). While in many EU and OECD countries MFA is already part of the official environmental statistics reporting system, MFA data are also available for an increasing number of emerging and developing countries (Giljum et al. 2014; Schandl and West 2010; West and Schandl 2013; WU Vienna 2016). Material flow-based indicators (for instance, Domestic Material Consumption (DMC) in relation to GDP) have already been integrated in various European indicator sets such as the "Roadmap to a Resource Efficient Europe" (European Commission 2011). But in response to the increasing demand, significant efforts have been devoted to establish methodologies for calculating footprint-type indicators (see above; Wiedmann et al. 2015; Giljum et al. 2015; Lutter and Giljum 2014).

Water accounts have already been included in statistical systems on the national level (for example, DESTATIS 2013), and water data are collected by EUROSTAT and the OECD via a Joint Questionnaire on Inland Waters (e.g., OECD and Eurostat 2008).

Also, the United Nations set up a standard for compiling water accounts in their System of Environmental-Economic Accounting for Water (United Nations 2012). These statistics mostly represent the domestic uptake of water, but do also allow for accounting for water actually consumed (i.e. water lost throughout the production process via incorporation into the product, evaporation, transpiration by plants, etc.). Despite the existence of suitable accounting structures, data availability is rather unsatisfying, which can be explained, for instance, by a lack of statistical expertise, scattering of data collection among too many institutions, or a lack of political relevance.

In contrast to territorial water indicators, water footprinting indicators account for the water input along production and consumption processes. As in the cases of raw materials or land, to calculate such indicators on the macro level, in top-down approaches water abstraction quantities are allocated to specific abstracting economic sectors. Then the flow of "virtual water" is followed through the world economy in parallel to economic trade interrelationships. In order to obtain a comprehensive picture of human impact on the hydrological system, the water footprints can further be broken down to the watershed level and compared to locally prevailing water scarcity levels (for instance, Lutter et al. 2016b).

Land cover accounts are generally established from satellite images. For example, the EU Corine Land Cover (CLC) system, which is used by the European Environment Agency (EEA), is based on satellite images using a 100 m × 100 m grid (EEA 2006). Besides describing the geographical patterns of different land cover types across a country or region, such systems aim at investigating the ways and drivers of change over time. Also the Land Use and Cover Area frame survey (LUCAS) by EUROSTAT assesses the state and the dynamics of changes in land use and cover in the European Union based on a point survey (EUROSTAT 2010). Indicators on land use illustrate the land area required to produce a product or service (micro level) or all the goods produced or consumed in a region or country (macro level). Particularly valuable are indicators, which illustrate the change of land cover and land use from one year to another (e.g. expansion of built-up land at the cost of agricultural land) (EEA 2010a). Also in the area of land use, there is increasing interest in quantifying the land embodied in internationally traded products and in derived indicators such as the land footprint (Yu et al. 2013; Weinzettel et al. 2013; Bruckner et al. 2015; Kastner et al. 2012b).

4.2.2 Methodologies to Calculate Upstream Resource Flows and Footprints

In the era of globalisation, supply chains are increasingly organised on the international level, thus disconnecting the location of production from final consumption (Liu et al. 2013). Various local environmental and social impacts in – often developing – countries, which extract and process raw materials or manufacture products, are therefore often driven by final consumption in rich countries, e.g. in Europe or North America.

Over the past 15 years, the amount of resources embodied in international trade augmented even faster than global resource consumption (Giljum et al. 2014). Today, already more than a third of globally extracted materials serve as direct and indirect inputs for the production of internationally traded products (Wiedmann et al. 2015). This implies that traditional production-oriented, national perspectives to account for resource use might no longer be sufficient, as important drivers are being missed. Furthermore, territorial indicators of a country's resource use are not robust against the displacement of environmental burden through the outsourcing of resource-intensive production stages via international trade (see above). In contrast, consumption-based – or "footprint" – indicators consider the resources embodied in internationally traded products (Wiedmann 2016) and are thus independent from the location of production.

Mathis Wackernagel and William Rees were the first to coin the term "footprint" when introducing their concept of the Ecological Footprint (EF) in the early 1990s (Wackernagel and Rees 1996). The EF aims to illustrate the amount of nature required to sustain the consumption of goods and services by an individual person, a city or a country. The method was developed independently of previous achievements in life cycle assessment (LCA), accounting for a country's product in- and outflows and converting them into corresponding land areas. In past years, the scientific and public footprint discourse was dominated by the carbon footprint, which is more in line with the tradition of LCA. In the beginning, well-established process data and bottom-up methods from the field of LCA were applied. However, these approaches were quickly followed by studies applying multi-regional input–output (MRIO) analysis.

Over the past 15 years, footprint-type indicators have been developed for a number of resources, including materials, energy, water and land (for an overview, see Hoekstra and Wiedmann 2014; Wiedmann 2016; Giljum et al. 2013; Čuček et al. 2012). In the following, we focus on economy-wide resource flow accounting methods operating on the level of geographical or political entities, particularly countries. For economy-wide applications, it is crucial to account for trade flows and, to the extent possible, to consider regional differences in technologies and resource intensities. Therefore, data on international trade in commodities and services are combined with information on production processes and related resource inputs. Three variants of trade and footprint accounting models can be distinguished: bottom-up, top-down and mixed approaches. The presentation of the main features of these three variants is followed by a discussion on the use of monetary versus physical units in footprinting methods.

4.2.2.1 Bottom-Up Approaches

The first established bottom-up footprint accounting system were the National Footprint Accounts used to calculate the Ecological Footprint. They comprise detailed data on bilateral trade and domestic production expressed in quantities (e.g. tonnes), and 'apparent consumption' of a country is derived by calculating production plus imports minus exports. This form of economy-wide inventory analysis was later also applied for material, water and land footprints. The quantities of each product consumed in a country are multiplied with coefficients reflecting the related upstream resource use. These coefficients quantify the resources required along a product's supply chain.

Examples of bottom-up methods include the coefficients approach for the calculation of material footprints developed by the Wuppertal Institute (Dittrich et al. 2013; Schütz and Bringezu 2008; Dittrich et al. 2012), the Water Footprint methodology developed at University of Twente (Hoekstra et al. 2011), and the various bottom-up land footprint methods including, among others, Erb (2004), Würtenberger et al. (2006), Von Witzke and Noleppa (2010), Fader et al. (2011), (Kastner et al. 2014, 2011a, b, 2012a), Bringezu et al. (2012), and Mayer et al. (2014).

Coefficients for the water and land footprint of primary products are usually gathered from agricultural statistics or crop models. Coefficients for the case of the material footprint and for the water and land footprint of derived products are primarily obtained from LCAs. LCA studies are technically detailed, and they are often limited in their spatial specification. As a consequence, if applied on the global level, the sum of all footprints from bottom-up calculations will inevitably differ from the sum of all resource inputs and emission outputs.

One major advantage of bottom-up approaches is that they are able to consider country-specific characterization factors. This is of specific relevance due to the important role of regional climatic and soil conditions, among other factors, in the case of land and water footprints. As a consequence of the availability of detailed global agricultural statistics as well as spatially explicit climatic and agro-ecological information water and land footprints can be estimated on a high level of spatial detail (Hoekstra and Mekonnen 2012; Pfister and Bayer 2014).

4.2.2.2 Top-Down Approaches

In contrast to bottom-up methods, top-down approaches build on input-output analysis (IOA) which focuses on the economic structure of a country in the form of matrices that depict inter-industry flows, i.e. input-output tables. Each vector of an IO table can be interpreted as an inventory of production inputs. The environmental data on resource use linked as extensions to an IO table can be considered an inventory of environmental inputs such as raw materials and land area.

In general two main types are distinguished – single-region and multi-regional input-output (MRIO) models. Single-region IO models assume that imported products are produced with the same technology as domestic products. In the case of MRIO models, country IO tables are linked together via bilateral trade data (Wiedmann et al. 2011), by that means considering different technologies applied in each country. MRIO analysis allows tracking product value chains and related resource use along all life cycle stages of all products and services from resource extraction to final consumption, considering specific environmental intensities across countries (Tukker et al. 2013). Environmental MRIO analysis, thus, simply redistributes global resource inputs from the point of extraction by a specific sector to the point of consumption (e.g. final demand by households and government). The sum of all footprints hence is always equal to the sum of all resource inputs and emission outputs.

A major disadvantage of IOA is the limited commodity and regional detail determined by the sector and region definitions of an IO model. Further, inaccurate results are produced by the assumption of homogenous environmental characteristics of all products within a product group; for example, when products with widely diverging mass-value-ratios are aggregated into one product group (de Koning et al. 2015).

IOA is widely used for the case of material flow accounting.[1] An increasing number of studies apply MRIO models extended by water data.[2] Lenzen et al. (2013) incorporate water scarcity into an assessment of global virtual water flows between nations using input-output analysis. Lutter et al. (2016b) go one step further and use the MRIO database EXIOBASE[3] extended by comprehensive water data on the watershed level to quantify indirect water uses and scarcity impacts on the subnational level. The land footprint has been calculated using several MRIO databases[4] including EXIOBASE, GTAP,[5] WIOD[6] and Eora.[7] For a review see Bruckner et al. (2015). The LANDFLOW model is the only existing top-down accounting model purely based on physical data (Prieler et al. 2013).

4.2.2.3 Mixed Approaches

Mixed or hybrid approaches aim at exploiting the advantages from IOA in combination with physical trade accounts or process-based coefficients. Depending on the processing stage and data quality and availability, a differentiated approach for the calculation of footprint indicators for different products is applied. Typically, physical trade accounts and environmental characterisation factors for raw materials and products with a low level of processing are used and detailed environmental properties considered for those products where variation within a product group is particularly high. Processed commodities and finished goods with more complex production chains are treated with IOA, which allows considering all indirect effects and thus all upstream resource requirements (Ewing et al. 2012).

---

[1] Giljum et al. (2016c, 2015), Wiedmann et al. (2015), Bruckner et al. (2012b), Schaffartzik et al. (2014), Wiebe et al. (2012), Eisenmenger et al. (2016), Tukker et al. (2016).

[2] Daniels et al. (2011); Dietzenbacher and Velázquez (2007); Steen-Olsen et al. (2012); Ewing et al. (2012); Arto et al. (2016).

[3] www.exiobase.eu

[4] Wilting and Vringer (2009); Lugschitz et al. (2011); Bruckner et al. (2012a); Arto et al. (2012); Tukker et al. (2013); Yu et al. (2013); Bruckner et al. (2014).

[5] Global Trade Analysis Project (GTAP) (www.gtap.agecon.purdue.edu).

[6] World Input-Output Database (WIOD) (www.wiod.org).

[7] www.worldmrio.com

Hybrid models are increasingly applied in all areas of resource flow accounting, acknowledging their respective strengths and capabilities. The combination of top-down and bottom-up methods is realised in various ways. Hybrid approaches for the calculation of consumption-based material flow indicators integrate detailed statistics in mass units into monetary input-output tables, thereby creating mixed-unit IO tables (Buyny et al. 2009; Schoer et al. 2012; Schoer et al. 2013). Some land footprinting approaches apply input-output analysis to derive land intensity coefficients for highly processed products to complement the physical land flow accounts (Meier and Christen 2012; Meier et al. 2014). Another type of hybrid accounting applied to the analysis of global land flows sets up physical accounts to model crop flows and related embodied land flows from agricultural production to the first use stage (Weinzettel et al. 2011; Ewing et al. 2012; Steen-Olsen et al. 2012; Weinzettel et al. 2013, 2014). Also in the water area hybrid approaches have been applied in the past (Ewing et al. 2012). The resulting information, i.e. environmental extensions representing the intermediate consumption of primary products distinguished by region of origin, is then allocated to the monetary IO model, which is used to cover all supply chains from the first processing step onwards.

4.2.2.4 Calculations Based on Physical and/or Monetary Units

While all available bottom-up approaches are based on physical trade flows, top-down methods mostly apply monetary data. From a conceptual point of view, the use of monetary or physical trade flows for tracking embodied resource use or impacts could be considered equal, if all flows of product groups specified in an IO model within and between countries were fully homogenous in terms of prices. However, current data availability is still limited. Quality and resulting price differences in product flows result in differences of value-based and mass-based allocation methods (for instance, Bruckner et al. 2015).

In addition to the area of international trade, the operating unit also influences the allocation of environmental effects with regard to joint and by-products. LCAs draw on a variety of different allocation methods, for example, in relation to the products' economic value, mass or energy content. In contrast, allocation in IOA is determined by the unit of the IO table. It is often argued that monetary values, as applied in MRIO models, would represent the economic incentives, and thus the drivers behind production activities and ultimately environmental impacts. Therefore, value-based allocation is suggested as the basic rule also for LCA.

From a conceptual point of view, the different economy-wide footprinting methodologies should come up with the same results and can be considered 'equal', as they aim at a full life-cycle wide coverage of upstream effects, account for them in the form of tree or matrix structures, which eventually are equivalent and interchangeable, and consider trade relations with other countries in addition to domestic production. A convergence of methods may be expected in the future, given the steadily increasing data collection. However, the status quo is still limited; therefore the selection of suitable methodologies depends on the level of application.

4.3 Empirical Trends

The rapid development of methodologies to analyse human resource use and the increasing availability of data on national and international levels as described in the previous chapter has led to a significant increase in empirical studies on the use of natural resources for economic purposes. This section provides a selection of recently published results, illustrating the empirical trends of resource use on the global, European and national levels across the three resource categories of raw materials, water and land. Figure 4.1 starts with an overview of raw material use on the global level.

Figure 4.1 shows that global material extraction, which equals global material consumption, more than doubled in the past 30 years, from around 36 bill t in 1980 to almost 85 bill t in 2013, an overall growth of 132%. Two distinct phases of growth in material extraction can be identified. A first period of modest growth between 1980 and 2002, with an average annual growth rate of 1.8%. In this time period, a temporal stagnation of raw material extraction could be observed from 1990 to 1993, caused by the collapse of the former Soviet Union and the following recessions in many Eastern European and ex-USSR countries.

From 2003 onwards, growth of global material extraction increased significantly, with the average annual growth rate more than doubling to 4.1% in the period up to 2013. The observed growth in material extraction was mainly driven by the rise of emerging economies in Asia, most notably in China and India. While Asia held a share of 47% in global material extraction in 2002, this share increased to 61% in 2013.

Growth rates were unevenly distributed among the main material categories, as visible from Fig. 4.1. In particular, the extraction of industrial and construction min-

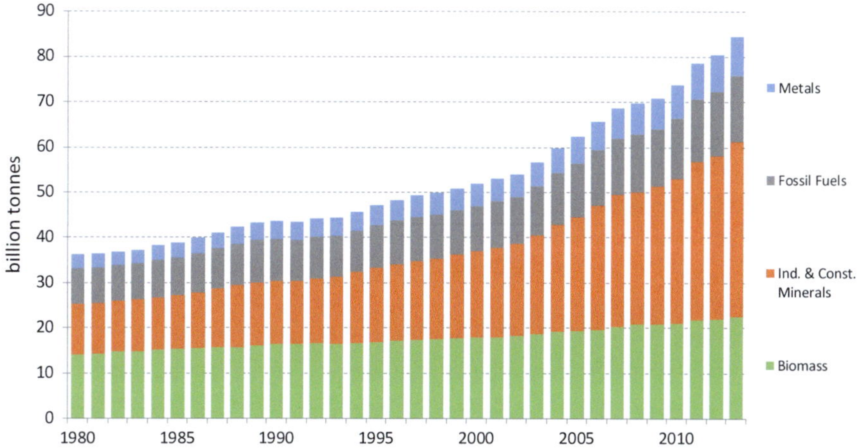

Fig. 4.1 Global extraction of raw materials, 1980–2013 (Source: www.materialflows.net; WU 2016)

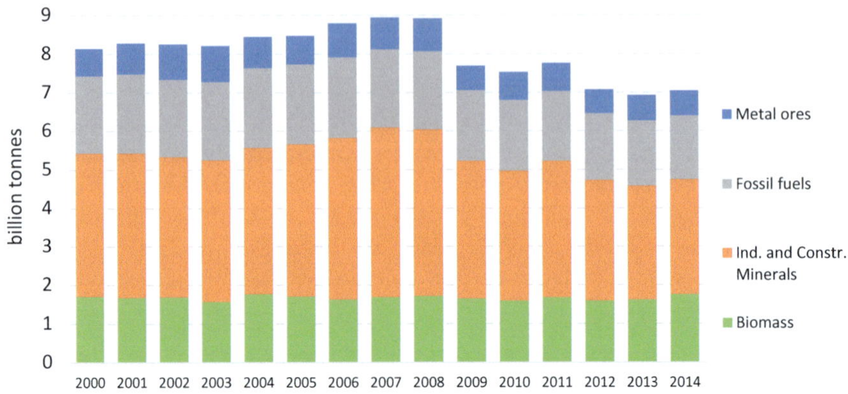

Fig. 4.2 Raw material consumption (RMC) of the EU-28, 2000–2014 (Source: EUROSTAT 2016)

erals increased significantly (by more than 240%), indicating the continued importance of this resource category for industrial development, in particular for the construction of housing, energy and transport infrastructure in emerging economies. Global extraction of metal ores increased by 183% and fossil fuels by 82% in the period of 33 years. Increases in biomass extraction amounted to 61%. The share of renewable resources in total resource extraction thus constantly decreased, from around 39% in 1980 to less than 27% in 2013.

In contrast to the global trend of continued growth, official statistics from EUROSTAT, the statistical office of the European Union, indicate that material consumption in the EU (measured from a footprint-perspective) has decreased in the past few years (Fig. 4.2).

Figure 4.2 illustrates a slightly rising trend of RMC between the years 2000 and 2008 – from around eight to around nine billion tonnes. However, with the start of the financial crisis and the following economic recession, RMC drastically declined, by more than 20% between 2008 and 2014. Looking more closely into the development of the different material categories, it can be observed that the category of 'industrial and construction minerals' was the category that accounted for most of the reduction. As EUROSTAT (2016) explains in its analysis of the trends, it was particularly the construction sector that was heavily hit by the economic crisis, with declines in gross value added of 11% from 2008 to 2010 and further 8% between 2010 and 2013. This stark decline in economic activities of the construction sector translated into a decrease in particular in the domestic extraction of construction minerals and thus leading to an overall decline in RMC. As Fig. 4.2 shows, the other material categories were hardly affected by the economic crisis.

With around 12.5 t RMC per capita in 2014, the EU still had a per capita consumption level above the ones observed for many other world regions. And this pattern is not only visible for materials, but also for other resource categories, such as water, land or carbon emissions (Tukker et al. 2016). The impacts of European consumption spread across the whole planet, as recent studies based on multi-regional input-output modelling have shown. In Fig. 4.3, one example regarding the global impacts of European consumption on the blue water footprint is illustrated.

Figure 4.3 shows the quantities of irrigation water from rivers, lakes and aquifers (so-called 'blue water') consumed in watersheds world-wide due to the production of products that are finally consumed in the EU-27, i.e. the EU blue water footprint. The darker the shading of the respective watershed, the higher the blue water consumption. The map illustrates that European consumption directly and indirectly (i.e. via international supply-chains) affects almost all watersheds world-wide. Large-scale irrigated agriculture in watersheds outside Europe take place in the Indus area in Pakistan and India with around 20 billion m^3 of irrigation water being spent for the cultivation of agricultural products that serve European consumption. Also in the Mississippi-related watersheds (mainly Mid-West US; ~9 billion m^3), the Nile (mainly Egypt and Sudan; ~5 billion m^3), and the Parana (mainly Argentina and Southern Brazil; ~4 billion m^3), significant amounts of irrigation water is being used, which forms part of the EU water footprint. Within Europe, the Danube River (mainly Hungary and Romania), the Guadalquivir (Southern Spain) and the Po (Northern Italy) contribute the largest irrigation water quantities for European final consumption.

Further advancement of methods and data have also allowed to perform assessments of the global land demand related to the final consumption in one country or world region. Figure 4.4 illustrates a recently published example on the global cropland footprint of the EU-28, i.e. cropland appropriated in various world regions to harvest products that directly and indirectly served EU-28 consumption.

The global land demand related to the EU's consumption of a large spectrum of products made of agricultural-based commodities slightly declined between 1995 and 2010, from around 170 million hectares to around 157 ha. Less cultivation of cropland within the EU is the main reason for the decreases in the EU cropland footprint, as net-trade, i.e. land embodied in imports and exports, remained stable across the time period. Agricultural commodities for livestock production held the highest share in all years, making up 49% of the EU's global cropland footprint in 1995 and 47% in 2010. Land use for production of commodities for food use also slightly declined, from 33% in 1995 to 31% in 2010. The main trend in the composition of the cropland footprint is a decreasing proportion of the use of food in favour of a higher proportion of the non-food sector, which increased its share from 14 to 18%. The increase in the crop-based non-food footprint is to a large extent due to vegetable oil and maize for production of biodiesel and ethanol respectively. One key aspect that distinguishes food from non-food use is the share of origin of the respective commodities. While the vast majority of cropland embodied in the EU's food consumption in 2010 stemmed from the EU itself (more than 86.5%), for non-food products only 35% were based on domestic land resources (Fischer et al. 2016). The rapidly growth of the non-food sector of Europe's bioeconomy is therefore significantly dependent on land resources outside Europe, in particular in Asia (Giljum et al. 2016b).

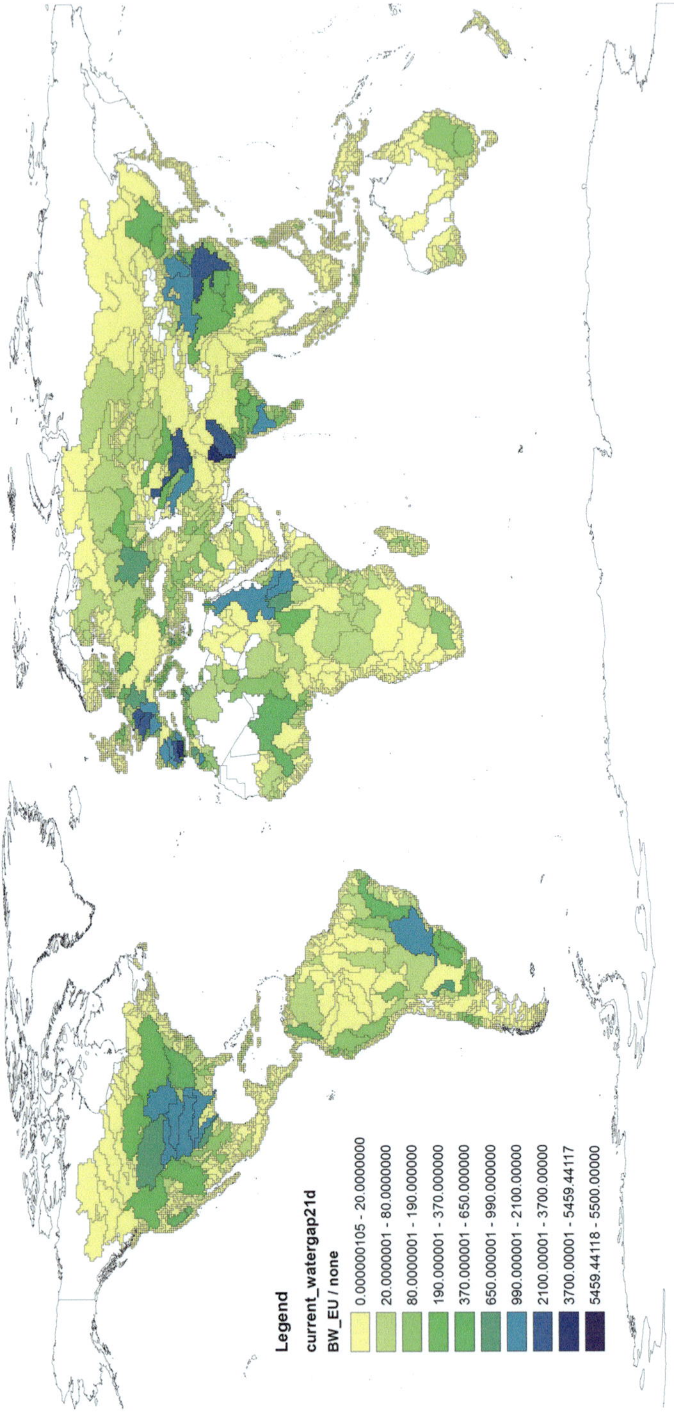

Fig. 4.3 EU-27 blue water footprint per watershed, 2007 (Source: Lutter et al. 2016b)

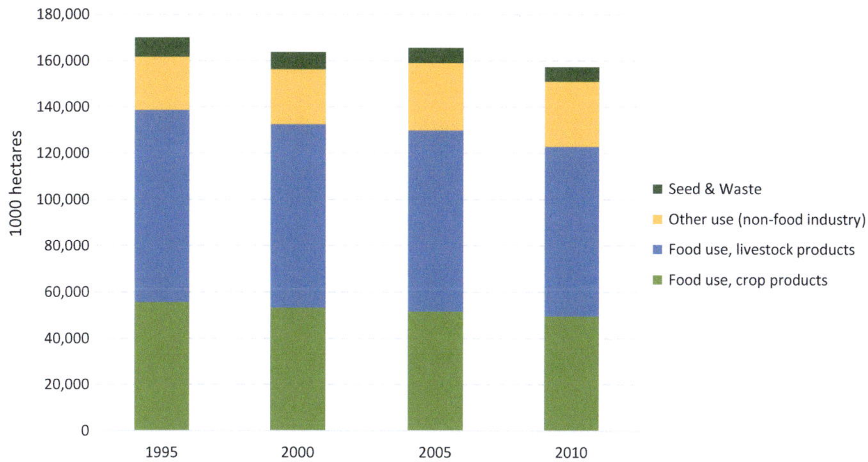

Fig. 4.4 EU-28 cropland footprint per major use categories, 1995–2010 (Source: Fischer et al. 2016)

4.4 Decoupling

Decoupling the use of natural resources and related environmental impacts from economic growth is a strategy that has been widely adopted in policy initiatives on the national, European and global level. Many regions and countries have developed strategies to substantially increase the material productivity of their economies and to reduce overall levels of natural resource use (Hinterberger et al. 2013). The EU, but also countries in other world regions, such as Japan and China, have started to implement policy frameworks and legislation that aim to accelerate resource efficiency improvements and guide investments into green sectors and technologies, supporting sustainable consumption and production (UNEP 2016).

In this chapter, we provide recent empirical evidence on decoupling trends on the global and European level, with a focus on the resource category of material flows. Figure 4.5 shows the global trends for GDP, population, material extraction and material intensity for the time period of 1980 to 2013. Material intensity is calculated by dividing material extraction by GDP in constant prices.

Figure 4.5 illustrates that between 1980 and 2013, global GDP grew by almost 150%. At the same time, global material extraction increased by around 130%. Across the whole time period, a slight relative decoupling could therefore be observed, i.e. average annual growth rates of constant GDP exceeding the average growth rates in material extraction.

However, the dynamic changed profoundly around the year 2002. Between 1980 and 2002, material intensity of the world economy decreased constantly, by an average of 1% per year, indicating a trend of relative decoupling. As a consequence, in the year 2002, around 20% less materials were required to produce one unit of global economic output compared to 1980. However, since 2002, the material inten-

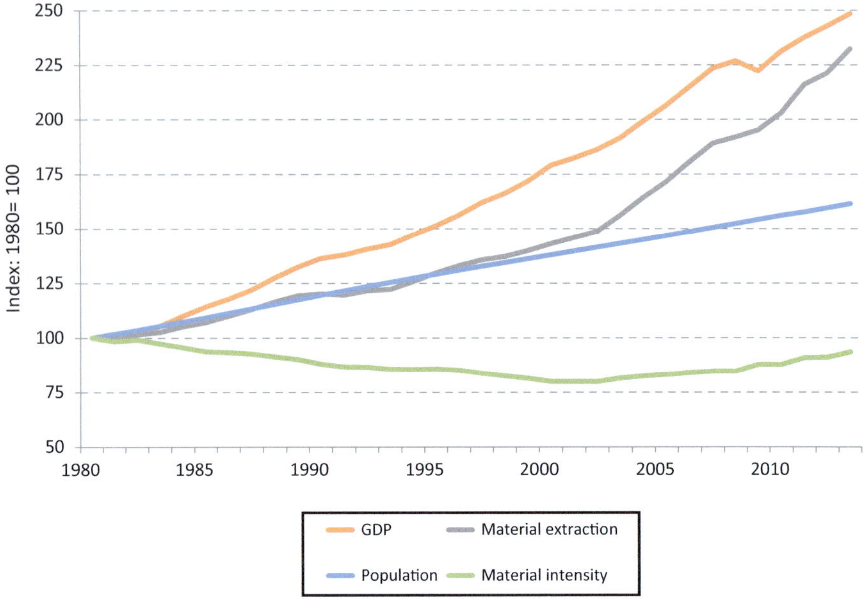

Fig. 4.5 Global trends of GDP, population, material extraction and material intensity (Source: www.materialflows.net; WU 2016)

sity curve moves upward, indicating that in the past 15 years, not even a relative decoupling has been achieved. In the period of 2002 to 2013, material intensity actually increased by an average factor of 1.4% per year. Due to the rapid expansion of material extraction activities in many world regions (see also Fig. 4.1 above), growth rates in extraction are thus above the global growth rates for GDP. Currently, the world economy is therefore on a path of re-materialisation and far away from any decoupling.

As with absolute levels of material consumption (see Fig. 4.2), different trends can be observed with regard to decoupling in the European Union. Figure 4.6 illustrates the trends, indexed to 1980, of the developments of GDP in volume terms, Raw Material Consumption (RMC) and related resource productivity, calculated as GDP over RMC.

EU-28 resource productivity, measured as GDP over RMC increased from around 1.5 Euro/kg in the year 2000 to 2.1 Euro/kg in 2014. However, as Fig. 4.6 illustrates, this development was not steady with a fundamental change happening at the beginning of the economic crisis. Resource productivity only increased modestly between the years 2000 and 2008, with an annual growth rate of around 1%. In contrast, from 2008 to 2014, resource productivity surged, with an average growth rate of more than 4%. This change can be explained by the developments of the underlying indicators. While RMC significantly declined after 2008, in particular due to a shrinking of the construction sector (see Fig. 4.2 above), GDP also declined between 2008 and 2009, however, started to get back on a growth path of

Fig. 4.6 Trends of GDP, RMC and resource productivity (GDP/RMC) for the European Union (Own calculations based on EUROSTAT 2016)

around 1% per annum by 2010. The slightly growing GDP together with a rapidly declining RMC caused the significant absolute decoupling that could be observed in the EU between 2008 and 2013. However, as Fig. 4.6 also illustrates, RMC was rising again in 2014, which might indicate a change in trends. Future data updates and analyses will reveal whether EU's absolute decoupling was only a temporal phenomenon linked to the financial crisis, or whether longer-term structural changes in the EU economy took place that will continue the trend of rising resource productivities.

4.5 Requirements and Outlook

As mentioned earlier, meaningful resource policies ideally build on solid official data on the appropriation of natural resources by societies. Such official data are compiled following internationally standardised formats. Some of the accounting methods for the different resource categories described above already exist in an internationally standardised format – for instance, material flow-based indicators on the product and the country level, the Carbon Footprint or Kyoto inventories of GHGs. The accounting method for other categories, in particular for water and land are currently under development. However, the measurement systems covering simultaneously various types of natural resources have been developed separately. As a consequence, further methodological harmonisation is still required, in order to improve the comparability of the results. In addition, aspects such as the definition of common system boundaries or the improvement of data collection have to be tackled (Giljum et al. 2016a).

As explained above, one important element for the calculation of consumption-based indicators are input-output tables to which environmental data are linked as so-called environmental extensions. These tables generally provide a high level of detail for the manufacturing and service sectors, which contribute the highest shares to GDP, but sectors more relevant for resource flows are often highly aggregated. A more detailed disaggregation of resource intensive sectors would help avoid errors due to aggregation of inhomogeneous products. Such a disaggregation should be combined with the integration of physical data into the MRIO system, as physical flows through the economic system can be reproduced better by using physical instead of monetary data (a so-called "mix-unit MRIO"). Finally, there is a strong need to improve the availability of input-output tables – especially in non-OCED countries, where a significant share of global material extraction takes place, and for a greater number of years including the extension of time series to the current year, or even "forecasting" data and indicators into the future.

When discussing the application of global multi-regional input-output models (GMRIO) in policy making, two major issues are at stake: (1) how can existing GMRIO databases be harmonized and become more aligned and (2) how can the acceptance of GMRIOs be increased among the statistical community? Both aspects are interconnected as the different GMRIOs available at the moment, produce – to a certain extent – different results for e.g. calculations of environmental footprints, which leads to problems in acceptance of GMRIOs.

Currently, most GMRIOs lack a 'statistical stamp', as they are produced by scientists. The only GMRIO that goes beyond this is the ICIO database of OECD which has the drawback of being far too aggregated to be of use for calculating water, materials, land and emissions footprints. However, the OECD ICIO database could be used as a starting point and then further detailed using the detailing procedures as developed particularly for GRMIOs as EXIOBASE or for Eora (see Sect. 4.2.2.2). For land and water footprint analyses, this for instance would imply further detailing the agricultural and food processing sectors. For material footprint analysis, the materials extraction sector would need detail. This would result in an officially accepted database readily available for any NSI or practitioner to be used.

With regard to *raw materials*, one important area of improvement is the availability of comprehensive and solid time series of material extraction data. While EUROSTAT only recently made material accounting obligatory, resulting in more comprehensive datasets provided by EU Member States at least the most recent years, longer time series are missing for many countries, in particular beyond the group of OECD countries. Hence, indicator calculations often depend on academic sources providing more extensive global databases compiled from various statistical datasets such as the database of WU Vienna (2016). There are ongoing efforts to further harmonize data and come up with one consistent worldwide dataset in the context of an ongoing project of the UNEP International Resource Panel (UNEP 2016).

Apart from material extraction data, also data on physical material trade are of high relevance for the calculation of footprint-type indicators – especially when applying hybrid approaches, but also in the context of linking national input-output tables to set up a multi-regional framework. The UN COMTRADE database (United Nations 2016) provides data from 1962 up to the most recent year, applying high credibility and transparency standards; however, data in physical units are incomplete and have to be estimated e.g. via average prices. Hence, improving the data situation as well as further research on the completion of patchy physical trade data is required (Lutter 2016).

In the area of *land use*, a major strength of top-down approaches is their global consistency, i.e. total global land use exactly equals the aggregated land footprint of all countries and regions worldwide. Moreover, linking a monetary MRIO model to the physical accounting approach allows for a detailed analysis of the land footprint for non-food products, which is the use category showing the highest growth rates and thus with particular policy relevance. This methodology therefore is suited to inform recent policy strategies focusing on the non-food use of biomass, such as bioeconomy strategies and initiatives launched on the national (German) and international (EU) level.

To improve the meaningfulness of such land accounting approaches, physical allocation models should be more detailed and apply a highly transparent supply and use structure. Moreover, the estimation of grassland utilization and footprints could be advanced by reliable estimates of extents of grassland used for grazing ruminant livestock, and more detailed reporting of the use of feed and fodder crops for specific livestock categories could replace the current estimation methods.

With regard to statistical requirements, FAO data on the production of fodder crops (e.g. grasses, forages and silages) have considerable improvement potential concerning completeness and robustness through the provision of consistent definitions of the physical resources involved (i.e. arable land or pasture land). In general, data from the FAO are the only option to realise global calculations, despite their incompleteness and poor quality in a number of areas. Most importantly, differences in classification between national approaches and the FAO tend to cause a shift especially between the categories of cropland and grassland. Hence, applying an international accounting approach using international data sources entails the risk of discrepancies with national statistical sources. This problem can be overcome by replacing data from FAO for a specific country of interest with data from official national trade and agricultural statistics, thus building a single-country national accounts consistent (SNAC) footprint accounting model. This is a highly recommendable step when implementing any kind of top-down trade and footprint model for monitoring or target setting purposes on the national level (Bruckner et al. 2015).

As explained earlier, a number of studies already aimed at applying approaches of global coverage to calculate footprint-type indicators in the area of *water appropriation*. Approaches like the one applied by Lutter et al. (2016b) used a very high level of detail in the area of agricultural production and managed to break down the results of agricultural water consumption data embodied in international trade to the watershed level. However, special attention should be given to a number of issues:

1. Data on water withdrawal and consumption on a high level of sectoral disaggregation only exists for the agricultural sector. Hence, there is the need to improve estimations of water consumption in industrial processes. Hydropower production has an especially large contribution (Mekonnen and Hoekstra 2011; Pfister et al. 2011).
2. With regard to water statistics, mainly modelled data are available in a level of detail appropriate for W-MRIO application. Ideally, consumption-based indicator calculations would be based on a comprehensive official dataset on water withdrawal and consumption on a detailed sector level. The foundations for this data compilation effort have been laid e.g. by the efforts of EUROSTAT to set up a system of water accounts (EUROSTAT 2014) which is itself based on the SEEA Water (United Nations 2012) or. Such a dataset would gain even more value if expanded by information on the precise – sub-national – spatial distribution of water requirements per industrial activity.
3. The quality of blue water scarcity assessments would be considerably improved with temporal disaggregation into monthly water consumption. Existing approaches include the criticality dimension by assuming the number of months experiencing at least moderate scarcity in each year (Lutter et al. 2016b; Hoekstra et al. 2012) which helps to identify hotspots by combining blue water scarcity and consumption in a single number for each watershed.

In Europe, two policies are of particular relevance with regard to water accounting – the Water Framework Directive (European Parliament and Council 2000) and the "Blueprint to Safeguard Europe's Water Resources" (European Commission 2012). Both focus on the watershed as the main management unit as well as on the measurement not only of water quality but also of quantities. The Blueprint also mentions the need to consider the global aspects of water use and water management related to European production and consumption patterns. Despite these two prominent policies, water statistics do not obtain too much attention in the European statistical system. Improving the collection of relevant water data and by that means increasing the availability and quality would be a prerequisite for comprehensive water use analyses.

4.6 Targets for Sustainable Resource Use?

The development of sound accounting frameworks for direct and indirect resource use is of specific relevance when it comes to the monitoring of progress towards an efficient, environmentally safe and socially fair use of global resources (Lutter 2016). However, it is argued by many researchers and policy makers that simply the measurement of resource use will not result in political or economic action towards an overall reduction. In contrast, targets set a clear orientation, providing concrete guidance and help to prioritise actions to achieve a policy objective. If properly enforced and supported by an appropriate mix of policy measures, they can be a powerful approach to addressing environmental issues (BIO IS et al. 2012). While the political discussion about specific targets on material resource use still appears to be somewhat languorous, scientific work has been advancing during the last decades.

In the case of *raw materials*, an important milestone was the publication of the Club of Rome report "Limits to growth" by Denis Meadows (Meadows and Club of Rome 1972), arguing that with continued trends of increasing raw material use humanity would soon run out of raw materials, causing at the same time irrevocable environmental damages. The discussion gained momentum through Schmidt-Bleek and colleagues from the Wuppertal Institute (for instance, von Weizsäcker et al. 1995, 1998; Schmidt-Bleek 1994, 1997) proposing a decrease in raw material use in industrialised countries by a factor of four, five or even up to ten, to stay within planetary boundaries and at the same time allow emerging economies to catch up with the "developed" world. In the following years, with regard to the identification of material use targets different indicators have been discussed – most prominently the DMC (BIO IS et al. 2012; Dittrich et al. 2012), RMC (Bringezu 2015), and TMC (Total Material Consumption; Bringezu 2011, 2015; Lettenmeier et al. 2014). To comprehensively calculate the DMC and RMC, methods and data exist. In contrast, as quantities of unused extraction are usually estimated and only available as averages for specific materials or countries, in the case of the TMC data availability is scarce (Lutter 2016).

As in the material case, so far specific *water use targets* have been discussed rather on the scientific level. This can be explained partly by the poor international data situation but also by the general water abundance within Europe. Having said that, in 2009, the European Environment Agency (EEA) proposed a target for the Water Exploitation Index (WEI), comparing available water resources with water abstractions (EEA 2010b). The EEA recommended to reduce the WEI below 20% until 2020 and below 10% until 2050 (EEA 2009). A Water Exploitation Index between 10 and 20% is considered as *"low water stress"*, WEI between 20 and 40% is indicating *"stress on water resources"* and above 40% represents *"severe water stress"*.

However, such recommendations tend to lose their relevance as soon as advances are made on the scientific level. These range from identifying the main criteria for target setting to designing a solid and comprehensive water accounting framework (United Nations 2012; Lutter forthcoming; EUROSTAT 2014). But especially with regard to the first aspect, coherence between the approaches of different stakeholders in this area is needed to come up with one approach to identify targets for sustainable water use, which can then be taken up by policy making.

Land area is increasingly exposed to growing demand by different types and intensities of land use. Monitoring and targeting domestic land demand requires a balanced share between different land use categories. Thereby, those land use categories should be preserved which have the smallest effects on land cover – e.g. wilderness areas but also forests (BIO IS et al. 2012). Conversely, the expansion of agricultural land should be strongly limited (compare Rockström et al. 2009 for their recommendation on limiting the percentage of global land cover converted to cropland to 15%).

On the global level, in addition to domestic land use aspects, distributional issues have to be considered. For instance, Europe is characterised by favourable biogeographic conditions (climate, soils) and a high availability of productive land per capita. As a consequence, Europe's demand for foreign land area (in particular productive land such as agricultural land) should be balanced. In relation to the request for stabilising biomass use and limiting the EU demand for limited land resources to a fair share, the target for zero net-demand of foreign land is considered well-balanced (BIO IS et al. 2012).

Setting targets can be considered a process of balancing different interests and perspectives. The cases of raw materials, water and land show that policy makers often hesitate to opt for specific targets, before science has proposed solid and agreed values. However, as data, accounting methods and resulting analyses advance and the environmental and social impacts related to resource use increase, the identification of reachable but courageous targets is seen as a necessary step towards concerted and effective action for an overall reduction of the use of natural resources.

References

Adriaanse A, Bringezu S, Hamond A, Moriguchi Y, Rodenburg E, Rogich D, Schütz H (1997) Resource flows: the Material Base of industrial economies. World Resource Institute, Washington

Arto I, Genty A, Rueda-Cantuche JM, Villanueva A, Andreoni V (2012) Global resources use and pollution:Vol. I, production, consumption and trade (1995–2008), EUR 25462. European Commission Joint Research Centre (Institute for prospective technological studies), Luxembourg

Arto I, Andreoni V, Rueda-Cantuche J (2016) Global use of water resources: a multiregional analysis of water use, water footprint and water trade balance. Water Resources and Economics 15:1–14

BIO IS, Institute for Social Ecology, Sustainable Europe Research Institute (SERI) (2012) Assessment of resource efficiency indicators and targets. Final report prepared for the European Commission. DG Environment, Brussels

Bringezu S (2011) Key elements for economy-wide sustainable resource management. Responsabilité & Environnement 61:78–87

Bringezu S (2015) Possible target corridor for sustainable use of global material resources. Resources 4(1):25–54

Bringezu S, O'Brien M, Schütz H (2012) Beyond biofuels: assessing global land use for domestic consumption of biomass: a conceptual and empirical contribution to sustainable management of global resources. Land Use Policy 29(1):224–232

Bruckner M, Giljum S, Lutz C, Wiebe KS (2012a) Materials embodied in international trade – global material extraction and consumption between 1995 and 2005. Glob Environ Chang 22(3):568–576

Bruckner M, Lugschitz B, Giljum S (2012b) Turkey's virtual land demand. A study on the virtual land embodied in Turkey's imports and exports of agricultural products. Sustainable Europe Research Institute (SERI), Vienna

Bruckner M, de Schutter L, Martinez A, Giljum S (2014) Consumption-based accounts of land use related greenhouse gas emissions for the European Union. In: BIO IS (ed) Resource efficiency policies for land use related climate mitigation. Final report prepared for the European Commission, DG CLIMA. Bio Intelligence Service at Deloitte, Paris

Bruckner M, Fischer G, Tramberend S, Giljum S (2015) Measuring telecouplings in the global land system: a review and comparative evaluation of land footprint accounting methods. Ecol Econ 114:11–21

Buyny Š, Klink S, Lauber U (2009) Verbesserung von Rohstoffproduktivität und Ressourcenschonung. Weiterentwicklung des direkten Materialinputindikators. Statistisches Bundesamt, Wiesbaden

Čuček L, Klemeš JJ, Kravanja Z (2012) A review of footprint analysis tools for monitoring impacts on sustainability. J Clean Prod 34:9–20

Daniels PL, Lenzen M, Kenway SJ (2011) The ins and outs of water use–a review of multi-region input–output analysis and water footprints for regional sustainability analysis and policy. Econ Syst Res 23(4):353–370

de Koning A, Bruckner M, Lutter S, Wood R, Stadler K, Tukker A (2015) Effect of aggregation and disaggregation on embodied material use of products in input–output analysis. Ecol Econ 116:289–299

DESTATIS (2013) Umweltnutzung und Wirtschaft. Bericht zu den Umweltökonomischen Gesamtrechnungen. Statistisches Bundesamt, Wiesbaden

Dietzenbacher E, Velázquez E (2007) Analysing Andalusian virtual water trade in an input-output framework. Reg Stud 41(2):185–196

Dittrich M, Giljum S, Lutter S, Polzin C (2012) Green economies around the world? The role of resource use for development and the environment, Vienna/Heidelberg

Dittrich M, Giljum S, Lutter S, Polzin C (2013) Aktualisierung von nationalen und internationalen Ressourcenkennzahlen, 07/2013. Umweltbundesamt, Dessau

EEA (2006) Land accounts for Europe 1990–2000. Towards integrated land and ecosystem accounting, 11/2006. European Environment Agency, Copenhagen

EEA (2009) Water resources across Europe — confronting water scarcity and drought, 2/2009. European Environment Agency, Copenhagen

EEA (2010a) The European environment. State and outlook 2010. Land use. European Environment Agency, Copenhagen

EEA (2010b) WEI – water exploitation index. European Environment Agency, Copenhagen

Eisenmenger N, Wiedenhofer D, Schaffartzik A, Giljum S, Bruckner M, Schandl H, Wiedmann TO, Lenzen M, Tukker A, Koning A (2016) Consumption-based material flow indicators — comparing six ways of calculating the Austrian raw material consumption providing six results. Ecol Econ 128:177–186

Erb KH (2004) Actual land demand of Austria 1926–2000: a variation on ecological footprint assessments. Land Use Policy 21(3):247–259

EREP (2014) Manifesto & Policy Recommendations. European Resource Efficiency Platform, Brussels

European Commission (2011) Roadmap to a Resource Efficient Europe. COM(2011) 571 final. European Commission, Brussels

European Commission (2012) A blueprint to safeguard Europe‘s water resources. European Commission, Brussels

European Parliament and Council (2000) Directive 2000/60/EC of the European Parliament and of the Council of 23 October 2000 establishing a framework for community action in the field of water policy

EUROSTAT (2001) Economy-wide material flow accounts and derived indicators. A methodological guide. Statistical Office of the European Union, Luxembourg

EUROSTAT (2010) LUCAS – a multi-purpose land use survey. Statistical Office of the European Communities, Luxembourg

EUROSTAT (2013) Economy-wide material flow accounts (EW-MFA). Compilation guide 2013. Statistical Office of the European Communities, Luxembourg
EUROSTAT. 2014. Personal communication with EUROSTAT, draft manual for physical water flow accounts (PWFA). Luxembourg
EUROSTAT (2016) Material flow accounts – flows in raw material equivalents. Available at: http://ec.europa.eu/eurostat/statistics-explained/index.php/Material_flow_accounts_-_flows_in_raw_material_equivalents. Statistical Office of the European Communities, Luxembourg
Ewing BR, Hawkins TR, Wiedmann TO, Galli A, Ertug Ercin A, Weinzettel J, Steen-Olsen K (2012) Integrating ecological and water footprint accounting in a multi-regional input–output framework. Ecol Indic 23(0):1–8
Fader M, Gerten D, Thammer M, Heinke J, Lotze-Campen H, Lucht W, Cramer W (2011) Internal and external green-blue agricultural water footprints of nations, and related water and land savings through trade. Hydrol Earth Syst Sci 15(5):1641–1660
Fischer G, Tramberend S, Bruckner M, Lieber M (2016) Quantifying the land footprint of Germany and the EU using a hybrid accounting model. German Federal Environment Agency (Umweltbundesamt), Berlin
Fischer-Kowalski M, Krausmann F, Giljum S, Lutter S, Mayer A, Bringezu S, Moriguchi Y, Schütz H, Schandl H, Weisz H (2011) Methodology and indicators of economy-wide material flow accounting. J Ind Ecol 15(6):855–876
Giljum S, Hinterberger F, Lutter S, Polzin C (2009) How to measure Europe's resource use. An analysis for Friends of the Earth Europe. Sustainable Europe Research Institute, Vienna
Giljum S, Lutter S, Bruckner M, Aparcana S (2013) State of play of national consumption-based indicators. A review and evaluation of available methods and data to calculate footprint-type (consumption-based) indicators for materials, water, land and carbon. Report for DG Environment of the European Commission. Sustainable Europe Research Institute (SERI), Vienna
Giljum S, Dittrich M, Lieber M, Lutter S (2014) Global patterns of material flows and their socio-economic and environmental implications: a MFA study on all countries world-wide from 1980 to 2009. Resources 3(1):319–339
Giljum S, Bruckner M, Martinez A (2015) Material footprint assessment in a global input-output framework. J Ind Ecol 19(5):792–804
Giljum S, Bruckner M, Gözet B, de Schutter L (2016a) Land under pressure. Global impacts of the EU bioeconomy. Friends of the Earth Europe, Brussels
Giljum S, Lutter S, Bruckner M (2016b) Measuring natural resource use from the micro to the macro level. In: Shmelev S (ed) Green economy reader. Lectures in ecological economics and sustainability. Springer, Berlin
Giljum S, Wieland H, Lutter S, Bruckner M, Wood R, Tukker A, Stadler K (2016c) Identifying priority areas for European resource policies: a MRIO-based material footprint assessment. Journal of Economic Structures 5(17):1–24
Hinterberger F, Giljum S, Omann I, Polsterer N, Stocker A, Burrell L, Campregher C, Fuchs D, Hartwig F (2013) green growth: from labour to resource productivity. Best practice examples, initiatives and policy options. Agence France de Developpment, Paris
Hoekstra AY, Mekonnen MM (2012) The water footprint of humanity. Proc Natl Acad Sci 109(9):3232–3237
Hoekstra AY, Wiedmann TO (2014) Humanity's unsustainable environmental footprint. Science 344(6188):1114–1117
Hoekstra AY, Chapagain AK, Aldaya MM, Mekonnen MM (2011) The water footprint assessment manual: setting the global standard. Earthscan, London
Hoekstra AY, Mekonnen MM, Chapagain AK, Mathews RE, Richter BD (2012) Global monthly water scarcity: blue water footprints versus blue water availability. PLoS One 7(2):e32688
Kastner T, Erb K-H, Nonhebel S (2011a) International wood trade and forest change: a global analysis. Glob Environ Chang 21(3):947–956

Kastner T, Kastner M, Nonhebel S (2011b) Tracing distant environmental impacts of agricultural products from a consumer perspective. Ecol Econ 70(6):1032–1040
Kastner T, Ibarrola Rivas MJ, Koch W, Nonhebel S (2012a) Global changes in diets and the consequences for land requirements for food. PNAS 109
Kastner T, Rivas MJI, Koch W, Nonhebel S (2012b) Global changes in diets and the consequences for land requirements for food. Proc Natl Acad Sci 109(18):6868–6872
Kastner T, Erb K-H, Haberl H (2014) Rapid growth in agricultural trade: effects on global area efficiency and the role of management. Environ Res Lett 9(3):034015
Kneese AV, Ayres RU, d'Arge RC (1970) Economics and the environment: a material balance approach. John Hopkins Press, Baltimore/London
Krausmann F, Weisz H, Eisenmenger N, Schütz H, Haas W, Schaffartzik A (2015) Economy-wide material flow accounting. Introduction and guide. Institute for Social Ecology, Vienna
Lenzen M, Moran D, Bhaduri A, Kanemoto K, Bekchanov M, Geschke A, Foran B (2013) International trade of scarce water. Ecol Econ 94:78–85
Lettenmeier M, Liedtke C, Rohn H (2014) Eight tons of material footprint—suggestion for a resource cap for household consumption in Finland. Resources 3(3):488–515
Liu J, Hull V, Batistella M, DeFries R, Dietz T, Fu F, Hertel TW, Izaurralde RC, Lambin EF, Li S (2013) Framing sustainability in a telecoupled world. Ecol Soc 18(2)
Lugschitz B, Bruckner M, Giljum S (2011) Europe's global land demand. A study on the actual land embodied in European imports and exports of agricultural and forestry products. Sustainble Europe Research Institute, Vienna
Lutter S (2016) Quantifying sustainable resource use within planetary boundariesthesis, Institute for Sustainable Economic Development, University of Natural Resources and Life Sciences (BOKU), Vienna, Austria
Lutter S (forthcoming) Developing a comprehensive framework to set targets for sustainable global water use. Ecol Indicators
Lutter S, Giljum S (2014) Demand-based measures of material flows. A review and comparative assessment of existing calculation methods and data options, ENV/EPOC/WPEI (2014), vol 1. OECD/Working Party on Environmental Information, Paris
Lutter S, Giljum S, Bruckner M (2016a) A review and comparative assessment of existing approaches to calculate material footprints. Ecol Econ 127:1–10
Lutter S, Pfister S, Giljum S, Wieland HP, Mutel C (2016b) Spatially explicit assessment of water embodied in European trade: a product-level multi-regional input-output analysis. Glob Environ Chang 38:171–182
Matthews E, Bringezu S, Fischer-Kowalski M, Huetller W, Kleijn R, Moriguchi Y, Ottke C, Rodenburg E, Rogich D, Schandl H, Schuetz H, van der Voet E, Weisz H (2000) The weight of nations. Material outflows from industrial economies. World Resources Institute, Washington
Mayer H, Flachmann C, Wachowiak M, Fehrentz P (2014) Nachhaltiger Konsum: Entwicklung eines deutschen Indikatorensatzes als Beitrag zu einer thematischen Erweiterung der deutschen Nachhaltigkeitsstrategie. Statistisches Bundesamt, Wiesbaden
Meadows DH, Club of Rome (1972) The limits to growth: a report for the Club of Rome's project on the predicament of mankind. Universe Books, New York
Meier T, Christen O (2012) Environmental impacts of dietary recommendations and dietary styles: Germany as an example. Environ Sci Technol 47(2):877–888
Meier T, Christen O, Semler E, Jahreis G, Voget-Kleschin L, Schrode A, Artmann M (2014) Balancing virtual land imports by a shift in the diet. Using a land balance approach to assess the sustainability of food consumption. Germany as an example. Appetite 74:20–34
Mekonnen MM, Hoekstra AY (2011) The water footprint of electricity from hydropower, 51. UNESCO-IHE, Delft
OECD (2007) Measuring material flows and resource productivity. The OECD guide ENV/EPOC/SE(2006)1/REV3. Environment Directorate. Organisation for Economic Co-operation and Development, Paris
OECD and Eurostat (2008) Joint questionnaire inland waters. OECD, Eurostat

Pfister S, Bayer P (2014) Monthly water stress: spatially and temporally explicit consumptive water footprint of global crop production. J Clean Prod 73:52–62
Pfister S, Saner D, Koehler A (2011) The environmental relevance of freshwater consumption in global power production. Int J Life Cycle Assess 16(6):580–591
Prieler S, Fischer G, Hizsnyik E, van Velthuizen H (2013) The LANDFLOW model: technical description of the LANDFLOW model. Annex A–H and chapters 3–4. In: VITO et al (eds) The impact of EU consumption on deforestation: comprehensive analysis of the impact of EU consumption on deforestation. DG ENV technical report – 2013 – 063. European Commission, Brussels
Rockström J, Steffen W, Noone K, Persson A, Chapin FS, Lambin EF, Lenton TM, Scheffer M, Folke C, Schellnhuber HJ, Nykvist B, de Wit CA, Hughes T, van der Leeuw S, Rodhe H, Sorlin S, Snyder PK, Costanza R, Svedin U, Falkenmark M, Karlberg L, Corell RW, Fabry VJ, Hansen J, Walker B, Liverman D, Richardson K, Crutzen P, Foley JA (2009) A safe operating space for humanity. Nature 461(7263):472
Schaffartzik A, Eisenmenger N, Krausmann F, Weisz H (2014) Consumption-based material flow accounting. Austrian trade and consumption in raw material equivalents 1995–2007. J Ind Ecol 18(1):102–112
Schandl H, West J (2010) Resource use and resource efficiency in the Asia–Pacific region. Glob Environ Chang 20(4):636–647
Schmidt-Bleek F (1994) Wie viel Umwelt braucht der Mensch ? MIPS – das Maß für ökologisches Wirtschaften. Birkhauser, Berlin/Basel
Schmidt-Bleek F (1997) MIPS and factor 10 for a sustainable and profitable economy. Wuppertal Institute, Wuppertal
Schoer K, Weinzettel J, Kovanda J, Giegrich J, Lauwigi C (2012) Raw material consumption of the European Union–concept, calculation method, and results. Environ Sci Technol 46(16):8903–8909
Schoer K, Wood R, Arto I, Weinzettel J (2013) Estimating raw material equivalents on a macro-level: comparison of multi-regional input–output analysis and hybrid LCI-IO. Environ Sci Technol 47(24):14282–14289
Schütz H, Bringezu S (2008) Resource consumption of Germany – indicators and definitions. 08/08. Federal Environment Agency, Dessau
Steen-Olsen K, Weinzettel J, Cranston G, Ercin AE, Hertwich EG (2012) Carbon, land, and water footprint accounts for the European Union: consumption, production, and displacements through international trade. Environ Sci Technol 46(20):10883–10891
Steffen W, Richardson K, Rockström J, Cornell SE, Fetzer I, Bennett EM, Biggs R, Carpenter SR, de Vries W, de Wit CA (2015) Planetary boundaries: guiding human development on a changing planet. Science 347:1259855
Tukker A, de Koning A, Wood R, Hawkins T, Lutter S, Acosta J, Rueda Cantuche JM, Bouwmeester M, Oosterhaven J, Drosdowski T (2013) EXIOPOL–development and illustratvie analyses of detailed global MR EE SUT/IOT. Econ Syst Res 25(1):50–70
Tukker A, Bulavskaya T, Giljum S, de Koning A, Lutter S, Simas M, Stadler K, Wood R (2016) Environmental and resource footprints in a global context: Europe's structural deficit in resource endowments. Glob Environ Chang 40:171–181
UNEP (2011) Towards a green economy: pathways to sustainable development and poverty eradication. UNEP, Nairobi
UNEP (2016) Global material flows and resource productivity. An assessment study of the UNEP international resource panel. Schandl H, Fischer-Kowalski M, West J, Giljum S, Dittrich M, Eisenmenger N, Geschke A, Lieber M, Wieland HP, Schaffartzik A, Krausmann F, Gierlinger S, Hosking K, Lenzen M, Tanikawa H, Miatto A, Fishman T. United Nations Environment Programme, Paris
United Nations (2012) SEEA-W. System of environmental-economic accounting for water. United Nations Statistics Division, New York
United Nations (2016) UN comtrade database: UN Trade Statistics Branch, Statistics Division

von Weizsäcker EU, Lovins A, Lovins H (1995) Faktor Vier. Doppelter Wohlstand – halbierter Naturverbrauch. Droemer Knaur, München

Von Weizsäcker EU, Weizsäcker EU, Lovins AB, Lovins LH (1998) Factor four: doubling wealth-halving resource use: the new report to the Club of Rome. Earthscan, London

Von Witzke H, Noleppa S (2010) EU agricultural production and trade: can more efficiency prevent increasing 'land-grabbing'outside of Europe? *Study commissioned by OPERA*

Wackernagel M, Rees WE (1996) Our ecological footprint – reducing human impact on the earth, the new catalyst bioregional series. New Society Publishers, Gabriola Island

Weinzettel J, Steen-Olson K, Galli A, Cranston G, Ercin E, Hawkins T, Wiedmann T, Hertwich EG (2011) Footprint family technical report: integration into MRIO model. OPEN-EU project report. NTNU, Trondheim

Weinzettel J, Hertwich EG, Peters GP, Steen-Olsen K, Galli A (2013) Affluence drives the global displacement of land use. Glob Environ Chang 23(2):433–438

Weinzettel J, Steen-Olsen K, Hertwich EG, Borucke M, Galli A (2014) Ecological footprint of nations: comparison of process analysis, and standard and hybrid multiregional input–output analysis. Ecol Econ 101:115–126

West J, Schandl H (2013) Material use and material efficiency in Latin America and the Caribbean. Ecol Econ 94:19–27

Wiebe C, Bruckner M, Giljum S, Lutz C, Polzin C (2012) Carbon and materials embodied in the international trade of emerging economies: a multi-regional input-output assessment of trends between 1995 and 2005. J Ind Ecol 16(4):636–646

Wiedmann T (2016) Impacts embodied in global trade flows. In: Taking stock of industrial ecology. Springer

Wiedmann T, Wilting HC, Lenzen M, Lutter S, Palm V (2011) Quo Vadis MRIO? Methodological, data and institutional requirements for multi-region input–output analysis. Ecol Econ 70(11):1937–1945

Wiedmann TO, Schandl H, Lenzen M, Moran D, Suh S, West J, Kanemoto K (2015) The material footprint of nations. Proc Natl Acad Sci 112(20):6271–6276

Wilting HC, Vringer K (2009) Carbon and land use accounting from a Producer's and a Consumer's perspective–an empirical examination covering the world. Econ Syst Res 21(3):291–310

WU (2016) Global material flow database (www.materialflows.net). Material extraction data, 1980–2013. Vienna University of Economics and Business (WU), Vienna

WU Vienna (2016) Global material flow database. 2016 version. Available at www.materialflows.net. Vienna

Würtenberger L, Koellner T, Binder CR (2006) Virtual land use and agricultural trade: estimating environmental and socio-economic impacts. Ecol Econ 57(4):679–697

WWF, Zoological Society of London, Global Footprint Network (2012) Living planet report 2012. Biodiversity, biocapacity and better choices. WWF, Gland

Yu Y, Feng K, Hubacek K (2013) Tele-connecting local consumption to global land use. Glob Environ Chang 23(5):1178–1186

Chapter 5
The Critical Raw Materials Concept: Subjective, Multifactorial and Ever-Developing

Jan Kosmol, Felix Müller, and Hermann Keßler

Abstract Criticality analysis has established itself as a multifactorial, action-oriented, socio-economic raw materials scarcity assessment method which is subject to continuous development. A raw material is critical when its supply is at risk and a company or economy is vulnerable to supply restrictions of that raw material. The binary labelling of raw materials as either critical or not delivers a strong message. However, each raw material has a characteristic risk profile which may not be described by an aggregated criticality score and a discrete treshold value. A differentiated interpretation allows for a deeper understanding of the raw material supply situation and for the adoption of appropriate measures. Criticality should be understood as a continuum, subjective to the raw material system in question. A harmonised criticality methodology presented in the industrial guideline on resource efficiency (VDI 4800-II) allows for a flexible application of the concept.

ÖkoRess, a research project of the German Environment Agency, examines why and how environmental aspects should be included into the criticality concept. A raw material is consequently environmentally critical if it exhibits a high overall environmental hazard potential and is at the same time of great importance for a company or economy. A high environmental hazard potential can indicate a future supply risk. The conclusions to be drawn, however, differ from the conclusions from conventional criticality analysis. Ecological criticality widens the focus to include measures used to foster responsible sourcing and mining practices, which until now have not been discussed in the context of criticality.

Keywords Criticality • Raw materials • Critical raw materials • Ecological criticality • Environmental hazard potential • Mining • Scarcity • Supply risk • Vulnerability • Raw material supply • Responsible sourcing • Responsible mining

---

J. Kosmol (✉) • F. Müller • H. Keßler
German Environment Agency (UBA), Wörlitzer Platz 1, 06844 Dessau-Roßlau, Germany
e-mail: jan.kosmol@uba.de

H. Lehmann (ed.), *Factor X*, Eco-Efficiency in Industry and Science 32,
https://doi.org/10.1007/978-3-319-50079-9_5

5.1 Introduction

A looming raw material scarcity is a key argument in favour of making our economy and actions resource-efficient and sustainable. Humankind has been concerned about the depletion of geological raw material resources since the early years of industrialisation. Many concepts to assess the threat of a physical, absolute raw material scarcity have been developed since (Bardi and Randers 2014; Hubbert 1956; Jevons 1865; May et al. 2012; Meadows and Meadows 1972; Sverdrup et al. 2017). All these theories are based on the intuitively plausible assumption that the earth's geological raw material reserves are finite, cannot be renewed within the reasonable human time scales and will be depleted over time. This narrative is reflected in the first edition of Germany's National Sustainability Strategy (The Federal Government 2002, p. 93):

> The Earth's stocks of raw materials are limited. Raw materials that we consume today are no longer available for future generations. Prudent and efficient use of scarce resources is therefore a key to sustainable development. Energy consumption is a central focus in this respect.

The concepts to assess absolute physical scarcity have been criticised for neglecting the factors describing the development of raw material supply and demand such as technical progress, and thus the variability of the raw material economic parameters "reserves" and "resources" (Giraud 2012; Rustad 2012). Criticism is directed in particular to the so called static lifetimes (e.g. reserves-to-production-ratios) that are misinterpreted as lifespans until depletion. They are raw material specific equilibrium values that reveal information on the global mean exploration periods (Wellmer 2008). The fact that the reserve range of copper is around 40 years, but lead's is only 20 years is not indicative of an alleged earlier exhaustion of global lead deposits (USGS 2016). They are not significant as a prognosis value for the lifetime of a material until exhaustion.

In the last 10 years, a scarcity assessment method has moved back into the focus of policy and science. This method does not look into the long-term depletion of geological resources but rather evaluates short- to medium-term availability of raw materials based on socio-economic drivers. The renaissance of this "criticality" can be explained as a response to unexpectedly strong price increases for some raw materials over the last decade which have largely been due to a sharp increase in demand from China. In addition, a growing market concentration in raw materials supply can be observed, for example due to high corporate and country concentration of production. Export restrictions were another reason for a strong imbalance in supply and demand. The criticality debate is characterised by the concern that a combination of various geological, technical, geopolitical and regulatory factors may jeopardise industrial and economic development by disruptive price rises or even interruptions in the supply of essential raw materials. Against this background, criticality analysis aims to identify the raw materials of a system (e.g. national economies, industry, enterprises) which uses them to fulfil essential functions for this system but whose supply is at risk. Criticality analysis enables the assessment of system vulnerability in the face of supply risks of specific raw materials causing supply disturbances. The criticality assessment does not correspond to the characterisation of total, but relative scarcities which occur when raw material demand of the reference system cannot be satisfied by the raw material supply in terms of time, space and organisation (VDI 2016).

5.2 Methods of Criticality Analysis

The work of the U.S. National Research Council (NRC 2008) and that of EU COM (2010) has recently directed attention to the criticality of raw materials. Both studies are the result of large, sovereign, interdisciplinary working groups. In their basic orientation, these studies still follow the tradition of national raw material committees and reports since the 1930s, where raw materials were evaluated based on their importance in the economic area investigated and the risk of an impaired raw material supply (Gandenberger et al. 2012; Paley et al. 1952). While "significance" mainly implied military considerations on national security in earlier works – the *strategic raw materials* concept was also introduced in this context (Wellmer 2012) – recent studies are motivated by economics and competitiveness.

The NRC performed pioneering methodical work and introduced a two-dimensional evaluation matrix based on the ISO 31000 risk analysis (Glöser et al. 2015; ISO 2009). Its structure was followed by numerous subsequent studies, which can now be regarded as constitutive for the definition of raw material criticality (Fig. 5.1) (Coulomb et al. 2015; Erdmann and Graedel 2011; Helbig et al. 2016). In

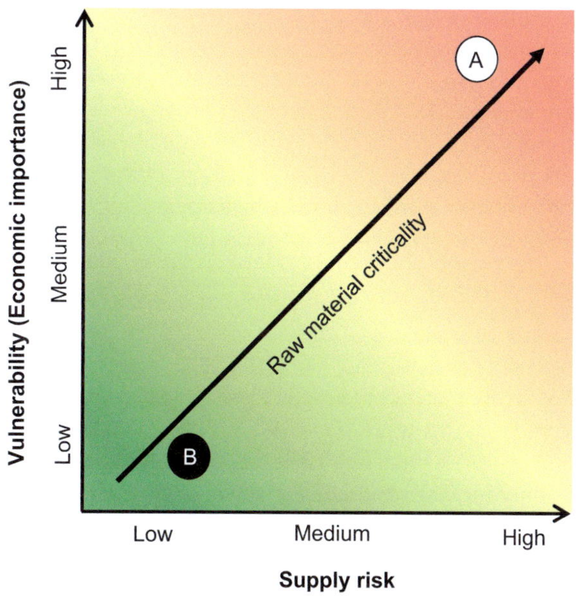

Fig. 5.1 Criticality matrix (The degree of criticality increases as one moves from the lower-left to the upper-right corner of the figure. In this example, raw material *A* is more critical than raw material *B*) (Source: own figure according to NRC 2008)

this context, the impact of supply restriction of raw materials for a system considered such as a national economy (vertical axis) is related to potential raw material supply risks (horizontal axis). The impact of supply restriction is assessed based on the size of the relevant demand sectors, substitution elasticity and the particular significance of raw materials for future technologies.

The supply risk is again described by criteria from five different perspectives of primary and secondary raw material availability: geologic (abundance), technical (extraction and processing), environmental and societal (environmentally and societally acceptable production), political (trade policies and actions), and economic (production cost and willingness to pay). The two-dimensional matrix yields three constitutive properties of criticality: (1) Critical raw materials are those that are both essential in use and subject to considerable supply risk. (2) Criticality is not a state of being or not-being, but a location on a multi-indicator continuum. (3) Criticality is not universal, rather it should be considered in terms of stakeholders and time since the characteristics of the two axes depend on these two determinants (Buijs et al. 2012; Graedel et al. 2014).

While the NRC study mainly focused on methodical principles and suggestions for better future information management on raw material criticality, a study was commissioned by the EU Commission in 2010 which not only aimed at identifying the critical raw materials for the European economy but also identified a further objective for criticality analysis: application as a guiding instrument of raw material policy (EU COM 2010). The analysis was designed to support the EU Raw Materials Initiative that encompasses measures in three areas to secure sustainable access from outside Europe, improve framework conditions for extracting minerals within Europe, and promote the recycling and resource efficiency of such materials (EU COM 2008). The investigation is also based on a two-dimensional criticality matrix from supply risk and economic significance. However, an additional third dimension is presented as a test criterion that relates to environmental risk, more precisely the risks that measures might be taken by countries with the intention of protecting the environment and by doing so endangering the supply of raw materials to the European Union (cf. Sect 5.4). This case manifested itself only a little later when the People's Republic of China, a leading producer of rare earths, announced a drastic reduction in production with the same justification, which, however, was seen as a violation of WTO law a few years later (The Guardian 2015).

As a result of EU COM (2010), 14 of 41 raw materials were evaluated as critical, with rare earths and the platinum metals considered as groups. The discrete list of critical raw materials sent out a strong message. However, the binary distinction between "critical" and "non-critical" counteracts the relative nature of the concept, i.e., a raw material is instead more or less critical than another raw material. At the very least, this was compensated for by the fast pace of some raw material markets and by the fact that reporting on critical raw materials for the EU was anchored as a permanent task. An update now containing 20 critical raw materials out of 54 was published in 2014 (EU COM 2014).

Complementing the two lead studies, a number of other economic-area-related criticality analyses were published simultaneously and subsequently for countries and German states (Länder). The studies either delineated critical raw materials in the two-dimensional matrix or contained an aggregated criticality index (Erdmann and Graedel 2011; Faulstich et al. 2014; Helbig et al. 2016; Morley and Eatherley 2008; Pfleger et al. 2009, 2012). Investigations have also been launched for individual technology fields such as energy and environmental technologies (Buchert et al. 2009; DoE 2010; Moss et al. 2011a; Reller 2011; Zepf et al. 2014).

In order to harmonise the criticality concept for various purposes, users, time horizons and organisational forms, Graedel et al. (2011) proposed a uniform method which could be adapted according to the subject of investigation. Particular attention is given to the use of indicators in the three dimensions of vulnerability to supply restrictions, supply risks and environmental implications, for which there is a transparent and updated database available. The approach was tested for various metal groups and an extensive indicator data pool was described (Graedel et al. 2015; Harper et al. 2015; Nassar et al. 2011b, Nassar et al. 2015a, b).

Inspired by this groundbreaking work, an expert group at the Association of German Engineers (VDI) has elaborated a harmonised criticality methodology as part of an industrial guideline on resource efficiency (VDI 4800-II) (VDI 2016). The guideline allows for a universal and flexible application of the concept on different levels from macro- to micro-scales, i.e. countries, economies, regions, industries, branches, sectors, technologies, companies, product lines, products and building blocks. This method is also action-oriented because it aims to identify particularly critical raw materials that must be treated as a priority. It also intends to develop measures that enable a reference system to overcome difficult raw material situations without sustaining impairment and to gain a strategic competitive advantage over other reference systems. Section 5.3 presents the VDI 4800 II criticality analysis in more detail and shows an example of its application.

5.3 Criticality Analysis According to VDI 4800 II

The two-dimensional approach clearly distinguishes between exogenous technical, socio-political and economic criteria that characterise the overall supply risks (Table 5.1) and subjective endogenous vulnerability criteria for the object in focus. The guideline describes **vulnerability** assessment for companies using indicators for the degree of concern and operational, tactical and strategic adjustment options. On the other hand, criteria and indicators of the **supply risks** represent a standardised approach that is applicable not only to mineral but also fossil raw materials and cultivation and extraction biomasses.

VDI 4800 (II) proposes three supply categories with a total of 13 criteria and representative indicators including scaling of value ranges (Table 5.1). The rationale of the respective criteria is explained below.

Table 5.1 Categories, criteria and indicators of the criticality dimension supply risk according to the guideline of the Association of German Engineers (VDI 4800 II) (VDI 2016)

Categories	Criteria	Indicator
Geological, technical and structural criteria	Static lifetime	Reserves to global annual production ratio
	Co-production/auxiliary production	Degree of co-production/ auxiliary production
	Recycling	Dissemination level of functional end-of-life recycling technologies
	Logistic restrictions	Economy of storage and transport
	Restrictions by natural phenomena	Distribution level of natural deposits/ cultivation areas
Geopolitical and regulatory criteria	Concentration of reserves by country	Herfindahl-Hirschman index of reserves
	Concentration of production by country	Herfindahl-Hirschman index of production by country
	Geopolitical risks of global production	Political country risk
	Regulatory situation for raw material projects	Regulatory country risk
Economic criteria	Enterprise concentration of global production	Herfindahl-Hirschman index of enterprises
	Global demand impulse	Degree of demand increase
	Substitutability	Technical feasibility and economic efficiency of substitutes in main applications
	Raw material price fluctuations	Annualised price volatility

5.3.1 Supply Risk: Geological, Technical and Structural Criteria

5.3.1.1 Static Lifetime

The ratio of reserves to global annual production is also referred to as "static lifetime". The ratio cannot be understood as an actual lifetime, but as a measure of the need to expand the supply through new exploration (expansion of reserves) or secondary raw materials (Wellmer 2008). For this reason, reserves is a variable quantity. The reserves for biotic raw materials should not be determined analogously since they are renewable raw materials. The total stock is interpreted as a reserve, and renewal rate is also taken into account to determine the annual production.

5.3.1.2 Co-production/Auxiliary Production

Many metals are present in association with other materials in their ores. By-products are present in comparatively lower concentrations, and their availability is strongly or exclusively coupled to the production capacity of one or more carrier elements, for example, gallium in bauxite ores to aluminium production, indium to zinc production. On the other hand, co-products have no clear carrier element. The metals concerned such as platinum group metals (PGM) and rare earths exist in defined mineralogical ratios, which require joint processing and refining to purify the individual elements. Both in the case of auxiliary and co-products, there is a high risk that an increased demand cannot be served within the short and medium term. The extension of the supply is structurally limited by the demand for the carrier elements and that for further co-elements (Nassar et al. 2015a).

5.3.1.3 Recycling

Recycling is both a strategy to conserve natural resources and to expand the supply base of primary raw materials with secondary raw materials. The latter usually has a positive impact on the supply situation. The spread of recycling is determined by physico-chemical (properties, applications) aspects on the one hand and by economic and technical (markets, technologies, infrastructure) ones on the other.

5.3.1.4 Logistical Restrictions

Logistical restrictions are above all relevant for biotic raw materials, building materials and industrial minerals when it comes to pricing and thereby pose a risk to supply. Logistical restrictions first include transport distances e.g. from the site of raw material extraction or harvest to the post-processor or end user. This can apply to reference radii that are economically and ecologically significant depending on the type of transport e.g. in the case of construction minerals. Second, logistical restrictions also include requirements and options of storage capability.

5.3.1.5 Restrictions Imposed by Natural Phenomena

Availability of biotic raw materials can in particular be influenced by natural phenomena (e.g. aridity, drought, flooding, pest infestation). Natural phenomena are short-term, hard-to-predict and therefore non-calculable, and often regionally restricted events that can critically affect global raw material availability. The prevalence of natural deposits/cultivation areas is considered the assessment criterion. The risk for regional biotic raw materials whose main cultivation areas are concentrated in specific climate regions on certain continents is estimated to be higher than that for raw materials produced globally.

5.3.2 *Supply Risk: Geopolitical and Regulatory Criteria*

5.3.2.1 Concentration of Reserves by Country

The risk of restrictive measures motivated by the supplier's trade policy to exploit market power is rather high when the reserves are concentrated in a few countries. Oligopolistic supply structures may have a significant impact on the amount of raw materials supplied and traded globally. However, if the reserves are distributed among a larger number of countries, the risk is significantly lower.

5.3.2.2 Concentration of Production by Country

If certain countries have a very high share of world production of a specific raw material, their market power enables them to dictate trading conditions and influence the quantity of the raw materials traded. A concentration of production by country may be the consequence of a concentration of reserves by country. However, even if reserves are distributed among many countries, raw material production may still be concentrated in a small number of countries. A concentration of raw material production can affect availability of raw materials in the short term (Buchholz et al. 2014).

5.3.2.3 Geopolitical Risks of Global Production

Access to raw materials is significantly influenced by political stability in the exporting countries and the regions of extraction (Kaufman et al. 2010). Prolonged political and military clashes in mostly repressive, undemocratic states can affect the availability of raw materials and lead to drastic interruptions of delivery. Raw material exploration and extraction can also be the driving forces for regional distribution conflicts and violent clashes. These aspects are increasingly affecting raw material policy by encouraging the development of certification mechanisms to prevent raw material shipments from conflict zones. Disclosure of systematic human rights violations in the producer countries can create veritable business risks in the added value chain by undermining corporate images and corporate social responsibility strategies (CSR strategies). This criterion is assessed using the governance indices of *Voice and Accountability* and *Political Stability and Absence of Violence*.

5.3.2.4 Regulatory Situation for Raw Material Projects

Government policies considerably influence raw material availability even in stable countries where political crises and violent conflicts are absent. Economic, tax and environmental policy conditions exert a great influence on the intensity and extent of the exploration of raw materials. Mining projects normally need long-term security for preparation and planning to justify large initial investments in the exploration of deposits. Societal, socio-economic and ecological conflicts of interest, for example, due to competition of use and occupational safety conditions, but also corruption and obstacles to infrastructure development, can severely limit mining projects despite significant raw material deposits (Kaufman et al. 2010). This criterion is assessed based on the governance indices of *Rule of Law, Regulatory Quality, Control of Corruption and Government Effectiveness*.

5.3.3 Supply Risk: Economic Criteria

5.3.3.1 Enterprise Concentration of Global Production

For competitive reasons, enterprises tend to expand their market shares against other companies to achieve economies of scale. If mining and processing enterprises are in the hands of a few firms, price risks are high due to their market power. Enterprises can significantly influence raw material supply and thus the price of raw materials. This applies not only to the extraction, but also to processing and refining in the further added value chain (Buchholz et al. 2014).

5.3.3.2 Global Demand Impulse

Population growth and economic development, especially in newly industrialising and developing countries, lead to significant increases in demand for a large number of raw materials (Schandl et al. 2016). The issue of price increase significantly depends on whether the supply can be expanded to the same extent and in the same period as planned. The more the demand increases and the more unstable it is, the greater the risk of high price increases. Special price risks are associated with surges in technological demand such as those triggered by future and environmental technologies (Marscheider-Weidemann et al. 2016). Such technologies, also considered as disruptive, involve rapid technological changes, large impact horizons in different areas of need, high economic production values and dramatic impacts on existing production and consumption patterns.

5.3.3.3 Substitutability

The extent to which raw materials can be substituted in their uses and applications is an economic criterion that affects supply risk. A substitution is to be assumed if the provision of a functional unit of use occurs through an at least partial exchange of a factor input. This can range from perfect substitutes to limitative production functions where substitution requires a completely new factor combination, mostly a new technological concept. The analysis of potential material substitutions enables estimation of potential quality and functionality losses as well as additional costs at the raw material or material level (Tercero Espinoza et al. 2013).

5.3.3.4 Raw Material Price Volatility

Price volatility is the fluctuation range of raw material prices over a certain period of time. The level of price volatility is the measure of the risk of a sharp change in price. This concerns the danger of sharp price increases as well as a clear price drop. Thus, price fluctuations can be interpreted as an indicator of high supply uncertainties on the raw material markets. The higher the price fluctuations, the stronger the restriction in planning security of systems (Mauss and Posch 2014).

5.3.4 Interpretation of the Results Using Phosphate as a Case Study

In the following, the application of the method according to VDI 4800 II is presented using phosphate as a case study.

5.3.4.1 Vulnerability

Phosphorus is an essential element for all forms of life and of existential importance for mankind's food supply. All agricultural production depends on sufficient phosphate supply. The possibility of substitution by other materials for phosphate does not exist in standard agricultural practice. However, this does not mean that the vulnerability of a system considered must per se be very high.

Vulnerability depends on the degree to which the system considered or stakeholders are affected by restricted availability and on their capability to react to reduce demand for supply and extract phosphorus from secure or alternative sources. For example, vulnerability may be high to very high for an entirely import-dependent economy [**A**] depending on whether or not phosphorus recovery structures exist, e.g. from wastewater, sewage sludge, incinerator ash, manure and animal meal.

Even if the degree of threat of an exporting country e.g. Morocco with its own primary production [**B**] is theoretically equally high because of agricultural interest, vulnerability is low because of its own reserves. Likewise, vulnerability for a farm [**C**] will be high depending on the diversification of production. In other sectors with technical phosphate applications, e.g. surface finishing, the degree of threat can also be very high but vulnerability still moderate, thanks to alternative methods or hedging instruments in the procurement process [**D**] (Fig. 5.3).

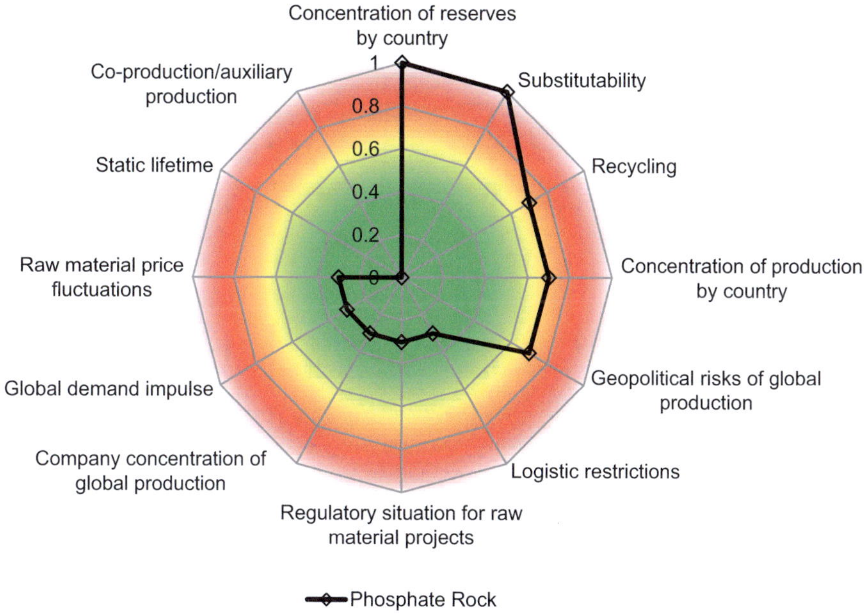

Fig. 5.2 Supply risk of phosphate (phosphate rock) (Source: own figure)

5.3.4.2 Supply Risk

Applying the above described assessment criteria on phosphate results in this broadly diversified characteristic supply risk profile (Fig. 5.2).

5.3.4.3 Interpretation

If aggregation is planned, a discrete value results for the dimension of supply risk. When an arithmetic mean and a scale of zero to one are used, 0.43 is obtained for phosphorus, where particularly problematic aspects (substitution, concentrations by country, recycling) have been levelled. However, the degressive addition method proposed by VDI 4800 II is rather different: it weights higher risks more strongly and a result of 0.81 is obtained, which suggests very high risks. Such a weighting is helpful in fulfilling an early warning function in multi-raw-material systems (VDI 2016).

Figure 5.3 shows the results in a matrix. It is clear that, given the same risk, the selection of the aggregation method contributes decisively to the localisation of phosphate in the more or less critical areas. Also, different vulnerabilities indicate the subjectivity of the concept: the configuration of the investigated system, its individual degree of threat and action alternatives become of prime significance.

However, the disaggregated level must be taken into consideration if targeted measures for the reduction of vulnerability based on criticality analysis are to

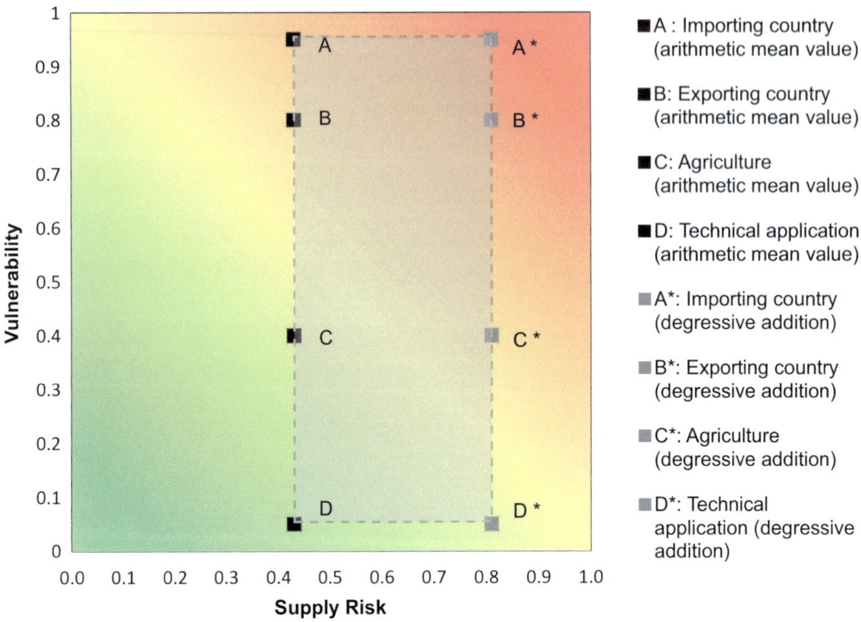

Fig. 5.3 Criticality case study for phosphorus (Source: Own figure)

be performed. This is because measures are generally taken in response to specific aspects of supply risk.

5.4 Further Development of the Environmental Dimension

5.4.1 Environmental Aspects – Why Should They Be Taken into Account? The Environmental Availability of Raw Materials

A focus of the current works on further development of the criticality concept and its methods of analysis is the consideration of environmental aspects in criticality analyses. Environmental aspects in studies of the renaissance of criticality analysis since 2008 were addressed in varying degrees and in various ways (cf. Sect 5.2) based mostly on the estimation that impacts of raw material extraction on the environment and subsequent societal reactions have significant and increasing effects on the availability of raw materials. The NRC identified in its study on critical raw materials for the US economy of 2008 "environmental and social availability" as one of five dimensions of availability for raw materials. This evaluation is described in the study with examples of conflicts with the local population through environmental impacts and land use competitions, however not with operationalised indicators in the criticality analysis (NRC 2008).

This logic, according to which the social reaction to environmental impacts could negatively affect the supply of raw materials and thus the supply situation, is found also in the first study of the European Commission on critical raw materials, published in 2010 (EU COM 2010, p. 32):

> Defining the criticality: To qualify as critical, a raw material must face high risks with regard to access to it, i.e. high supply risks or high environmental risks, and be of high economic importance.

In addition to the supply risk aggregated indicator, an aggregated indicator to the "Environmental Country Risk" was developed that was juxtaposed with the economic importance for the EU economy in order to analyse the criticality. This environmental governance indicator should depict for each raw material the potential that a raw-material-producing country adopts environmental policies that could limit the access to deposits or the supply of raw materials (EU COM 2010, p. 5):

> Two types of risks are considered: a) the 'supply risk' [...] and b) the 'environmental country risk' assessing the potential for environmental measures that may restrain access to deposits or the supply of raw materials.

The German Environment Agency communicated, in the public consultation of the report, criticism of possible raw materials policy conclusions that could be drawn from this analysis (UBA 2010, p. 1):

> This interpretation of environmental risks must not create an EU interest to keep environmental standards of raw material exporting countries on a low level. An enduring raw material supply, also for future generations in the sense of a sustainable materials management needs sound environmental standards world-wide.

Apart from the conclusions that are undesirable from the environmental policy point of view, the indicator describes quite well the phenomenon that was identified by NRC (2008) as "environmental availability": socially undesirable effects of mining activities on the environment can cause a social reaction that can lead to the polluter being burdened with the costs for the damage to ecosystems and human health, in that he/she is obligated to take measures to avoid these undesirable effects. In this way the production costs can rise and lead to a decrease in the economically and technically recoverable share of a deposit, the so-called reserves. They can also lead to investors' decisions on an investment in a mine project being dependent on the environmental hazard potential and on the environmental standards in force at the site. High environmental standards can make the exploitation of new deposits more expensive or even unprofitable. This can also lead to a relocation of production of raw materials from minerals containing pollutants to countries with weaker environmental governance.

Therefore the higher the environmental hazard potential of extraction of a raw material and the weaker the environmental governance in a producing country, the higher the probability that there will be in the future a scarcity in the supply of raw materials through a tightening of environmental standards. In addition to geological, technical and socio-political factors, the availability of raw materials is consequently also influenced by environmental factors. The raw-material markets react first however to this "new scarcity" when the environmental factors become economic factors through effectively implemented environmental standards. Herein lies a difference to the supply risk indicators in the relevant criticality analyses, as for example the concentration of production by country, which describe the current situation and thereby directly depict an increased probability for price surges or even interruptions of the raw material supply. The commodity-related environmental hazard potential as a scarcity indicator is linked on the other hand to an expected trend. Environmental hazard potentials only indicate a supply risk in the real economy when one assumes an increasing internalisation of external costs through voluntary or compulsory environmental standards. According to current evaluations by various experts, this expectation is entirely plausible (Angerer et al. 2016; Birshan et al. 2015; ICMM 2012).

5.4.2 Environmental Aspects – How Should They Be Taken into Account? The ÖkoRess Methodology for Environmental Hazard Potentials

The significance of externalities for the analysis of raw material criticality is also a subject of discussion in a current study of the OECD, however without being operationalised in the selected method of evaluation (Coulomb et al. 2015). Environmental hazard potentials on the basis of lifecycle inventory data and end point models are considered in various works on raw material criticality (Buchert et al. 2017; Graedel et al. 2011; Graedel et al. 2015). However, the extent to which available life cycle inventory data can fully map the relevant environmental hazard potentials of mining and, characteristically, of global raw material production, must be critically checked. Current studies on the environmental hazard potentials of raw material extraction from mines indicate that the impacts of the mining and processing of extracted raw materials reflect poorly or even not at all in the relevant lifecycle inventory data (Dehoust et al. 2017a, b; International Institute for Industrial Environmental Economics (IEEE) 2016). This applies for example to the case of acidification and heavy metal mobilisation due to auto-oxidation of mineral processing residues (Acid Mine Drainage), the risk of tailings dam failure or penetration of mining into protected areas.

In order to fill in these gaps in the environmental analysis of raw materials, the German Environment Agency has commissioned a research project (ÖkoRess) with the purpose of developing a methodology for assessing environmental hazard potentials of abiotic raw materials (Dehoust et al. 2017a). The analysis methodology is outlined below.

Acknowledging the variability and complexity of environmental challenges in the mining sector, a bottom-up approach was chosen and a site-specific analysis method was developed based on 40 case studies on mining projects world-wide.

The case studies indicated that the environmental impacts of raw material extraction depend significantly on

- geological conditions (e.g. geochemical composition),
- the mining and processing technology deployed (e.g. open-pit mining/underground mining, type of ore processing),
- as well as the environmental/locational factors (water stress, natural risk of incidents, ecosystem sensitivity).

Based on these findings, an evaluation grid with 11 indicators was developed with each one assigned to an environmental objective (e.g. avoidance of hazardous emissions). Each indicator is analysed based on measuring instructions with diverse handouts on a three-stage scale so that even without detailed data a multifactorial environmental hazard profile of mining projects can be generated in order to obtain an initial assessment.

Based on this, a commodity-related analysis approach was developed with the goal of establishing a basis for the analysis of the above-described "environmental availability of raw materials" and to achieve a contribution to the further development of criticality metrics. The analysis refers – unlike in life cycle assessment (LCA) – not to a unit of mass of a raw material or a basic material, but by analogy with supply risk analysis to the world production of a raw material or a basic material. This has the advantage that environmental hazard potentials are assessed also in terms of their global magnitude, but brings with it the disadvantage that no comparison is possible of the defined quantities of raw materials and that the analysis results cannot be used for the life cycle assessment according to ISO 14040 (ISO 2006). The claim to depict the current situation of the world production of each raw material puts particular requirements on the analysis because there is no recourse to comprehensive statistics for the totality of all producing mines.

In all three assessment levels generalisations are possible and acceptable, knowing full well that there can also be exceptions to the rule for some raw materials. Nevertheless, typical types of deposits and conditions of formation can be gleaned for each raw material from the technical literature related to deposits, from which commodity-related, geological environmental hazard potentials can be deduced, e.g. potential for Acid Mine Drainage or the release of heavy metals and radionuclides. Commodity-related generalisations in the area of mining and processing technology are likewise possible based on the standard methodology documented in technical literature on mining engineering. Comparable deposits are mined and their products processed using comparable technology on the basis of the worldwide competition in the mining sector. In such a manner it can for example be determined for each raw material whether it is mined predominantly underground or in open pits or whether chemical auxiliary agents are used during the ore processing from which a technology-related environmental hazard potential can be derived. The analysis assumes that with site-specific characteristics the environmental hazard potential depends strongly on local conditions. In this manner, for example, incidents (e.g. dam failures) due to natural phenomena are more likely in areas with an elevated risk of earthquakes, floodings or landslides. Usage competitions for

Table 5.2 Commodity-related environmental hazard potentials evaluation grid (Dehoust et al. 2017a)

	Goal	Indicator
Geology	Avoidance of contamination risk	1. Requirements for Acid Mine Drainage (AMD)
		2. Associated heavy metals
		3. Associated radioactive substances
Technology	Limitation of the intervention in a natural environment	4. Extraction methods
	Avoidance of contamination risk	5. Application of auxiliary agents
Natural environment	Avoidance of risk of incidents relative to nature	6. Risk of incidents through floodings, earthquakes, storms, landslides
	Avoidance of water-usage competition	7. Water stress index, WSIand desert areas
	Protection/preservation of high-quality ecosystems	8. Identified protected areas and AZE sites
Social environment	Implementation of standards	9. Environmental governance in the most important producing countries
Value chain	Limitation to the total extent of interventions	10. Cumulative raw material demand of world production (CRD_{global})
	Limitation to the primary energy demand	11. Cumulative energy demand of world production (CED_{global})

water are more likely in areas with high water stress and the damage to sensitive ecosystems is more likely if mining occurs in protected areas. In contrast to the other analysis levels, the environmental hazard potential can be assessed on the basis of a statistical evaluation by means of databases on the world-wide geographical distribution of deposits, country-related extraction volumes and geodata on water stress, protected areas and risk of failure through extreme natural events (Dehoust et al. 2017a).

The overview of the low, medium or high environmental hazard potentials in the evaluation grid for each of the 11 indicators determined (Table 5.2) permits it to assess how high the probability is of its occurrence and the extent of its negative environmental impacts for extraction and processing of a raw material (Dehoust et al. 2017a).

5.4.3 Aggregation/Merger

Criticality analyses in accordance with the matrix concept work with aggregated indicators for the vulnerability and supply risk dimensions (cf. Sects. 5.3 and 5.4). In order for the analysis of the environmental hazard potential to apply to criticality analyses, a merger is required accordingly of the 11 individual results of a raw material-specific environmental hazard profile to a qualitative or quantitative overall result.

An aggregation does indeed elevate the communicability and political applicability of the results but is inevitably associated with a loss of information and the application of subjective value judgments. Because the analysis of the individual indicators, except for indicators 10 and 11 that serve as scaling parameters for the magnitude of the environmental impacts, do not already occur on an ordinal scale, a method for the qualitative merger of the results was developed that also generates a three-stage ordinally scaled overall result for reasons of a preferably extensive scientific objectivity of the results.

This method is based on a method proposed by the German Environment Agency in 1999 on impact analysis in life cycle assessment, which is presently being revised (Schmitz and Paulini 1999): For this purpose the indicators are grouped at first according to common environmental objectives, such as e.g. avoidance of contamination risk and the so-called influential frame conditions (e.g. raw material extraction in the Arctic) and the analyses are merged according to specified rules. The environmental objectives are thereby arranged hierarchically according to their environmental significance. The evaluations of Rockström et al. on exceeding planetary boundaries are thereby used as standards (Rockström et al. 2009; Steffen et al. 2015). Finally, the merged environmental hazard potential for three environmental objectives and the merged influential frame conditions lead to a classification of raw materials according to an overall environmental hazard potential that can be represented as an independent dimension of criticality in line with an environmental availability of raw materials to picture the environmental criticality (Fig. 5.4).

5.5 Conclusion

Criticality analysis has established itself as a multi-layered and complex, action-oriented socio-economic raw materials assessment method to deal with scarcities (VDI 2016). The first studies that correspond to today's logic of criticality analysis were motivated by a raw material security policy and date back to the 1970s with the basics dating back even to the 1930s (Buijs et al. 2012; Gandenberger et al. 2012). Numerous new criticality metrics have been developed since 2008 (cf. Sect. 5.2). Various raw materials are classified as critical depending on the criteria used within the metrics, reference systems and up-to-datedness of the analyses. In the trend of studies, raw materials for environmental and future technologies are classified as particularly critical. These include, for example, essential raw materials for thin-film photovoltaic technologies such as indium, tellurium, gallium or germanium. In addition to special metals, precious metals such as platinum, rhodium and gold, or refractory metals required in significantly larger amounts such as tungsten, molybdenum, tantalum, niobium and chromium are also classified as critical. The labelling of discrete critical raw materials delivers a strong message. However, each raw material has a characteristic risk profile as pointed out for the case of phosphate

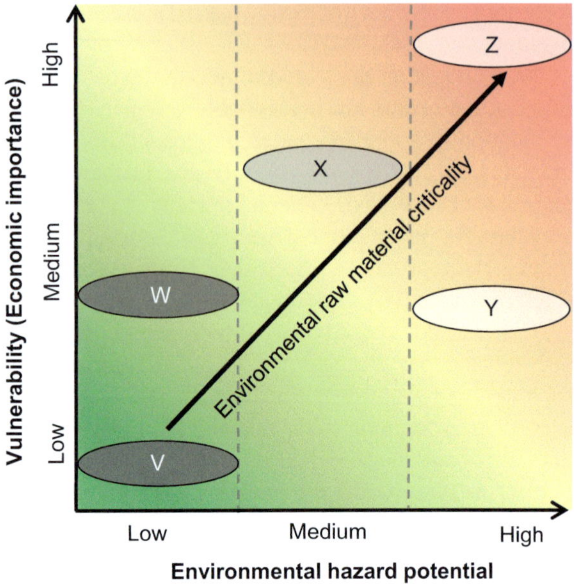

Fig. 5.4 Vulnerability vs. environmental hazard potential: environmental criticality (The degree of criticality increases as one moves from the lower-left to the upper-right corner of the figure. In this example, raw material *Z* is more critical than raw material *V*) (Source: own figure)

rock (cf. Sect. 5.3). These aspects are not described by an aggregated, weighted reduction to a discrete criticality threshold value. A differentiated interpretation of the individual indicators not only allows for a deeper understanding of the raw material supply situation but also for the adoption of appropriate measures to overcome tough supply situations. Criticality of raw materials therefore should be understood as a continuum, subjective to raw material systems in question. A harmonised criticality methodology recently presented in the industrial guideline on resource efficiency (VDI 4800-II) allows for a flexible application of the concept for various practitioners on different levels from macro- to micro-scales (VDI 2016).

The criticality concept is subject to a continuous development that reflects the manifold challenges of a safe and more sustainable raw material supply over time (Walz et al. 2016). A focus of the current works on the further development of the criticality concept is the consideration of environmental aspects (cf. Sect. 5.4). A raw material is consequently environmentally critical if it exhibits a high overall environmental hazard potential and is at the same time of great importance for the raw material use reference system and this system would therefore be particularly vulnerable to a manifestation of environmental hazard potentials (Fig. 5.4).

Even though a high environmental hazard potential can indicate a future supply risk through globally increasing environmental standards in the mining sector (cf. Sect. 5.4), the conclusions to be drawn differ in part from those of a conventional criticality analysis (Graedel et al. 2014). This is mainly because environmental and development policy goals must not be jeopardized. Increasing material efficiency and recycling are options that can help to increase supply security as well as decrease environmental pressure. The ÖkoRess method, however, does not allow to derive substitution measures for various reasons: Substituting a raw material with higher ecological criticality with a raw material with lower ecological criticality can lead to a ban of raw materials from developing countries with weak environmental governance, which is undesirable from a development policy perspective. Furthermore, an assessment of the environmental effects of raw material substitution requires an analysis along the entire product life-cycle using quantitative life-cycle-inventory data relating to a mass unit of material. Warehousing operations of environmental critical raw materials or entering into long term supply contracts would be options to increase supply security but might jeopardize environmental policy goals.

Despite its relevance as an indicator for future scarcities, an elevated ecological criticality indicates a higher probability for serious environmental impacts and subsequent conflicts for raw materials upon which companies or economies depend. Therefore, the focus is widened to measures in the areas of Corporate Social Responsibility (CSR) and sustainable supply chain management, which until now were not discussed in the context of criticality. Possible courses of action include the application of due diligence and certification schemes as well as direct engagement of downstream countries and companies with upstream raw material producers in order to foster responsible mining practices. Environmental criticality can be seen in this sense as a tool for the prioritisation of starting points for private sector or government measures for the strengthening of entrepreneurial responsibility in the raw materials supply chain (Fairphone 2017).

Disclaimer This paper does not necessarily reflect the opinion or the policies of the German Federal Environment Agency.

References

Angerer G et al. (2016) Rohstoffe für die Energieversorgung der Zukunft: Geologie – Märkte – Umwelteinflüsse. München

Bardi U, Randers J (2014) Extracted: how the quest for mineral wealth is plundering the planet; a reportnto the Club of Rome/Ugo Bardi. VT Chelsea Green Publishing, White River Junction

Birshan M, Decaix G, Ferreira N, Robinson H (2015) Is there hidden treasure in the mining industry? Low equity prices may offer important M&A opportunities for the mining industry. McKinsey & Company

Buchert M, Schueler D, Bleher D (2009) Critical metals for future sustainable technologies and their recycling potential. UNEP/Öko-Institut e.V, Paris/Darmstadt

Buchert M, Behrendt S, Degreif S, Bulach W, Schüler D (2017) SubSkrit: Substituting critical raw materials in environmental technologies (FKZ: 3714 93 316 0). Öko Institut e.V – Institute for applied ecology. IZT – Institut for Future Studies and Technology Assessment

Buchholz P, Huy D, Liedtke M, Schmidt M (2014) DERA-Rohstoffliste 2014 – Angebotskonzentration bei mineralischen Rohstoffen und Zwischenprodukten – potenzielle Preis- und Lieferrisiken.. DERA Rohstoffinformationen. DERA – Deutsche Rohstoffagentur in der Bundesanstalt für Geowissenschaften und Rohstoffe, Berlin

Buijs B, Sievers H, LAT E (2012) Limits to the critical raw materials approach. Proc ICE Waste Resour Manag 165:201–208

Coulomb R, Dietz S, Godunova M, Nielsen TB (2015) Critical minerals today and in 2030. OECD Publishing, Paris. doi:10.1787/5jrtknwm5hr5-en

Dehoust G, Manhart A, Vogt R, Kämper C, Giegrich J, Priester M, Rechlin A (2017a) ÖkoRess: environmental limits, environmental availability and environmental criticality of primary raw materials (FKZ: 3713 93 302). Öko Institut – Institute for applied ecology; ifeu – Institute for Energy and Environmental Research; projekt-consult
Dehoust G et al. (2017b) ÖkoRess II: further development of policy options for an ecological raw materials policy (FKZ: 3715 32 310 0). Öko Institut – Institute for applied ecology; ifeu – Institute for Energy and Environmental Research; projekt-consult; adelphi research
DoE U (2010) Critical materials strategy. U.S. Department of Energy, Washington
Erdmann L, Graedel TE (2011) Criticality of non-fuel minerals: a review of major approaches and analyses. Environ Sci Technol 45:7620–7630. doi:10.1021/es200563g
EU COM (2008) The raw materials initiative — meeting our critical needs for growth and jobs in Europe. Communication from the Commission to the Council and the European Parliament. COM (2008) 699 final, 4 November 2008. EU COM, Brussels
EU COM (2010) Critical raw materials for the EU. European Commission, Brussels
EU COM (2014) Report on critical raw materials for the EU. European Commission, Brussels
Fairphone (2017) Fairer materials – a list of the 10 we're focusing on Fairphone. https://www.fairphone.com/en/2017/01/26/fairer-materials-a-list-of-the-next-10-were-taking-on/. Accessed 6 march 2017
Faulstich M, Franke M, Mocker M, Kaufhold T, Kroop S (2014) Kritische Rohstoffe für Baden-Württemberg : Grundlagen und Empfehlungen im Rahmen der Landesstrategie Ressourceneffizienz. Umweltwirtschaftsforum: uwf; die betriebswirtschaftlich-ökologisch orientierte Fachzeitschrift 22:133–137
Gandenberger C, Glöser S, Marscheider-Weidemann F, Ostertag K, Walz R (2012) Die Versorgung der deutschen Wirtschaft mit Roh- und Werkstoffen für Hochtechnologien – Präzisierung und Weiterentwicklung der deutschen Rohstoffstrategie. TAB – Büro für Technikfolgen-Abschätzung beim Deutschen Bundestag, Berlin
Giraud P-N (2012) A note on Hubbert's thesis on mineral commodities production peaks and derived forecasting techniques. Procedia Eng 46:22–26. doi:10.1016/j.proeng.2012.09.441
Glöser S, Tercero Espinoza L, Gandenberger C, Faulstich M (2015) Raw material criticality in the context of classical risk assessment. Resour Policy 44:35–46. doi:10.1016/j.resourpol.2014.12.003
Graedel TE et al (2011) Methodology of metal criticality determination. Environ Sci Technol 46:1063–1070. doi:10.1021/es203534z
Graedel T, Gunn G, Espinoza LT (2014) 1. Metal resources, use and criticality. In: Critical metals handbook. Wiley, Chichester, pp 1–19
Graedel TE, Harper EM, Nassar NT, Nuss P, Reck BK (2015) Criticality of metals and metalloids. Proc Natl Acad Sci 112:4257–4262. doi:10.1073/pnas.1500415112
Harper EM et al (2015) Criticality of the geological zinc, tin, and lead family. J Ind Ecol 19:628–644. doi:10.1111/jiec.12213
Helbig C, Wietschel L, Thorenz A, Tuma A (2016) How to evaluate raw material vulnerability – an overview. Resour Policy 48:13–24. doi:10.1016/j.resourpol.2016.02.003
Hubbert MK (1956) Nuclear energy and the fossil fuels. Paper presented at the drilling and production practice, New York, 1 January 1956
ICMM (2012) Trends in the mining and metals industry. International Council on Mining and Metals (ICMM), London
International Institute for Industrial Environmental Economics (IEEE) (2016) Circle of light. The impact of the LED lifecycle. Lund University, Lund
ISO (2006) ISO 14040:2006 environmental management – life cycle assessment – principles and framework vol ISO 14040:2006. DIN Deutsches Institut für Normung e.V, Berlin
ISO (2009) ISO 31000:2009 risk management – principles and guidelines vol ISO 31000:2009. DIN Deutsches Institut für Normung e.V, Berlin
Jevons WS (1865) The coal question: an enquiry concerning the progress of the Nation, and the probable exhaustion of our coal-mines. Macmillan, London

Kaufman D, Kray A, Mastruzzi M (2010) The worldwide governance Indicators: methodology and analytical issues. Draft policy research working paper. Brookings Institution/World Bank
Marscheider-Weidemann F, Langkau S, Hummen T, Erdmann L, Espinoza LT, Angerer G (2016) Rohstoffe für Zukunftstechnologien 2016 – Raw materials for emerging technologies 2016. German Mineral Resources Agency DERA at the Federal Institute for Geosciences and Natural Resources BGR, Berlin
Mauss R, Posch PN (2014) Marktpreisrisiken rohstoffintensiver Unternehmen – Identifikation und Management. In: Kausch P (ed) Strategische Rohstoffe – Risikovorsorge. Springer, Berlin, p 39
May D, Prior T, Cordell D, Giurco D (2012) Peak minerals: theoretical foundations and practical application. Nat Resour Res 21:43–60. doi:10.1007/s11053-011-9163-z
Meadows D, Meadows D (1972) The limits to growth. Earth Island Ltd., London
Morley N, Eatherley D (2008) Material security: ensuring resource availability for the UK economy. C-Tech Innovation Limited, Chester
Moss R, Tzimas E, Kara H, Willis P, Kooroshy J (2011) Critical metals in strategic energy technologies. JRC-scientific and strategic reports, European Commission Joint Research Centre Institute for Energy and Transport, Luxembourg
Nassar NT et al (2011) Criticality of the geological copper family. Environ Sci Technol 46:1071–1078. doi:10.1021/es203535w
Nassar N, Graedel T, Harper E (2015a) By-product metals are technologically essential but have problematic supply. Sci Adv 1:1–10
Nassar NT, Du X, Graedel TE (2015b) Criticality of the rare earth elements. J Ind Ecol:1–11. doi:10.1111/jiec.12237
NRC (2008) Minerals, critical minerals, and the US economy. National Academies Press, Washington, DC
Paley W, Brown G, Bunker A, Hodgins E, Mason E (1952) Resources for freedom: a report to the President. Vol Bd. 1–5. U.S. Government Printing Office, Washington DC
Pfeifer S, Franke M, Mocker M, Faulstich M (2012) Resource strategies for Hesse with special consideration on secondary raw materials. Wasser und Abfall 14:10–14. doi:10.1365/s35152-012-0152-2
Pfleger P, Lichtblau K, Bardt H, Reller A (2009) Rohstoffsituation Bayern: Keine Zukunft ohne Rohstoffe. VBW (Vereinigung der Bayerischen Wirtschaft), München
Reller A (2011) Criticality of metal resources for functional materials used in electronics and microelectronics. Phys Status Solidi (RRL) – Rapid Res Lett 5:309–311. doi:10.1002/pssr.201105126
Rockström J et al (2009) A safe operating space for humanity. Nature 461:472–475
Rustad JR (2012) Peak nothing: recent trends in mineral resource production. Environ Sci Technol 46:1903–1906. doi:10.1021/es203065g
Schandl H et al (2016) Global material flows and resource productivity. An assessment study of the UNEP international resource panel. United Nations Environment Programme, Paris
Schmitz S, Paulini I (1999) Valuation as an element of life cycle assessments - German Federal Environmental Agency method for impact indicator standardization, impact category grouping (ranking), and interpretation in accordance with ISO 14042 and 14043. Federal Environment Agency, Berlin.
Steffen W et al (2015) Planetary boundaries: guiding human development on a changing planet. Science 347. doi:10.1126/science.1259855
Sverdrup HU, Ragnarsdottir KV, Koca D (2017) An assessment of metal supply sustainability as an input to policy: security of supply extraction rates, stocks-in-use, recycling, and risk of scarcity. J Clean Prod 140:359–372. doi:10.1016/j.jclepro.2015.06.085
Tercero Espinoza L et al (2013) Critical raw materials substitution profiles. CRM InnoNet Substitution of Critical Raw Materials, Brussels
The Federal Government (2002) Perspectives for Germany: our strategy for sustainable development; abridged version. The Federal Government, Berlin
The Guardian (2015) China scraps quotas on rare earths after WTO complaint. Guardian News and Media Limited. https://www.theguardian.com/world/2015/jan/05/china-scraps-quotas-rare-earth-wto-complaint. Accessed 5 March 2017

UBA (2010) Statement of the German Environment Agency (UBA) on the "critical raw materials for the EU". German Environment Agency (UBA), Dessau-Rosslau

USGS (2016) Mineral commodity summaries 2016. U.S. Geological Survey, Reston. http://dx.doi.org/10.3133/70140094

VDI (2016) VDI 4800 Blatt 2 (Draft): Ressourceneffizienz – Bewertung des Rohstoffaufwands. Association of german engineers (VDI)

Walz R, Bodenheimer M, Gandenberger C (2016) Kritikalität und Positionalität: Was ist kritisch für wen – und weshalb? In: Exner A, Held M, Kümmerer K (eds) Kritische Metalle in der Großen Transformation. Springer, Berlin/Heidelberg. doi:10.1007/978-3-662-44839-7

Wellmer FW (2008) Reserves and resources of the geosphere, terms so often misunderstood. Is the life index of reserves of natural resources a guide to the future? Z dt Ges Geowiss 159:575–590

Wellmer FW (2012) Was sind wirtschaftsstrategische Rohstoffe? In: Thomé-Kozmiensky KJ, Goldmann D (eds) Recycling und Rohstoffe, Band 5, vol 5. TK-Verlag, Neuruppin, pp 525–535

Zepf V, Simmons J, Reller A, Ashfield M, Rennie C (2014) Materials critical to the energy industry – an introduction. BP p.l.c, London

Chapter 6
Equitable, Just Access to Natural Resources: Environmental Narratives during Worsening Climate Crises

Patrick Bond

Abstract The emergence of distinct resource management narratives—based on principles, analysis, strategies, tactics and alliances—is a natural signal of a maturing field. The speed of this process has been hastened since the 1990s thanks to concern about climate change. What had existed as a relatively minor area of environmental economics that can be termed 'ecological modernisation' suddenly was put into practice—e.g. in water commoditisation and carbon markets—by the time of the 2002 World Summit on Sustainable Development. Simultaneously a discourse about society-nature relations known as environmental justice emerged as a result of differential race impacts of pollution. By the 2000s, this approach had also given birth to 'climate justice' politics. The divergences associated with these arguments are evident in global climate governance, where the 2015 Paris Climate Agreement and G7 deal a few months prior reveal very different standpoints. Now the Trump regime's withdrawal from international obligations amplifies why a new approach that reconciles the ecological modernisation, sustainable development and climate justice perspectives is vital. One useful case is the gradual adoption of 'natural capital accounting' techniques, which in Africa should justify an anti-extractivist politics. The bottom line, though, is that the current neoliberal regime which relies upon market signals is not appropriate for natural resource management, especially in addressing the most vital problem: climate change.

Keywords Climate change • Climate justice • Copenhagen Accord • Ecological Modernisation • Extractivism • International non-governmental organisations • Paris Climate Agreement • Sustainable development

---

P. Bond (✉)
University of the Witwatersrand School of Governance,
2 St David's Place, Parktown, 2050 Johannesburg, South Africa
e-mail: patrick.bond@wits.ac.za

H. Lehmann (ed.), *Factor X*, Eco-Efficiency in Industry and Science 32,
https://doi.org/10.1007/978-3-319-50079-9_6

6.1 Introduction

Never before has the struggle over access to natural resources been as urgent, as at present, what with climate-related crises emerging and leaving behind vast swathes of environmental and social destruction. Water is the most obvious, given the acute droughts and heat waves that swept southern Africa in 2015–2016 and remain in east Africa. These events led to thousands of human and non-human fatalities in the Horn from Kenya to Somalia to Sudan and Ethiopia, not to mention Chile, India and California (where an estimated 100 million trees died). But such droughts—followed in some cases by extreme storms and flooding whose run-off is amplified by the baked ground—are increasingly common. Rising sea levels, shrinking polar ice cover, ocean acidification, massive methane emissions from tundra, coral bleaching and so many other manifestations of the climate crisis populate our daily news reports.

For the sake of equitable, just access to resources—sustainably conserved as if generational stewardship mattered—these crises compel a much more serious world-scale examination of the historic responsibility for ecological damage. Mitigation, adaptation and—in some ways most importantly as an unresolved conflict—compensation should be the urgent tasks of those who have benefited thanks to the over-consumption of fossil fuels. Indeed, the natural resources that society most desperately needs to restrict access to, while at the same time radically improving access to in poor societies, are fossil fuels. As climate activists properly insist, 'leave the oil under the soil, the coal in the hole, the tar-sand in the land and the fracking shale gas under the grass!'

The word 'Anthropocene' has come into vogue to indicate humankind's destructive role in changing our physical terrain at the global scale. Some blame anthropomorphic experiences entirely, a very tempting strategy at the scale of the individual human being, in view of the rise of Donald Trump (2016) and his 'America First Energy Plan':

> We're going to rescind all the job-destroying Obama executive actions including the Climate Action Plan… We're going to save the coal industry [and] Keystone Pipeline. We're going to lift moratoriums on energy production in federal areas. We're going to revoke policies that impose unwarranted restrictions on new drilling technologies… We're going to cancel the Paris Climate Agreement and stop all payments of US tax dollars to UN global warming programs.

But a more scientific way to frame the damage which Trump's fossil-fuel allies and financiers—and also their neoliberal opponents—are imposing on the planet is the 'Capitalocene' as Jason Moore (2013) puts it. For this is not simply a maniacal climate denialist at work; the entire system—world capitalism—externalises the costs of production wherever possible. The standard orthodox remedy is the 'internalisation of those externalities,' typically through a taxation system that should provide a disincentive to pollution or other eco-social costs. But from the standpoint of fair resource allocation, dilemmas of efficiency, tax incidence and tax-price elasticity immediately arise:

- will a fix such as a carbon tax allocate the polluting activities efficiently?;
- will the tax be passed along to consumers without affecting the polluter's balance sheet?;
- if the balance sheet is affected, what level of penalty will compel a change?; and
- will that change be simply incremental (merely adjusting the existing system at the margins) or could it be transformational (e.g. a shift to an entirely new production or emissions regime)?

The contestation of such narratives is increasingly vital, in the wake of the British electorate's rejection of the European Union and the US electorate's curious selection of Trump as President of the United States, and with right-wing movements in other European countries (and Turkey, Hungary, the Philippines and other sites) also now becoming a collective threat to human survival. These right-wing risings further signal limits to the global-scale strategies of neoliberalism, which have allowed right-populist narratives a new lease on life. But in the same breath, the contradictions have also given birth to genuine countervailing politics from the left, now imminent in the rising ferment of mass protest. Yet the right-wing version of populism has the initiative in part because a *unifying* narrative, which the left is still lacking. (Such a narrative, I argue below, must tackle the question of access to resources by grappling honestly with the legacies of ecological modernisation and sustainable development.)

Next, one must overlay upon extremely divergent and dangerous political tendencies the failure of global governance, especially in relation to nature. The impression exists in much of the world that a viable global governance arrangement was achieved in Paris in December 2015, as a result of a supposedly inclusive process leading to the deal, including the 2015 G7 conference in Elmau and two prior path-breaking world climate summits: Copenhagen (2009) and Durban (2011). Irrevocable steps were (allegedly) being taken toward economic decarbonisation, so as to save the planet from catastrophic climate change and solve this aspect of the broader crisis of social reproduction. The reforms would come from within, 'bottom-up' in the case of each country's voluntary 'Intended Nationally Determined Contributions' (INDCs) for the mitigation of greenhouse gas emissions, with wildly unrealistic objectives for limiting temperature rise to 1.5° but no realistic way to do so.

Opposing this elite consensus, one of the most visible International Non-Governmental Organisations (INGOs), Friends of the Earth International (2015), as well as most of what are considered climate justice (CJ) movement components, condemned what they regard as *decaying* global climate governance. However, given the power balance prevailing in Paris, such opposition made no difference whatsoever, and so several debilitating features of the neoliberal strategy were locked in as a result:

- no legally-binding responsibilities and no accountability mechanisms;
- inadequate stated aspirations for lowering global temperatures (the INDCs would leave the earth 3° warmer) (Intergovernmental Panel on Climate Change 2015);
- no liabilities for past greenhouse gas emissions (hence an unpaid 'climate debt');

- renewed opportunities to game the emissions-reduction system through state-subsidised carbon trading and offsets (soon moving from the European Union and North America to the emerging markets led by China); and
- a pass given to emissions emanating from the military, air transport and shipping.

These are fatal flaws. But in any case, it appears that Trump will scuttle Paris, by backing out to the benefit of US-based polluters—whether formally by leaving the United Nations Framework Convention on Climate Change (UNFCCC) or informally by simply not complying with promised emissions cuts. The CJ answer to both Trump and the top-down policy regime—one overwhelmingly favourable to the United States, from where the strategy emanated—appears to be two-fold: an intensification of bottom-up strategies that aim to weaken GHG-emitting state and corporate targets through both direct action (disruptions) and financial divestment. Activists can at least take heart from a November 2016 poll by Yale and George Mason Universities (2016) which found that 69% of US registered voters endorsed Paris (only 13% were opposed) and 78% supported taxes or regulations against greenhouse gas emissions (with 10% opposed).

If Trump keeps his promise to abrogate US participation in the overarching UN climate convention, the CJ strategy appears prescient: to undo the damage at local and national scales in the process of accumulating sufficient power to go global. This strategy was witnessed in the last prior successful attack on global neoliberalism: the acquisition of generically-produced AIDS medicines, achievable in a just manner only when in 2001 intellectual property protections were removed at the World Trade Organisation's Doha ministerial summit. Extraordinary increases in life expectancy resulted, e.g., in South Africa from 52 years in 2004 to 62 years in 2015.

For the foreseeable future, to be sure, the *global* balance of forces appears extremely adverse—especially with the general rise of right-wing populism within neoliberal austerity, and the decline of the Latin American centre-left regimes—and so contrary to Paris propaganda, no system-saving change appears possible at that scale. This adverse situation for global reformism may therefore permit a decisive shift of orientation by INGOs towards the CJ approach, especially because of the potential for unity against Trump at sites like Standing Rock, North Dakota and the April 29, 2017 mass mobilisation on climate in Washington, DC. Not only CJ activists, but INGOs such as Greenpeace and 350.org have taken up direct action and divestment strategies, respectively, during this rapidly-closing window to address climate change and related eco-system breakdowns effectively and *fairly*.

6.2 Intersectional Climate Politics

To achieve justice in resource distribution will require unprecedented alliances. The most acute case is climate change. 'This changes everything', as Naomi Klein (2014) puts it. Intersectional linkages have become even more vital during the presidency of

Donald Trump, just after the British electorate's rejection of the European Union; these events and others to come confirm the demise of global governance, especially in relation to climate policy. With that comes the certainty of runaway climate change. The political turn of 2016 sets the stage not only for similar right-wing populist movements gathering pace in Europe, joining dangerous authoritarian leaders in Turkey, Hungary and the Philippines, but also an excuse for worsening pollution from the Brazil, Russia, India, China and South Africa (BRICS) bloc.

What can be done? Is the new political situation appropriate for renewed attention to social-movement resistance, especially in the form of CJ? After all, the elite strategy associated with climate policy gambles at the June 2015 G7 Summit hosted by Angela Merkel in Elmau and later that year in Paris at the UNFCCC climate conference would attempt to change the trajectory of the world. But this was not due to decisive action, but rather *the opposite:* failure to grapple with climate change as an existential crisis for humanity (Bond 2016). In both sites, the assembled world leaders' economic, geopolitical, technical, ideological and media powers were dedicated to what they presumed was an irreversible, logical proposition: *marginal, market-driven changes augmented by a slight degree of state regulatory assistance and technological breakthroughs will decarbonise the world's energy, land-transport and production systems, and also protect forests*.

Moreover, nothing of substance was offered at either summit to reduce climate change caused by air transport, shipping, the military, corporate agriculture, overconsumption and methane-intensive disposal sources. The self-confidence of those signing the Paris Climate Agreement was a reflection of how far from reality global climate governance had roamed—and how quickly they would be given an unprecedented reality check. The flaws in the elites' logic would lead to two reactions: an initial leftist critique of the Paris Agreement's reliance upon capitalism's self-correction mechanisms and hence the downplaying of climate justice, on the one hand; and then, on the other, a revival of climate denialism along with the rise of extreme petro-military complex power within the country most guilty of historic greenhouse gas pollution. Starting at Copenhagen's 15th UNFCCC Conference of the Parties (COP) in 2009 and gaining momentum in Durban 2 years later (Bond 2012b), the US State Department's chief climate negotiator, Todd Stern, successfully drove the UN negotiations away from the four essential principles that will be required in a future global governance regime to achieve climate justice:

- ensuring emissions-cut commitments are sufficient to halt runaway climate change;
- making the cuts legally binding with accountability mechanisms;
- distributing the burden of cuts fairly based on responsibility for causing the crisis; and
- offering adequate financial compensation to repair weather-related 'loss and damage' occurring directly because of that historic liability (Bond 2012a).

Then in 2015, the Elmau G7 goal was set: 'net zero carbon emissions' by 2100, *50 years too late*. And instead of full decarbonisation, the G7 endorsed 'net' strategies (these are based not on direct cuts but instead on offsets, emissions trading,

'Reducing Emissions through Deforestation and Forest Degradation' known as REDD and carbon sequestration) (Reyes 2015). As for the rest of the world, including the high-pollution emerging markets—especially the BRICS—the so-called 'bottom up' pledge-and-review strategy that Stern imposed in Copenhagen was once again endorsed by the major new emitters (Brazil, South Africa, India and China—termed the 'Basic' countries in 2009). Six months after Elmau, at the COP21, the 'Intended Nationally Determined Contributions'—i.e. *voluntary* pledged cuts—agreed upon by Paris signatories were so low that even if achieved, they would collectively raise the temperature for 2100 to above 3°, likely catalysing runaway climate change (Bond 2016).

Given the extreme dangers to civilisation and planet Earth's species represented by Trump regression and by the Paris and Elmau deals, it is vital that a civil society movement aimed at world policy-makers emerges more forcefully, and that such a movement prevails against both climate denialism and the Paris climate policy within the next decade at the latest. But how will this counter-movement arise? The critical question is whether either or both forces will muster the oppositional power necessary to reverse Trump's petro-military politics and the 'marketisation' of climate policy. Can *global* civil society generate the second part of Karl Polanyi's 'double movement', corresponding to Burawoy's (2013) update of Polanyi (albeit Western-centric yet still conceptually valuable)? If not, both the Trump withdrawal from climate governance and the gambles made in Elmau and Paris will likely result in ecological catastrophe for as climate scientists point out, greenhouse gas emissions lag times mean that market reactions are too little, too late.

How might INGOs or CJs or some combination move the world economy and society off the current trajectory? As shown below, civil society forces currently appear bogged down in interminable conflict over principles, analysis, strategies, tactics and alliances (the 'pasta' problem). The former include the most active Climate Action Network (CAN) members—Worldwide Fund for Nature (WWF) and Greenpeace (2015)—but also one notable self-exiled group from CAN, the environmental justice movement, Friends of the Earth International (FOEI), which typically allies closely with grassroots movements. The INGOs are much more open to alliances with politicians and in some cases corporations and green business federations. And to make matters more complicated, the leaders of the CAN's US chapter have embraced climate justice with gusto. The two most savvy INGOs, 350.org and Avaaz, have become known largely through highly creative social media campaigning, and some—like 350.org's Bill McKibben, Greenpeace International's 2009–2015 director Kumi Naidoo and Greenpeace USA's Annie Leonard—have well-recognised, visionary leadership. In contrast, CJ groups are committed to global critique but essentially local-level solutions, to militant strategies and to 'direct action' tactics described by their best-known proponent, Klein (2014), as 'Blockadia'. To the extent they tackle corporate power at its financial Achilles Heel, they support the divestment strategy catalysed by 350.org. But their strength, especially in the wake of Paris, is in the use of a disruptive repertoire to defend land, water and air against polluters.

The peak moment of Blockadia was probably the defence of North Dakota Treaty Land that the Dakota people had won generations ago, and that is now threatened by the Dakota Access Pipe Line (DAPL). At Standing Rock, popular opposition to DAPL in late 2016 was formidable, and Barack Obama's Army Corps of Engineers backed down. In February 2017, however, DAPL opponents were routed by the Trump regime and forced to leave the land. The anti-DAPL movement had carefully shaped their narrative to highlight water and land protection, as well as the spiritual identity of Native Americans, and hence their politics fused social protection and emancipation.

6.3 Climate Justice Contested

Blockadia activists (depending upon circumstances) point out how the success of their local battles against oil, gas, coal and major greenhouse gas emitters will also benefit humankind and the planet. But the local climate activist movement is so broad—as witnessed in the diversity of signs that appeared at the 400,000-strong New York Peoples March on Climate in September 2014—that all manner of interventions qualify as climate activism (Klein 2013). However, an authentic CJ nomenclature relies in part upon the Climate Justice Now! network's 2007 launch at the Bali COP13 (in opposition to CAN which was seen as too market-oriented). There were five founding principles:

- reduced consumption;
- huge financial transfers from North to South based on historical responsibility and ecological debt for adaptation and mitigation costs paid for by redirecting military budgets, innovative taxes and debt cancellation;
- leaving fossil fuels in the ground and investing in appropriate energy-efficiency and safe, clean and community-led renewable energy;
- rights-based resource conservation that enforces indigenous land rights and promotes peoples' sovereignty over energy, forests, land and water; and
- sustainable family farming, fishing and peoples' food sovereignty.

Some high-profile climate advocates—such as Mary Robinson (a supporter of carbon trading)—soon appropriated the CJ concept for use in a manner inconsistent with these demands, and other strategies for equity also came into dispute. Proponents of Greenhouse Gas Development Rights and 'Contraction and Convergence' approaches also advocated the sale of surpluses on the markets (such as Russia's 'Hot Air' emissions rights under the Kyoto Protocol), which has the tendency to turn a ceiling into a floor. Other concepts such as Common but Differentiated Responsibilities between national states and Converging Per Capita Emissions were, however, in the spirit of CJ as defined in Cochabamba. Most importantly, as Edgardo Lander (2010) explained in his review, the Cochabamba approach to CJ brought together the main contemporary struggles: 'justice/equality/war/militarisation, free trade, food sovereignty, agribusiness, peasants' rights, struggles against patriarchy, defence of indigenous peoples' rights, migration, the critique of the dominant Eurocentric/colonial patterns of knowledge, as well as struggles for democracy.'

The stereotypical premise here is that the INGOs are pragmatic and hence their normative approach is deal-making within existing UNFCCC constraints. In contrast, CJs are principled, radical and unbending in their opposition to compromise on a matter as vital as climate change, and are increasingly unwilling to countenance the kinds of compromises that the December 2015 Paris UNFCCC COP 21 summit represented. This is a simple dichotomy, one that begins to break down somewhat upon closer examination (e.g. Greenpeace's direct actions). However, in the field of climate politics, conditions are becoming so desperate that the more militant, localised approach may be judged by future generations *as the more pragmatic step required for basic civilisational reproduction*, especially if the alternative is what can be termed 'neoliberal nature' or 'ecological modernisation', a conceptual framing implicitly adopted by both world elites and many INGOs.

In what may be its most advanced form of such self-correction within neoliberal capitalism, Deutsche Bank's Pavan Sukhdev initiated 'The Economics of Ecosystems and Biodiversity' (TEEB) within the UN Environment Program to 'make nature's values visible' and thus 'help decision-makers recognise the wide range of benefits provided by ecosystems and biodiversity, demonstrate their values in economic terms and, where appropriate, capture those values in decision-making.' TEEB's search for optimal resource use emphasises 'low-hanging fruit' that can achieve the least costly form of market-facilitated environmental management.

To understand the way CJ engages these contested narratives, consider next, the wider terrain of neoliberal nature. There we find groups which adopt insider positions in relation to global power structures. They broadly agree with the conceptual premises behind global incremental change—following market principles—on the one hand, versus, on the other hand, CJ movements that work locally and reject market strategies. And finally, if we consider how climate policy analyses, strategies, tactics and alliances emerge to lend themselves to this dichotomy, we see INGOs and CJs in conflict over markets and technicist solutions—or 'false solutions' as CJs argue—and that in turn allows us to reframe both INGO and grassroots CJ argumentation in a way that might lead to a different outcome. The rise of Trump makes the search for unity all the more urgent, but also more feasible if a world divestment movement picks up momentum.

6.4 The Wider Terrain of Struggle: Neoliberal Nature

The very different climate framing narratives and the policy strategies that follow them do not represent a brand new debate: distinctions in scale politics and the degree of political pragmatism date back decades within environmentalism. Andrew Jamison's (2001) book *The Making of Green Knowledge* identified a distinct division between first, a mode of thinking and practice he termed 'green business', which co-opted environmentalism into the nexus of capital accumulation and flexible regulatory regimes, while deploying rhetoric of sustainable development and the 'Triple Bottom Line'. The green business ontology is grounded in faith in science and technology, instrumental rationality and market democracy.

In contrast, second, 'critical ecology movements' resist the greening of business by invoking environmental justice, demanding stronger laws and enforcement and engaging in campaigns against corporations and states which despoil the environment. Jamison posited four types of environmentalisms: civic work on campaigns and social ecology; professional interventions based upon science and law; militant direct action and personal environmentalism. Each of these has either reformist or revolutionary currents. Regardless, their politicisation of ecology runs counter to green business in virtually all issues and processes, as will be explored below.

Green business networks have been around for decades, and prominent ones today include the UN Global Compact, World Business Council on Sustainable Development and World Forum on Natural Capital. In sector after sector, they continue to promote the notion that profit can be reconciled with environmental stewardship, e.g., in the Marseille-based World Water Council in which commercialisation of the most basic element of life is firmly promoted so as to achieve more efficient, sustainable management of the resource. These networks are dedicated to natural capital accounting (up to a point), Payment for Ecosystem Services, cleaner production, green products and environmental management systems.

One revealing example of a market-friendly strategy that continues to divide the environmental movement is carbon trading (Bond 2012a). Misgivings first arose about its pilot in the form of lowering US sulphur dioxide emissions in Southern California, which were slower and less effective than command-and-control strategies adopted in Germany's Ruhr Valley during the early 1990s. Nevertheless, large environmental INGOs endorsed the idea when presented with it as a deal-breaking demand by US vice president Al Gore at the COP3 in Kyoto. Gore promised that Washington would sign the Kyoto Protocol if it included carbon markets as an escape hatch for companies that polluted too much and then desired the right to purchase other companies' pollution permits. The US Senate had already voted 95-0 against endorsing Kyoto. Even though Gore won this critical concession, there was no change in attitude on Capitol Hill, so the US never ratified the Kyoto Protocol. Yet carbon markets later became one of the most important wedge issues dividing INGOs from the CJ movement (Lohmann 2006; Leonard 2009; Bond 2012a; Bond et al. 2012). Aside from FOEI, INGOs like Greenpeace sought reforms not abolition of the carbon markets, with WWF strongly endorsing such markets along with renewables investment and innovation (Bryant 2016, 12–13). At some point, the failure of carbon markets should be forcefully addressed by INGOs and their justification for ongoing futile reform advocacy reconsidered.

But this is not the only aspect of neoliberal nature that splits global from local CJ activists; there are other 'false solutions' to the climate and other environmental crises. Many more continue to emerge from private-sector Dr. Strangeloves, some in alliance with the business-oriented INGOs, including:

- controversial forms of so-called 'cleaner energy' such as nuclear fission, 'clean coal', fracking shale gas, hydropower and hydrogen;
- biofuels, biomass, biochar; and
- other geoengineering gambles such as Carbon Capture and Storage, Genetically Modified trees and other biomas, sulfates in the air to shut out the sun, iron filings in the sea to create algae blooms, artificial microbes to convert plant biomass into fuels, chemicals and products and large-scale solar reflection such as industrial-scale plastic-wrap for deserts.

Several very small INGOs with a decidedly CJ orientation—the ETC Group, EcoNexus, the African Biodiversity Network, Gaia and Biofuelwatch (2010)—confirm their opposition to false solutions: 'The shift from petroleum to biomass is, in fact, worsening climate change, increasing deforestation and biodiversity loss, degrading soils and depleting water supplies. Further, the new "bio-based" economy threatens livelihoods, especially in the global South where it encourages "land grabs".' As Kathy McAfee (2012) puts it, 'Compensating the poor and other land users for practices that maintain healthy, "service-producing" ecosystems may be an important part of strategies for sustainable and equitable development. Serious problems arise, however, when such compensation schemes are framed as markets.'

If the 'net' emissions reduction strategy is not questioned, not only will carbon trading and offsets potentially revive (with all their intrinsic problems unresolved), but a panoply of false solutions will be financed by the UNFCCC's Green Climate Fund (GCF) (advertised as soon reaching $100 billion/annum—so long as major donors such as Washington do not default on their obligations). Yet even when INGOs with a CJ orientation get involved in global technicist advocacy, debilitating problems emerge due to adverse power relations, as the GCF has already demonstrated. Sarah Bracking (2015) criticises *both* the mainstream INGOs and CJ participants who 'invested resources and energy into a process that distracts from other types of politics and issue framing' required to address climate finance.

There is just one case, however, in which a neoliberal nature strategy may have appeal to those with a CJ orientation: in contesting extraction of fossil fuels (and other raw materials). This can easily be done in sites where it can be demonstrated that drilling for oil or coal does not make sense *economically*—not just in terms of pollution and environmental (including climate) damage, social dislocation and disrupted spiritual values that are normally the basis for opposition. The main economic argument is that by calculating natural resource depletion associated with extraction, and comparing the outflow of those values with the inflow of retained profits and reinvestment made by the corporations which do the extraction, the overall impact is net negative.

Even though the World Bank has traditionally endorsed extraction, including of fossil fuels, several World Bank staff in the group studying Wealth Accounting and the Valuation of Ecosystem Services (WAVES) annually calculate 'adjusted net savings' as an augmentation of national economic accounting. WAVES' results are extremely disturbing. For example, the World Bank's 2014 *Little Green Data Book*

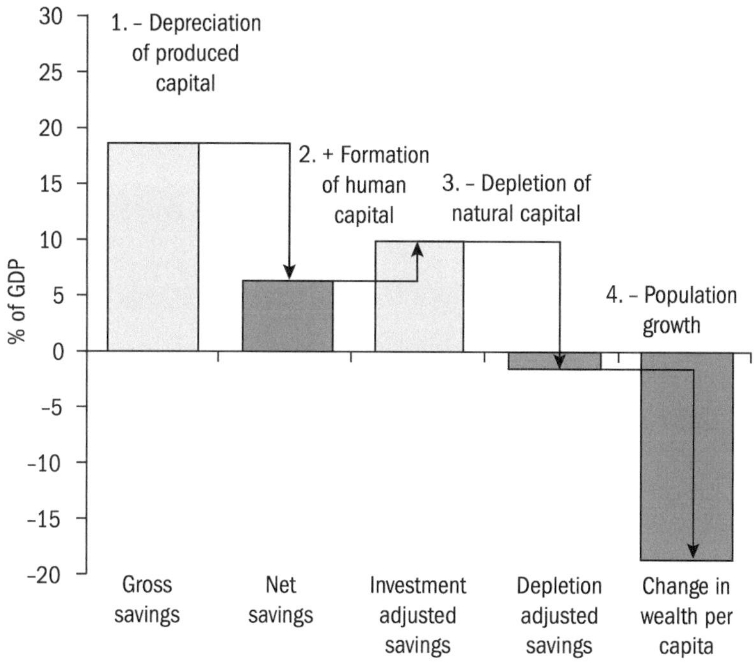

Fig. 6.1 World Bank (2014, vi) decomposes change in wealth per capita, Sub-Saharan Africa, 2010 (Source: public domain)

conceded that '88% of Sub-Saharan African countries were found to be depleting their wealth in 2010', with a 12% decline in per capita wealth that year attributed to the extraction of minerals, energy and forest products (natural capital) (World Bank 2014, 8) (Fig. 6.1).

The adjusted net savings measure is the most ambitious attempt to comprehend changes in wealth incorporating nature. Sub-Saharan Africans had the world's second most dramatic loss between gross and adjusted savings. For North Africa and the Middle East, gross savings were 27.9% but adjusted savings were 8.1% thanks mainly to energy depletion being 12.4% of GNI. In contrast, resource-rich wealthy countries—including Canada, the United States, Australia and Norway—witness sufficient reinvestment by (home-based) corporations, that their natural capital depletion was outweighed by new physical capital, leaving a net positive outcome (World Bank 2015, 12).

Why might CJ groups dedicated to decommodification of life tolerate such counting exercises, given that their premise is the monetary valuation of natural resources (Sullivan 2013)? Paul Robbins and Sarah Moore (2015) also ask whether, 'Amidst—and despite—its deep-seated rejection of technocratic fixes, can political ecology reconcile itself with eco-modernism?' They answer cautiously in the affirmative: 'We suggest that we join together to render eco-modern political ecology a therapeutic empirical project. Rather than become entrenched in an ongoing battle over the dysfunction of the other group's phobic attachments, then, we would instead explicitly engage them, working together to pose specific questions, open to productive exploration' (one of which might be whether natural capital accounting can be deployed to negate most existing forms of extractivism).

What CJ-informed opponents of natural capital accounting have most trouble in criticising, ultimately, is the need to punish polluters by considering the formal monetary liabilities—or some approximation since nature is priceless—so that reparations to environment and affected peoples is sufficiently financed, and in the process an incentive is generated not to pollute in the future. This is the reason to make at least a rough monetary case for 'ecological debt' payments in courts of law. For example, of Nigeria's $11.5 billion claim against Shell for a 2011 oil spill, more than half is meant to compensate the local, small-scale fishing industry. The liability owed to silicosis-afflicted mineworker victims of Anglo American and other gold mining houses has begun to reach payment stage. The South African firms Gencor and Cape PLC had to pay $65 million a decade ago to settle South African asbestos lawsuits after they lost their last appeal in the UK House of Lords. Similar arguments should be made against the MNCs most responsible for what the UN calls loss and damage due to climate change. Ideally, over time, this strategy would develop as 'fine-and-ban', so that as a corporation makes an egregious error, it is fined punitively for the damage done, and then sent packing.

To be sure, there is a danger that if 'fine-and-ban' is not the local state policy, then natural capital accounting will lead, instead, to a 'fee' for pollution, with the damage continuing, alongside ongoing payment. That would be the result if a formal market emerged, such as the EU ETS, and naturally CJ activists beginning with the Durban Group for Climate Justice firmly rejected these in 2004 (Lohmann 2016). The distinction should thus be clear, between valuing nature for ecological debt payment purposes (a fine and ban) on the one hand, and on the other pricing nature for market-making (a fee). As Vandana Shiva put it in a 2014 South African talk, 'We should use natural capital as a red light to destruction, not as a green light' (Bond 2014). The 'red light' strategy is an example of a potential rapprochement between INGO and CJ framing strategies, emphasising technicist analysis in the ecological modernisation tradition as well as being useful to anti-extractivist campaigners who want an economic argument against fossil fuel depletion.

In sum, natural capital accounting is potentially one bridge narrative between INGOs and CJ, especially in making the economic argument to leave resources in the ground, especially fossil fuels. Instead of extracting non-renewable resources when they demonstrably lead to much lower adjusted savings, is there scope for a different narrative addressing the climate debt that ideally should be paid to those who suffer climate change and who are also residents of fossil fuel reserve sites? This has been one route taken by Oilwatch members to implore national leaders in places like Ecuador (the Yasuni case) and Nigeria (Ogoniland) to leave fossil fuels untouched (Bond 2012a). To arrive at that narrative requires one more detour through the philosophies of environmental management: sustainable development.

6.5 The Scorched Earth of Sustainable Development Narratives

If there is an alternative worldview to neoliberal nature, most INGO and CJ narrative shapers and strategists would immediately point to the phrase 'sustainable development'. The 1987 United Nations World Commission on Environment and Development (1987) led by Gro Harlem Brundtland offered a definition still worth returning to. Not only does it contain the intergenerational requirement so well known in the first clause of her definition: 'Sustainable development is development that meets the needs of the present without compromising the ability of future generations to meet their needs'. Consider, as well, her next two sub-clauses in the definition, which first observe 'the concept of "needs", in particular the essential needs of the world's poor, to which overriding priority should be given', hence generating grounds for social justice advocacy. Second, 'the idea of limitations imposed by the state of technology and social organisation on the environment's ability to meet present and future needs' repudiates pro-growth assumptions of those who use the words sustainable development in public relations greenwashing.

The idea gained popularity in 1972 with the first Earth Summit in Stockholm and in *The Limits to Growth* (Club of Rome 1972), culminating in the Brundtland Commission and 1992 Rio Earth Summit. But soon, co-opted by corporations during the 1990s, sustainability was downgraded in favour of neoliberal ideologues' advocacy of export-led growth and the commodification of nature. It poked its head up once again at a 2002 UN earth summit in Johannesburg, which unfortunately fused the UN's strategy with the for-profit agendas of privatisers, carbon traders and mega-corporations which supported the UN Global Compact (which was mostly a fund-raising exercise for a beleaguered institution). Then in 2012 at the next Rio Earth Summit, sustainability was fused with 'Green Economy' rhetoric, biodiversity off-setting and market-centric climate change policy. Sustainability had again flowered, but now with a much more direct relationship to neoliberal nature (Büscher et al. 2014). For the 2015–2030 period, Sustainable Development Goals are now the mantra of the UN and many other multilateral agencies, in spite of extensive critique of the realities they elide (such as by the scholar-activist network TheRules.org [2015]).

But even if this weak version of the sustainability narrative is contested by CJ critics—and also still attacked by the most pollution-intensive fractions of capital—there is no questioning the problem of rampant socio-environmental unsustainability as the world hits what the Club of Rome (1972) had long warned would be 'limits to growth.' These were later termed by the Stockholm Resilience Centre 'planetary boundaries,' of which the most serious threat is diminished carrying capacity for greenhouse gases that cause climate change, and in turn, ocean acidification. There are others: biodiversity loss, stratospheric ozone depletion (abated by the 1987 Montreal Protocol that phased out CFCs by 1996—but leaving atmospheric aerosols as a danger), crises in the biogeochemical nitrogen and phosphorus cycles, other resource input constraints, chemical pollution, freshwater adulteration and evaporation and shortages of arable land (Magdoff and Foster 2011).

Seeking sustainability, many INGOs believe in a 'green capitalism' strategy based on arguments by Gore (2009) and Paul Hawken et al. (1999) (for a critique see Tanuro 2014). But as Ariel Salleh (2010) argues, a serious consideration of externalised costs should include at least three kinds of surplus extractions, both economic and thermodynamic, never comprehensively incorporated by reformers: (1) the social debt to inadequately paid workers; (2) an embodied debt to women family caregivers and (3) an ecological debt drawn on nature at large. The more conservative INGOs have simply ignored the logical trajectory of 'polluter pays' externalisation in the sense that Salleh points out. In contrast, concepts of the left stress sustainability as achieved through distributional equity, non-materialist values and a critique (and transcendence) of the capitalist mode of production. They include the environmental justice vision that African-American activists in North Carolina began to articulate in the 1980s (Bullard 2000); 'anti-extractivism' and the 'rights of nature' articulated by Ecuadorean and Bolivian activists and constitutions (even if not in public policy, as pointed out by Accion Ecologica [2014]) along with the Andean indigenous peoples' versions of *buen vivir* (living well) and allied ideas (Council of Canadians et al. 2011); 'degrowth' (*décroissance*) (Latouche 2004); post-GDP 'well-being' national accounting (Fioramonti 2014), such as Bhutan's Gross National Happiness, which emphasises sufficiency; 'the commons' (Linebaugh 2008) and eco-socialism (Kovel 2007). Strategies for transitioning to genuinely sustainable societies and economies are also hotly debated (Swilling and Annecke 2012; Scoones et al. 2015).

With such creative options flowering—albeit in sometimes reformist mode harking back to indigenous conservation, mere accounting reforms and the slowing (not ending) of capitalism—genuine sustainability does ultimately depend on the nature of the critique of unsustainability. Perhaps the most popular systemic analysis comes from Annie Leonard's (2007) *Story of Stuff* film and book which link the spectrum of extraction, production, distribution, consumption and disposal. Klein's (2014) *This Changes Everything* puts the onus on capitalism for climate change. But expressed most bluntly, Martinez-Alier and Spangenberg (2012) explain what is truly at stake: 'Unsustainable development is not a *market failure* to be fixed but a *market system failure*: expecting results from the market that it cannot deliver, like long-term thinking, environmental consciousness and social responsibility.'

6.6 Conclusion: From Duelling Narratives to Practical Fusions

Hope springs from several sources in play within climate politics. On the one hand, there is underlying humility in the current generation of INGO leaders. And on the other, there is a profound organic intelligence on the part of local CJs, who have the potential to take their perspective onto what at first blush appears to be extremely hostile terrain of ecological modernisation: natural capital accounting. Moreover, the CJ movement's clear grassroots principles are often intimidating to the INGOs. Indeed, more than 30 INGO leaders recently explained why CJ and similar grassroots forces *are holding the INGOs to account*, in an extraordinarily frank and refreshing confessional: 'A new and increasingly connected generation of women and men activists across the globe question how much of our energy is trapped in the internal bureaucracy and the comfort of our brands and organisations. They move quickly, often without the kinds of structures that slow us down. In doing so, they challenge how much time we—you and I—spend in elite conferences and tracking policy cycles that have little or no outcomes for the poor' (Civicus et al. 2014).

It is that cringe-worthy honesty that opens the door to alliances with CJ groups which want, as the Civicus et al. INGO leaders put it, to 'challenge the business as usual approach. Prioritise a local community meeting rather than the big glitzy conferences where outcomes are pre-determined.' It is here that even in the most difficult circumstances, the United States, new linkages of women, Muslims, Latinos, African-Americans, immigrants, indigenous Native Americans, other minorities, the LGBTQ community, poor people, trade unionists, environmentalists and social justice activists are increasingly common, offering a 'social self-defence' that activist Jeremy Brecher (2017) identifies in his survey of anti-Trump struggles at the time of the inauguration. Trump's giant ego, public relations gaffes and inevitable allegations of business misconduct draw his firms towards such chaos like a moth to a flame. As a result, high-profile personal critiques will continue, e.g., via the #grabyourwallet boycott of 75 Trump-related firms, which was quickly successful in removing Ivanka Trump's lines from Nordstrom, Nieman Marcus and Belk.

Trump company targets will continue to be relevant, especially because the new president refuses to divest any of his holdings. But it is in attacking Trump's policies where CJ and other climate activists can find unprecedented unity, using the social protection and emancipation bases and interlinking these with radical environmentalism. To illustrate, Trump's (2016) economic plan is partly an 'Autobahn-Keynesianism': to build filthy infrastructure (fossil-fuel pipelines, airports, roads and bridges); cancel international climate obligations; retract shale gas restrictions and the ban on the Keystone oil pipeline; encourage drilling; defund renewable energy and public transport; and destroy the Environmental Protection Agency (EPA). His choices for the main climate-related Cabinet positions left no room for doubt: former ExxonMobil chief executive Rex Tillerson as Secretary of State; Scott Pruitt as EPA Director (based on his Oklahoma career attacking EPA) and former Texas governor Rick Perry as Secretary of Energy. Tillerson was not only a major contributor to climate policy inertia over several decades as an ExxonMobil leader. More recently his contract for a massive $500 billion Siberian oil drill earned him the Russian 'Order of Friendship' from Vladimir Putin in 2013. A year later, the deal was postponed due to sanctions that followed Putin's invasion of the Crimea.

The climate critique of Trump is also the basis for divestment, e.g., from firms associated with Trump's cabinet and top officials: such as Goldman Sachs bank, ExxonMobil oil, Koch Industries oil, Lockheed Martin military, General Dynamics military and Wells Fargo bank (which is also facing a Standing Rock pipeline-divestment campaign). A broader world divestment movement—drawing from both social protection and emancipation currents—would build on conceptual tools that have been around for years and that immediately came to life after Trump's election (Bond 2017):

- A decade earlier, Joseph Stiglitz had argued that 'unless producers in America face the full cost of their emissions, Europe, Japan and all the countries of the world should impose trade sanctions against the US.'
- Klein reacted to Trump's election: 'We need to start demanding economic sanctions in the face of this treaty-shedding lawlessness.'
- Representing French business, conservative ex-president Nicolas Sarkozy threatened, 'I will demand that Europe put in place a carbon tax at its border, a tax of 1–3%, for all products coming from the US, if the US doesn't apply environmental rules that we are imposing on our companies.'
- The *New York Times* quoted a leading Mexican official at the UNFCCC COP22 summit in Marrakesh: 'A carbon tariff against the US is an option for us,' a stance echoed by a Canadian official.

Some INGOs are already playing a major role in these sorts of crucial battles. Even while it had become obvious that the carbon trading strategy countenanced by Greenpeace had failed, the impact of the group's attacks on Shell Oil was formidable in 2015, far outweighing the failed EU ETS reforms in strategic importance. Any institutional cost-benefit analysis of the INGOs' emissions market advocacy (e.g. the astonishing $200 million spent during 2009–2010 US congressional lobbying for cap-and-trade legislation) would logically place Blockadia strategies far ahead in the benefits category although not without considerable costs (Shell's legal threats against Greenpeace plus the Portland court's fines for blocking its bridge access in mid-2015, for instance). Similarly, 350.org's commitments to direct action grow more vibrant, the more the frustrations rise about the slow pace of state and corporate decarbonisation. In late 2016, this was evident at the Standing Rock showdown where several INGOs assisted Native Americans in fighting (and initially defeating) the DAPL in a manner suffused with respect and local ownership. (Partly as a result, the framing of 'water protectors' rather than climate warriors was emphasised.)

INGO visionaries are fully aware of the limitations of their structural location. For example, African anti-extractive activists—ranging from faith-movement progressives to ActionAid—have responded vigorously to challenges made by Farai Maguwu and Christelle Terreblanche to the 'Alternative Mining Indaba' (AMI), held every February in Cape Town to coincide with the African Mining Indaba of major corporate and state attendees. Instead of being resolutely committed to fighting mining—especially coal, which is increasingly destructive across a range of constituencies—the AMI tends towards mild-mannered reforms. The dispute recorded in Maguwu et al. (2016)—including several duelling op-ed articles in March 2016 in the main African ezine, *Pambazuka*—is one example of the INGO-CJ tensions noted above.

Another example is a reform network of capital and the state, the Extractive Industries Transparency Initiative (EITI), which in 2016 witnessed a legitimacy challenge from the (INGO) 'Publish What You Pay' movement of Soros-funded NGOs (some of which have grassroots CJ connections) when ecause EITI imposed a 'civil society' representative in their decision-making processes through a dubious process.

In other words, more connectivities between these differently-located and philosophically-divergent types of civil society—INGOs and CJs—may unearth the frictions, especially if CJ continues to draw both social protection and emancipation under its ideological umbrella. It is incumbent now upon the better-resourced INGOs to take up the challenge made by Civicus et al. (2014) to provide not just auto-critique but new modes of operation sensitive to the (often more radical) grassroots agenda.

That means, second, a complementary move by CJ groups might be considered, both to scale up their critique so they can offer concrete *global scale* analysis and start networking properly, and then gather sufficient confidence to take on INGO rhetoric, much of which was learned in struggles within the system. This is the argument made by one of the world's leading contemporary historical materialists, David Harvey (1996, 401), who insists that local activists must become more forward-looking, to more decisively link up with each other, and to take the broadest terrain as their mandate (including cultural and spiritual features of ecological and social life):

> The reinsertion of 'rational ordering' indicates that such a movement will have no option, as it broadens out from its militant particularist base, but to reclaim for itself a non-coopted and non-perverted version of the theses of ecological modernisation. On the one hand that means subsuming the highly geographically differentiated desire for cultural autonomy and dispersion, for the proliferation of tradition and difference within a more global politics, but on the other hand making the quest for environmental and social justice central rather than peripheral concerns. For that to happen, the environmental justice movement has to radicalise the ecological modernisation discourse.

To radicalise ecological modernisation requires that in search of a just distribution of natural resources, environmental activists should not boycott the neoliberal nature thesis but instead engage. To be sure, Erik Swyngedouw (2010) warns of the 'post-political' quagmire, when it comes to combating climate change and the corporations and states behind it, insofar as carbon markets and the Green Climate Fund represent 'the predominance of a managerial logic in all aspects of life [and] the reduction of the political to administration where decision-making is increasingly considered to be a question of expert knowledge and not of political position.'

The knowledge sources that undergird such efforts are typically divided into the technical disciplines of green business, the political traditions of eco-social justice and the transcendental experiences of the eco-socialist project. But discovering the way battles over resource allocation unfold in this era of climate crises is surely a matter of debate through praxis in the months and years ahead.

Acknowledgement An earlier version of this paper was presented at the Ernst Strüngmann Forum in Frankfurt, whose participants are thanked for very useful inputs.

References

Bond P (2012a) Politics of climate justice: paralysis above, movement below. University of KwaZulu-Natal Press, Pietermaritzburg

Bond P (2012b) Durban's conference of polluters, market failure and critic failure. Ephemera 12(1–2):42–69

Bond P (2014) Can natural capital accounting come of age in Africa? TripleCrisis, 9 July. http://triplecrisis.com/can-natural-capital-accounting-come-of-age-in-africa-part-2/

Bond P (2016) Who wins from 'climate apartheid'? African climate justice narratives about the Paris COP21. New Politics 60:122–129

Bond P (2017) Tripping up Trumpism through global boycott divestment sanctions. Counterpunch, 20 Jan. http://www.counterpunch.org/2017/01/20/tripping-up-trumpism-through-global-boycott-divestment-sanctions/

Bond P, Allen F, Amisi B, Brunner K, Castel-Branco R, Dorsey M, Ernsting A, Gambirazzio G, Hathaway T, Nel A, Nham W, Sharife K (2012) The CDM in Africa cannot deliver the money. University of KwaZulu-natal Centre for civil society and the Dartmouth College Climate Justice Research Project, published by the Environmental Justice Organisations, Liabilities and Trade Project, Autonomous University of Barcelona. http://www.ejolt.org

Bracking S (2015) The anti-politics of climate finance: the creation and performativity of the green climate fund. Antipode 47(2):281–302

Brecher J (2017) Social Self-Defense: Protecting People and Planet against Trump and Trumpism. Labor4Sustainability, January, http://www.labor4sustainability.org/uncategorized/social-self-defense-protecting-people-and-planet-against-trump-and-trumpism/

Bryant G (2016) The politics of carbon market design: rethinking the techno-politics and post-politics of climate change. Antipode 48:2

Bullard R (2000) Dumping in Dixie: race, class and environmental quality. Westview Press, New York

Burawoy M (2013) Marxism after Polany. In: Satgar V, Williams M (eds) Marxisms in the 21st century. Wits University Press, Johannesburg, pp 34–52

Büscher B, Dressler W, Fletcher R (eds) (2014) Nature™ Inc.: environmental conservation in the neoliberal age. The University of Arizona Press, Tucson

Civicus et al (2014) An open letter to our fellow activists across the globe: building from below and beyond borders. Rustler's Valley, South Africa, 6 August. http://blogs.civicus.org/civicus/2014/08/06/an-open-letter-to-our-fellow-activists-across-the-globe-building-from-below-and-beyond-borders/

Club of Rome (1972) The limits to growth. Universe Books, New York

Council of Canadians, Global Exchange and Fundacion Pachamama (2011) The rights of nature: the case for a universal declaration on the rights of mother earth. Ottawa

ETC Group, EcoNexus, the African Biodiversity Network, Gaia and Biofuelwatch (2010) Biofuels, bioenergy and biochar: false solutions lead to land-grabbing. Press statement, Cancun, Mexico, 10 December. http://www.gaiafoundation.org/blog/bioeconomy-new-threat-livelihoods-and-biodiversity

Fioramonti L (2014) How numbers rule the world. Zed Books, London

Friends of the Earth International (FOEI) (2015) Call to action! Mobilising around COP21—what, where, when? Email from Dipti Bhatnagar Maputo, 12 August

Gore A (2009) Our choice. Rodale, New York

Greenpeace (2015) Greenpeace responds to climate progress at today's G7 meeting. Amsterdam, 12 June. http://www.greenpeace.org/usa/en/media-center/news-releases/Greenpeace-Responds-to-Climate-Progress-at-Todays-G7-Meeting/. Accessed 12 June 2015

Harvey D (1996) Justice, nature and the geography of difference. Basil Blackwell, Oxford

Hawken P, Lovins A, Hunter Lovins L (1999) Natural capitalism. Little, Brown, Boston

Intergovernmental Panel on Climate Change (2015) Synthesis report on the aggregate effect of intended nationally determined contributions. New York. https://unfccc.int/resource/docs/2015/cop21/eng/07.pdf

Jamison A (2001) The making of green knowledge: environmental politics and cultural transformation. Cambridge University Press, Cambridge
Klein N (2013) Conversation. Earth Isl J, September. http://www.earthisland.org/journal/index.php/eij/article/naomi_klein/
Klein N (2014) This changes everything. Simon & Schuster, New York
Kovel J (2007) The enemy of nature: the end of capitalism or the end of the world. Zed Books, London
Lander E (2010) Reflections on the Cochabamba climate summit. Transnational Institute, Amsterdam, 29 April. https://www.tni.org/es/art%C3%ADculo/reflexiones-sobre-la-cumbre-del-clima-en-cochabamba?content_language=en
Latouche S (2004) Degrowth economics. Le Monde Diplomatique 11. https://mondediplo.com/2004/11/14latouche
Leonard A (2009) Story of cap and trade. Story of Stuff Project, Berkeley. http://www.storyofcapandtrade.org
Linebaugh P (2008) The magna carta manifesto: liberties and commons for all. University of California Press, Berkeley
Lohmann L (2006) Carbon trading: a critical conversation on climate change, privatisation and power. Development Dialogue 48
Magdoff F, Foster JB (2011) What every environmentalist needs to know about capitalism: a citizen's guide to capitalism and the environment. Monthly Review Press, New York
Maguwu F, Terreblanche C, Rutledge C, Capel J, Lorgat H (2016) Alternative mining indaba debate. University of KwaZulu-Natal Centre for Civil Society, February–March. http://ccs.ukzn.ac.za
Martinez-Alier J, Spangenberg J (2012) Green growth. http://www.ejolt.org/2015/09/green-growth/
McAfee K (2012) Nature in the market-world: ecosystem services and inequality. Development 55:25–33
Moore J (2013) Anthropocene or Capitalocene? On the origins of our crisis. Part 1: Excerpt from ecology and the accumulation of capital. https://jasonwmoore.wordpress.com
Reyes O (2015) Leaders' declaration G7 summit (climate section) annotated. Institute for Policy Studies, Washington, 9 June. http://genius.com/G7-leaders-declaration-g7-summit-climate-section-annotated/
Robbins P, Moore S (2015) Love your symptoms: a sympathetic diagnosis of the eco-modernist manifesto. Entitlefellows, 19 June. https://entitleblog.org/2015/06/19/love-your-symptoms-a-sympathetic-diagnosis-of-the-eco-modernist-manifesto/
Salleh A (2010) From metabolic rift to metabolic value: reflections on environmental sociology and the alternative globalization movement. Organ Environ 23(2):205–219
Scoones I, Leach M, Newell P (2015) The politics of green transformation. Routledge, London
Sullivan S (2013) Should nature have to prove its value? Green Economy Coalition, London, 13 July. http://www.greeneconomycoalition.org/know-how/should-nature-have-prove-its-value
Swilling M, Annecke E (2012) Just transitions: explorations of sustainability in an unfair world. University of Cape Town Press, Cape Town
Swyngedouw E (2010) Apocalypse forever? Theory Culture Soc 27(2–3):213–232
Tanuro D (2014) Green capitalism: why it can't work. Fernwood Publishing, Toronto
TheRules.org (2015) SDGs. London. http://therules.org/tag/sdgs/
Trump D (2016) An America First Energy Plan. New York, 27 May. https://www.donaldjtrump.com/press-releases/an-america-first-energy-plan
World Bank (2014) Little green data book 2014. Washington, The World Bank Group
World Bank (2015) Little green data book 2015. Washington, The World Bank Group
World Commission on Environment and Development (1987) Our common future. Oxford University Press, Oxford. Climate Change and the American Mind, New Haven: Yale Program on Climate Change Communication (climatecommunication.yale.edu) and Fairfax: the George Mason University Center for Climate Change Communication. climatechangecommunication.org

Part II
Implementation Strategies

Chapter 7
Circular Economy: Origins and Future Orientations

Riina Antikainen, David Lazarevic, and Jyri Seppälä

Abstract The circular economy is increasingly attracting the attention of various actors in Europe and globally. It refers to closing material loops and prolonging the lifetime of materials; and, as such, presents a radically different socio-technological future compared to the unsustainable conventional 'take-make-dispose' economic model. The concepts underpinning the circular economy are not new, and ecological economics, environmental economics and industrial ecology have been highlighted as its significant antecedents. The circular economy requires involvement of all the societal actors: companies, which bring new circular economy business models; consumers, who create the demand for products and services that apply circular economy principles and decision makers, who support the transition with 'better' policy instruments and governance. The circular economy is expected to bring multiple benefits to the environment and the economy, but only a few examples have demonstrated the circular economy's potential economic benefits for industrial actors. This chapter provides an overview of the concepts, principles, expectations, strategies, business models, indicators and future trends connected to the circular economy.

Keywords Circular economy • Theories • Strategies • Indicators • Business models • Future trends

7.1 Introduction

The circular economy is increasingly attracting the attention of governments, businesses and non-governmental organizations (NGOs). It is proposed as a response to the unsustainable conventional 'take-make-dispose' economic model (Ellen MacArthur Foundation 2013) and as a concept that articulates a socio-technological future radically different to the one that exists today (Lazarevic et al. 2016).

R. Antikainen (✉) • D. Lazarevic • J. Seppälä
Finnish Environment Institute, Mechelininkatu 34A, 00250 Helsinki, Finland
e-mail: riina.antikainen@ymparisto.fi; david.lazarevic@ymparisto.fi; jyri.seppala@ymparisto.fi

H. Lehmann (ed.), *Factor X*, Eco-Efficiency in Industry and Science 32,
https://doi.org/10.1007/978-3-319-50079-9_7

The origin of the term has been ascribed to many authors, and although descriptions include a range of meanings and associations, they generally entail representations of cyclical closed-loop systems (Murray et al. 2015). Today, the most prominent definition is provided by the Ellen MacArthur Foundation (2013, 7), which describes the circular economy as

> an industrial system that is restorative or regenerative by intention and design. It replaces the 'end-of-life' concept with restoration, shifts towards the use of renewable energy, eliminates the use of toxic chemicals, which impair reuse, and aims for the elimination of waste through the superior design of materials, products, systems, and, within this, business models.

As the circular economy is a concept that is still in the making, it has been used to denote different modes of material and organisational coordination. For instance, in China, the concept of the circular economy has shifted slowly from a narrow focus on waste recycling to a wider focus on closed material loops in production, distribution and consumption (Su et al. 2013). In Europe, the transition to a circular economy has been framed in broader terms, concerning the systemic changes required in regards to technologies, organisation, society, consumer behaviour, finance methods and policies; including elements such as product-service systems and the collaborative economy (European Commission 2014). For others, such as Stahel (2013), moving toward a circular economy is essentially a question of economics and profit maximisation brought about by stock optimisation. This chapter provides an overview of the concepts, principles, expectations, strategies, business models, indicators and future trends connected to the circular economy.

7.2 Background and Concepts

7.2.1 Origins

The concepts underpinning the circular economy are not new. For instance, the idea of using waste as a resource for economic activity at the firm level can be traced back to the nineteenth century (e.g. Simmonds 1862). Moreover, before the industrial revolution, municipal waste management facilities in Western Europe were described as "material recovery facilities that used manual sorting and sieving to produce secondary raw materials" (Velis et al. 2009, 1283).

The circular economy concept can be seen as the next potential phase in the evolution of waste management systems. Following the need for resource-efficiency, many countries and governments have implemented initiatives guided by ambitious waste management policies. Germany was a pioneer in this field with the *Closed Substance Cycle and Waste Management Act* (Kreislaufwirtschafts- und Abfallgesetz), enacted in 1994, which aimed at promoting closed substance cycle waste management to conserve natural resources and to ensure the environmentally compatible disposal of waste. Japan has also been highlighted as a forerunner after it initiated a legal framework in 2002 through the *Basic Law for Establishing a*

Recycling-Based Society, leading to a fundamental plan for establishing a sound material-cycle society. However, China was the first country to use the term 'circular economy' in terms of waste and resource policy, see Sect. 7.4.

7.2.2 Conceptual Underpinnings

Several attempts have been made to trace the conceptual and theoretical origins of the circular economy (see Murray et al. 2015; Ghisellini et al. 2016). In these efforts, ecological economics, environmental economics, and industrial ecology have all been highlighted as significant antecedents.

Boulding's (1966) seminal *Spaceship Earth* essay espoused the notion that a closed earth and a closed sphere of human activity—whose primary concern should be stock maintenance—would require all outputs from consumption to be constantly recycled. Ecological economists continued this line of thought. For instance, Georgescu-Roegen (1971) proposed a fourth law of thermodynamics where matter, like energy, becomes progressively unavailable. Although controversy has surrounded this proposition—the application of the law of entropy to matter (Ayres 1981)—the message that economic systems should consist of the maximum amount of recycling and renewables possible remains valid. Furthermore, Andersen (2007) suggests that the circular economy is often justified by environmental economics; which contends that the environment provides amenity values, a resource base for the economy, a sink for residual flows and a life support system, and that unpriced or underpriced services should be internalised in the economy.

Beyond the conceptual underpinnings of ecological economics and environmental economics, industrial ecology has been suggested to have perhaps the greatest practical influence on the development of the circular economy (Andersen 2007). Industrial ecology is defined as "the study of material and energy flows resulting from human activities", providing the basis for "developing approaches to close cycles in such a way that the ecological impact of these activities is minimized" (Boons and Howard-Grenville 2009, 13). The primary concerns of industrial ecology are to: improve the metabolic pathways of industrial processes and material use, create closed-loop industrial ecosystems, dematerialise industrial output and systematise patterns of energy use (Ehrenfeld 1997). Thus, the circular economy radically departs from the conventional modes of coordination in the current linear economy.

The Ellen McArthur Foundation also recognises the contribution of more recent ideas in their view of the circular economy, such as: cradle-to-cradle (Braungart and McDonough 2002), the performance economy (Stahel 2013), biomimicry (Benyus 1997), natural capitalism (Hawken et al. 2013), the blue economy (Pauli 2010), and regenerative design (Lyle 1996).

The circular economy is also closely connected to the concepts of the green economy and the bioeconomy. The green economy is perceived as a pathway to sustainability by addressing financial and climate change problems as well as other environmental impacts (OECD 2011; UNEP 2011). The bioeconomy includes all economic activities that are linked to the development and the use of biological products and processes (OECD 2009). Often, the green economy is considered to be a wider umbrella concept when compared to the circular or bioeconomy, in addition to many other more practical concepts and approaches enhancing more sustainable economies (Loiseau et al. 2016). The appearance of these three concepts in scientific literature has been increasing rapidly in recent years, with much of the circular economy research emanating from China (D'Amato et al. 2016).

7.2.3 Organisational Articulations of the Circular Economy

A useful lecture-key to understand the potential material forms of the circular economy is provided by Stahel and Clift (2015). Drawing on references to capital stock (natural, cultural, human, manufactured and financial) and flows, three interpretations of the circular economy are outlined.

The *loop* economy focuses on material flows, whereby product materials are not lost from the economy as waste and instead recycled for the same use (Stahel and Clift 2015). Loops consist of reuse, repair, remanufacturing and material reprocessing, all taking place within local economies, regional and global supply chains. Material ownership typically changes with each loop. Examples include global bulk material recycling, 'high quality' material recycling in the European Union (EU) and the reuse of products in local economies.

Utilising the same loops, the *lake* economy has a primary focus on optimising and managing the use of stock (not flows) and value preservation without changes to ownership (Stahel and Clift 2015). Examples include the operational leasing of vehicles, construction and medical equipment. Maintaining ownership of resource stocks may provide economic advantages through reuse and remanufacturing strategies.

The *performance* economy goes one step further and focuses on optimising the value obtained from using stock and is operationalised through business models that sell goods or molecules as 'services' (Stahel and Clift 2015). It is related to the objective of creating "the highest possible use value for the longest possible time while consuming as few material resources and energy as possible" (Stahel 2015, 128). Strategies for the performance economy are based on the need to reduce the volume and slow the flow of materials through the economy. Examples include selling tire use by the kilometre (e.g., Michelin), power by the hour (e.g., Rolls-Royce turbines) and pay-per-copy office printing (e.g., Xerox). Such an approach reflects the move to product-service systems, defined as "a mix of tangible products and intangible services designed and combined so that they are jointly capable of fulfilling final customer needs" (Tukker 2015, 76).

7.3 Circular Economy Principles and Expected Benefits

Several underlying principles have been suggested to coordinate activity in a circular economy. Stahel (2013) identifies five principles with a focus on the quality of stock in the economy: (1) the smaller the loop (activity-wise and geographically), the more profitable and efficient it is; (2) loops have no beginning and no end; (3) the speed of circular flows is crucial: the efficiency of managing stock in the circular economy is increased with a decreasing flow speed; (4) continued ownership is cost-efficient: reuse, repair and remanufacture without a change of ownership save double transaction costs; and (5) a circular economy needs functioning markets. For the Ellen McArthur Foundation, the circular economy rests on the following three principles that address both resource and system challenges: (1) preserve and enhance natural capital by controlling finite stocks and balancing renewable resource flows; (2) optimise resource yields by circulating products, components and materials at the highest utility at all times in both technical and biological cycles; and (3) foster system effectiveness by revealing and designing out negative externalities.

By adhering to such principles, the circular economy is expected to bring multiple benefits to the environment and the economy. The potential environmental benefits can be categorized, for example, as follows: less input and use of natural resources, increased share of renewable and recyclable resources and energy, reduced emissions, fewer material losses and residuals and keeping the value of products components and materials in the economy (EEA 2016).

From an economic perspective, based on an extensive literature review and summary by Ghisellini et al. (2016), the adoption of circular economy principles is expected to deliver economic benefits for companies and municipalities, due to a decrease of the problem of waste management, a reduction in environmental externalities (lower pollution) and new jobs, opportunities and increased welfare for low income households. In Europe, it has been estimated that by taking advantage of rapid technological development the circular economy could create resource productivity growth of up to 3% annually, and following the circular economy scenario would bring a 7% increase in the gross domestic product (GDP) as compared to current development, with additional benefits for employment (Ellen McArthur Foundation et al. 2015). On a national level, circular economy benefits have been estimated for Dutch, Finnish, French, Spanish and Swedish economies (Wijkman and Skånberg 2015). The assessment identified significant carbon emission reductions, job creation and improvement in the trade balance for all five countries if resource efficiency scenarios are followed. However, whilst recent analyses and discussions about circular economy, and its development, have mainly focused on resource scarcity and environmental impacts, only a few examples have demonstrated the circular economy's potential economic benefits for industrial actors (Lieder and Rashid 2016).

7.4 National Circular Economy Strategies

7.4.1 China's Circular Economy Initiatives

China's engagement with the circular economy started as far back as 1998, and the approach was formally accepted by the central government in 2002 (Yuan et al. 2006). However, this initial approach was driven by a 3R's approach (i.e. reduction, reuse, recycle) (Yuan et al. 2006), which was adopted in the waste management policy of several Western European countries much earlier—such as the Netherlands, in 1978 and Germany, in 1986 (Buclet and Godard 2000). China's Circular Economy Promotion Law is intended to promote the development of the circular economy, improve the resource utilization efficiency, protect and improve the environment and realise sustainable development. It defines the circular economy as "a generic term for the reducing, reusing and recycling activities conducted in the process of production, circulation and consumption" (People's Republic of China 2008). In China, the circular economy is promoted as a top-down national political objective, in contrast to the bottom-up approaches adopted by areas and countries such as EU, Japan and the USA (Ghisellini et al. 2016).

7.4.2 The European Union's Circular Economy Package

In the EU, the discussion on the circular economy kicked off in full force in July 2014 when the European Commission published its communication *Towards a circular economy: a zero waste programme for Europe* (European Commission, 2014). In December 2014, the new Commission withdrew the communication. However, at the same time, the Commission committed to present a new 'more ambitious' package by the end of 2015, which would cover the full economic cycle and not just waste reduction targets. A public consultation was held between May and August 2015, and the *Closing the loop—An EU action plan for the Circular Economy* package was released in December 2015 (European Commission, 2015).

The EU aims at maintaining the value of products, materials and resources in the economy for as long as possible, minimising the generation of waste and developing a sustainable, low carbon, resource efficient and competitive economy. The circular economy is expected to boost the EU's competitiveness, help to create jobs and new business opportunities as well as more efficient and innovative ways of producing and consuming. Revised legislative proposals set targets for the reduction of waste and a long-term path for waste management and recycling (see more detailed description in Chap. 18 in this book).

Following the activities of the EU, many countries including the Netherlands, Scotland and Finland have launched initiatives on the circular economy, as described below.

7.4.3 The Dutch Circular Economy Programme

In the Netherlands, a government-wide programme called *A circular economy in the Netherlands by 2050* was launched in September 2016 (The Ministry of Infrastructure et al. 2016). Its vision of a future-proof, sustainable economy and a liveable earth for future generations requires efficient use and recycling of raw materials as well as obtaining them in sustainable manner. It also necessitates fewer raw materials due to more efficient products and services, thereby helping to reduce the pressures on the living environment and public health.

The programme aims at a completely circular Dutch economy by 2050. The first milestone, in 2030, is a 50% reduction in the use of raw materials like minerals, fossil-based fuels and metals. Five chains and sectors have been given priority in the transition: biomass and food, plastics, manufacturing, construction and consumer goods. To accelerate the transition to a circular economy, the Dutch government plans to draw up 'transition agendas' in these areas, so that by 2050 they will only be using sustainably produced, renewable or generally available raw materials and be generating as little residual waste as possible. To support the high-quality recycling of products, smart return and collection systems are planned. A national raw materials agreement will be concluded with societal partners including the business community, government authorities and NGOs. Additionally, 27 million euros will be earmarked for improved waste separation and to fund new innovations aimed at improving the recycling capability of products.

7.4.4 A Circular Economy Strategy for Scotland

Scotland has launched a circular economy strategy—*Making things last: A Circular Economy Strategy for Scotland*—that builds on the country's progress on the zero waste and resource efficiency agendas and targets a broad set of business and industry opportunities (The Scottish Government 2016). Transition towards more circular economy is expected to benefit the environment, the economy and the communities by cutting waste and carbon emissions and reducing reliance on scarce resources, improving productivity, opening up new markets and improving resilience, and providing more low cost options to access goods with opportunities for social enterprise.

Actions will be taken in four priority areas, namely food and drink and the broader bio-economy, remanufacturing, construction and the built environment and energy infrastructure. The strategy also includes a new food waste reduction target—to cut food waste by a third by 2025—being the first of its kind in Europe. Additionally, a new approach to producer responsibility will be explored through a single framework for all product types that drives choices for reuse, repair and remanufacture. In addition to existing producer responsibility schemes (for batteries, electronic equipment, end of life vehicles and packaging), tires, furniture and mattresses will be taken under consideration. Furthermore, Scotland aims at building a better evidence base to create more understanding on how products and materials flow through the economy, both to measure progress and to identify opportunities. This will be supported by refreshing the key indicators and a mandatory use of the electronic 'edoc' system for waste in Scotland.

7.4.5 The Finnish Circular Economy Roadmap

The world's first circular economy roadmap, *Leading the cycle—Finnish road map to a circular economy 2016–2025*, was published in Finland in September 2016 (Sitra 2016). It was drafted under the direction of the Finnish Innovation Fund Sitra, in co-operation with the Ministry of the Environment, the Ministry of Agriculture and Forestry, the Ministry of Economic Affairs and Employment the business sector and other key stakeholders. Following the roadmap, Finland intends to be a pioneer in the circular economy over the next 5–10 years. The roadmap is based on the idea of maximizing the use of materials and retaining their value in the loop for as long as possible, having the foundation for earnings in services and intelligence-based digital solutions. The transition is seen to require co-operation across the sectoral and industrial borders, and the largest potential for new circular economy solutions is expected to appear in the areas between traditional sectors.

The first steps will be taken by means of five focus areas, based on Finland's traditional strengths, namely sustainable food system; forest-based loops; technical loops; transport and logistics; and common action between legislators, companies, universities and research institutes, consumers and citizens and regions in order to achieve systemic change.

The implementation of the roadmap is organised into three measures: policy measures, key projects and pilots. Selected key projects include: regional co-operation bringing sustainable local food to everyday life, utilizing public procurement and promoting nutrient recycling, recovery of valuable and rare materials contained in electrical and electronic devices by development of a demonstration plant, Finland hosting the World Circular Economy Forum 2017 in Helsinki, open data creating low-carbon and smart transport for development work in the Helsinki metropolitan area, forest industry bioproducts moving from labs to trials, testing the replacement of fossil fuels and companies using production and community side streams, promoting industrial symbiosis concept.

7.5 Circular Economy in Business

New types of business and revenue models are expected to play a significant role in the transition to a circular economy, with the focus placed on the provision of solutions for identified business and consumer customer needs as opposed to the production and sale of products. Business has a key role as the provider of products and services, with product design being one of the main factors affecting the lifetime, material composition, energy efficiency, repairability, decomposability and recyclability of products (Bocken et al. 2016). New business models may significantly decrease the material intensity of the economy, whereby renting, leasing and sharing are enabled by the design of product-service systems.

The transition to a circular economy also means changes in the roles of a broad range of societal and economic actors. Consumers will play a major role in creating the demand for products and services that apply circular economy principles. The role of consumers may also change. The idea of 'prosumers'—individuals who at the same time consume and provide or produce assets—places citizens at both the end and beginning of the supply chain. For example, in local energy markets, prosumerism is a growing trend. People who have installed solar panels can sell the excess energy to the grid. Similar models are also emerging in the sharing economy, e.g., the sharing of private cars and homes.

In the literature, there are several propositions of how to categorise business models; most of them are very similar and use the criterion of the source of value creation (Lewandowski 2016). The Ellen MacArthur Foundation (2015a) presents a broad *ReSOLVE* categorization: *Regenerate*, *Share*, *Optimise*, *Loop*, *Virtualise*, *and Exchange*. *Regenerate* refers to shifting to renewable energy and materials. *Share* denotes the sharing and recycling economy as well as prolonging the life of products. *Optimise* refers to increased efficiency, waste minimization and utilization of information and communications technology (ICT). *Loop* is defined as closing the technical and biological material cycles. *Virtualise* deals with direct and indirect dematerialization. Finally, *Exchange* calls for the utilization of novel materials and technologies. Circular economy business model types are diverse and range from the conversion of non-recyclable waste materials into useable heat, electricity or fuel, e.g., by anaerobic digestion to product leasing models and new production technologies such as 3D printing (Lewandowski 2016).

Even though a wide range of examples of circular economy business models and their implementation exists, the scaling-up of these activities is a challenging task given the path decencies and lock-in in prevailing institutional structures, material infrastructure and actor networks. A large-scale change requires radical change in business mind-sets and management commitment but also legislation and policy and support for infrastructure and social awareness (Lieder and Rashid 2016).

7.6 Assessing the Impacts of Circular Economy

In order to plan progress towards a more circular economy and implement cost-effective actions at product-service system and national levels, it is important to have an ex-ante analysis of the effects of policy measures and circular economy business models. Several micro and macro-level assessment methods exist to study the three dimensions of sustainability; each with their own strengths and limitations that should be recognized in order to utilize them in the appropriate contexts and to allow for the proper interpretation of results.

Circular economy business models will change the interactions between industry sectors and customers along the value chain and the whole economy. The consequences of introducing new business models can be analysed with the help of life cycle based tools on the product/service, company or sector level. Life cycle assessment (LCA) is a standardized method for assessing the environmental considerations of a product and service throughout its entire life cycle (ISO 2006a, b). A complete life cycle includes all product or service phases, i.e., raw material extraction, processing, transportation, manufacturing, distribution, use, re-use, maintenance and recycling as well as final disposal. LCA has been applied in a wide variety of cases, including various waste management assessments from the early 1990s (Björklund et al. 2010). To include economic aspects into the analysis, Swarr et al. (2011) has published a code of practice for environmental life cycle costing, being a consistent framework with the LCA ISO standards. With social LCA (Benoît and Mazijn 2009) a comprehensive three-pillar product or service life cycle study can be performed.

In addition to life cycle based tools, environmental impact assessment, strategic environmental assessment, cost-benefit analysis, risk assessment, material flow accounting or analysis, substance flow analysis, environmental management system and their combinations can be used for the economic and environmental assessments of different circular economy measures (see Finnveden et al. 2007). However, these tools may have too narrow a scope for many decision situations. Therefore, tools such as environmentally extended input-output (EEIO) are more suitable for producing environmental and economic information to support decision-making concerning the circular economy. EEIO tables can give insights into the environmental interventions, employment and value of economic transactions between different sectors in an economy, including output for exports, capital formation and final government and private consumption (Eder et al. 2006). For example, Wijkman and Skånberg (2015) used an EEIO-based model to analyse GDP, employment and greenhouse gas emissions caused by three different circular economy policy scenarios for Dutch, Finnish, French, Spanish and Swedish economies.

EEIO cannot be used to draw very detailed conclusions from a macro-level study because of their quite limited numbers of sectors (typically 60–150 sectors). Furthermore, EEIO is based on static accounting, which makes it less suitable for long-term trend analysis. For future outlooks, models such as computable general equilibrium (CGE) models are more suitable (e.g. Dixon and Jorgenson 2013). On the other hand, CGE models can be integrated into EEIO models (Eder et al. 2006).

Indicators are often considered as a good tool to demonstrate the progress in certain areas; although they have also been recognized to have large weaknesses, and their use in decision support can be challenging (Rosenström 2009). The Ellen MacArthur Foundation (2015b) has proposed circularity indicators to measure the effectiveness of a company's transition from linear to circular models. There is also a range of indicators measuring some elements of circular economy, including indicators related to resource use and resource efficiency, the data for which is collected by many organisations, e.g., the Organisation for Economic Co-operation and Development (OECD) and Eurostat (Resource efficiency scoreboard). However, existing indicators and indicator sets have their limitations, and development work around circular economy indicators continues for example by the European Commission.

7.7 Future Trends in Circular Economy

The transition to a circular economy is a multi-level governance challenge. However, Hobson (2016, 89) notes that current "academic, policy and business-led analyses frame transformations towards the CE as predominantly issues of innovation, technical systems, fiscal and business incentives, and reformulated business models." In doing so, governance by corporate business (Hisschemöller et al. 2006) is the dominant paradigm in the clear majority of pathways to the circular economy, with the role of the state relegated to ensuring 'better'(less) regulation (Lazarevic et al. 2016). There is currently a significant knowledge gap in terms of governance options at the EU, Member State, local authority, private sector and citizen levels. This is not a trivial matter as the circular economy calls for a wholesale transformation of the conventions underpinning the current economic systems.

The role of research is important to determine the appropriate forms of circular economy that decision-makers, business and citizens may want to act upon. This requires research to assess the ecological impacts, costs and benefits of potential circular economy products and services. It is also important to identify possible negative side effects caused by a circular economy transition and to identify the potential actors that may be negatively affected.

Technological innovations are still needed to recycle and reuse contaminated and mixed material flows. In addition to new technologies, societal and organizational innovations related to ownership, fashion and emotion are needed to increase both demand and supply of circular economy business models. In addition to 'better' regulation, governments can support the development other policy instruments, including fiscal and other economic instruments. Furthermore, in a more urbanized world, the role of cities in supporting the circular economy is significant. This means, among other things, how land use planning, smart technologies and virtual platforms, logistics and collection systems enable the closure of material loops.

Communication and information strategies are needed to raise the awareness of manufacturers and the public about responsibility for products throughout their service lives (Stahel 2016). In addition, consumers should be provided with information on the consequences of their choices. It is important that behaviour changes not lead to negative rebound effects; hence, if monetary resources are saved from the purchase of fewer products, the reallocation of funds becomes an important question.

7.8 Conclusions

The move towards a more circular economy is a widely acknowledged need. However, major challenges face this economic transformation. One aspect is that due to the number of overlapping concepts, confusion among stakeholders and decision-makers may lead towards discussion at the general level of definitions instead of action on the ground. Overlapping concepts may also cause competition in policy agendas, whereas promoting the fulfilment of more sustainable economies in general would benefit from creating synergies from various concepts and practical work being implemented under these concepts. Furthermore, the current framing of the circular economy, as a pathway to large scale transformative change, often neglects the roots and origins of the issues it claims to remedy, and, as such, has been suggested to be far from the radical change that it is promoted to be (Hobson and Lynch 2016).

Many countries are incorporating the circular economy in their strategic policy agendas, and similarly, more and more companies are implementing circular practices as their core business models. However, only small minority of companies worldwide have adopted a true circular mindset as a core business strategy, and most business still aim at selling more materials and physical products. New circular economy businesses often require new business logic on earnings and return on investment, which can be a challenging task for existing business and new entrepreneurs.

So far, circular economy discussions have more focused on defining the concepts, bringing up the overall and business benefits and creating a political agenda. Areas such as the role of (smart) cities and consumers in context of circular economy have achieved less attention. However, their role in the transformation is crucial and without them the modes of economic coordination cannot be changed.

References

Andersen MS (2007) An introductory note on the environmental economics of the circular economy. Sustain Sci 2(1):133–140. doi:10.1007/s11625-006-0013-6

Ayres RU (1981) Eco-thermodynamics: economics and the second law. Ecol Econ 26:189–209

Benoît C, Mazijn B (eds) (2009) Guidelines for social life cycle assessment of products. United Nations Environment Programme (UNEP), Paris

Benyus JM (1997) Biomimicry: innovation inspired by nature. William Morrow, New York

Björklund A, Finnveden G, Roth L (2010) Application of LCA in waste management. In: Christensen TH (ed) Solid waste technology & management. Wiley, Chichester, pp 137–160

Bocken NMP, de Pauw I, Bakker C, van der Grinten B (2016) Product design and business model strategies for a circular economy. J Ind Prod Eng 33(5):308–320. doi:10.1080/21681015.2016.1172124

Boons F, Howard-Grenville JA (eds) (2009) The social embeddedness of industrial ecology. Edward Elgar Publishing, England

Braungart M, McDonough W (2002) Cradle to cradle: remaking the way we make things. North Point Pr, London

Buclet N, Godard O (2000) Municipal waste management in Europe: a comparative study in building regimes. Kluwer Academic Publishing, Dordrecht
D'Amato D et al (2016) Green, circular, bio-economy: a comparative analysis. Manuscript
Dixon P, Jorgenson DW (eds) (2013) Handbook of computable general equilibrium modeling, vol 1A and 1B. North Holland, Amsterdam
Eder P, Delgado L, Neuwahl F (eds), Arnold T, Huppes G, Van Oers L, Heijungs R (2006) Environmentally extended input-output tables and models for Europe. Technical Report Series Report EUR 22194 EN. Joint Research Centre (DG JRC), Institute for Prospective Technological Studies, Seville
EEA (2016) Circular economy in Europe. Developing the knowledge base. EEA Report No 2/2016. European Environment Agency, Publications Office of the European Union, Luxembourg
Ehrenfeld JR (1997) Industrial ecology: a framework for product and process design. J Clean Prod 5(1–2):87–95. doi:10.1016/S0959-6526(97)00015-2
Ellen MacArthur Foundation (2013) Towards the circular economy: economic and business rationale for an accelerated transition. Available: https://www.ellenmacarthurfoundation.org/assets/downloads/publications/Ellen-MacArthur-Foundation-Towards-the-Circular-Economy-vol.1.pdf. Accessed 2 Feb 2017
Ellen MacArthur Foundation (2015a) Towards a circular economy: business rationale for an accelerated transition. Available: https://www.ellenmacarthurfoundation.org/assets/downloads/TCE_Ellen-MacArthur-Foundation_9-Dec-2015.pdf. Accessed 2 Feb 2017
Ellen MacArthur Foundation (2015b) Circularity indicators. An approach to measuring circularity project overview. Available: https://wwwellenmacarthurfoundationorg/assets/downloads/insight/Circularity-Indicators_Project-Overview_May2015pdf. Accessed 2 Feb 2017
Ellen MacArthur Foundation, SUN, McKinsey Center for Business and Environment (2015) Growth within: a circular economy vision for a competitive Europe. Available: https://wwwellenmacarthurfoundationorg/assets/downloads/publications/EllenMacArthurFoundation_Growth-Within_July15pdf. Accessed 2 Feb 2017
European Commission (2014) Communication from the Commission to the European Parliament, the Council, the European Economic and Social Committee and the Committee of the Regions: towards a circular economy: a zero waste programme for Europe. Brussels.
European Commission (2015) Communication from the Commission to the European Parliament, the Council, the European Economic and Social Committee and the Committee of the Regions—closing the loop—an EU action plan for the Circular Economy. Brussels
Finnveden G, Björklund A, Ekvall T, Moberg Å (2007) Environmental and economic assessment methods for waste management decision-support: possibilities and limitations. Waste Manag Res 25(3):263–269
Georgescu-Roegen N (1971) The entropy law and the economic process. Harvard University Press, Lincoln
Ghisellini P, Cialani C, Ulgiati S (2016) A review on circular economy: the expected transition to a balanced interplay of environmental and economic systems. J Clean Prod 144:11–32. doi:10.1016/j.jclepro.2015.09.007
Hawken P, Lovins AB, Lovins LH (2013) Natural capitalism: the next industrial revolution. Routledge, London
Hisschemöller M, Bode R, van de Kerkhof M (2006) What governs the transition to a sustainable hydrogen economy? Articulating the relationship between technologies and political institutions. Energy Policy 34(11):1227–1235. doi:10.1016/j.enpol.2005.12.005
Hobson K (2016) Closing the loop or squaring the circle? Locating generative spaces for the circular economy. Prog Hum Geogr 40(1):88–104
Hobson K, Lynch N (2016) Diversifying and de-growing the circular economy: radical social transformation in a resource-scarce world. Futures 82:15–25
ISO (International Organization for Standardization) (2006a) ISO 14040: environmental management—life cycle assessment—principles and framework. International Organization for Standardization, Geneva

ISO (International Organization for Standardization) (2006b) ISO 14044: environmental management—life cycle assessment—requirements and guidelines. International Organization for Standardization, Geneva
Lazarevic D, Valve H, Kautto P (2016) What makes the circular economy? Expectations and visions of a European circular economy. In: Paper presented at the 7th international sustainability transitions conference, Wuppertal, 6–9 Sept 2016
Lewandowski M (2016) Designing the business models for circular economy—towards the conceptual framework. Sustainability 8(1):43. doi:10.3390/su8010043
Lieder M, Rashid A (2016) Towards circular economy implementation: a comprehensive review in context of manufacturing industry. J Clean Prod 115:36–51. doi:10.1016/j.jclepro.2015.12.042
Loiseau E, Saikku L, Antikainen R, Leskinen P, Pitkänen K, Droste N, Hansjürgens B, Kuikman P, Thomsen M (2016) Green economy and related concepts: an overview. J Clean Prod 139:361–371. doi:10.1016/j.jclepro.2016.08.024
Lyle J (1996) Regenerative design for sustainable development. Wiley, New York
Murray A, Skene K, Haynes K (2015) The circular economy: an interdisciplinary exploration of the concept and application in a global context. J Bus Ethics. doi:10.1007/s10551-015-2693-2
OECD (2009) The Bioeconomy to 2030: desingning a policy agenda. OECD Publishing, Paris.
OECD (2011) Towards Green Growth: Monitoring Progress. OECD Indicators, OECD Green Growth Studies, OECD Publishing, Paris
Pauli G (2010) The blue economy. Paradigm Publications, Taos
People's Republic of China (2008) Circular economy promotion law of the People's Republic of China
Rosenström U (2009) Sustainable development indicators: much wanted, less used? Monographs of the Boreal Environment Research 33. Finnish Environment Institute, Helsinki. http://urn.fi/URN:ISBN:978-952-11-3414-2
Simmonds PL (1862) Undeveloped substances: or, hints for enterprise in neglected fields. Robert Hardwicke, London
Sitra (2016) Leading the cycle Finnish road map to a circular economy 2016–2025. Sitra studies 121. http://wwwsitrafi/julkaisut/Selvityksiä-sarja/Selvityksia121pdf. Accessed 2 Feb 2017
Stahel WR (2013) Policy for material efficiency-sustainable taxation as a departure from the throw-away society. Philos Transact A Math Phys Eng Sci 371:20110567. doi:10.1098/rsta.2011.0567
Stahel WR (2015) The performance economy: business models for the functional service economy. In: Misra KB (ed) Handbook of performability engineering. Springer, London, pp 127–137
Stahel WR (2016) Circular economy. Nature 531:435–438
Stahel WR, Clift R (2015) Stocks and flows in the performance economy. In: Clift R, Druckman A (eds) Taking stock of industrial ecology. Springer, Heidelberg
Su B, Heshmati A, Geng Y, Yu X (2013) A review of the circular economy in China: moving from rhetoric to implementation. J Clean Prod 42:215–227. doi:10.1016/j.jclepro.2012.11.020
Swarr TE, Hunkeler D, Klöpffe W, Pesonen HL, Ciroth A, Brent AC, Pagan R (2011) Environmental life cycle costing: a code of practice. Society of Environmental Chemistry and Toxicology (SETAC), Pensacola
The Ministry of Infrastructure, the Environment and the Ministry of Economic Affairs, also on behalf of the Ministry of Foreign Affairs and the Ministry of the Interior and Kingdom Relations (2016) A circular economy in the Netherlands by 2050. Available: https://wwwgovernmentnl/binaries/government/documents/policy-notes/2016/09/14/a-circular-economy-in-the-netherlands-by-2050/17037+Circulaire+Economie_ENPDF. Accessed 2 Feb 2017
The Scottish Government (2016) Making things last—a circular economy strategy for Scotland. Available: http://wwwgovscot/Resource/0049/00494471pdf. Accessed 2 Feb 2017
Tukker A (2015) Product services for a resource-efficient and circular economy—a review. J Clean Prod 97:76–91. doi:10.1016/j.jclepro.2013.11.049
UNEP (2011) Towards a green economy: pathways to sustainable development and poverty eradication. United Nations Environment Programme, Nairobi

Velis C, Wilson DC, Cheeseman CR (2009) 19th century London dust-yards: a case study in closed-loop resource efficiency. Waste Manag 29(4):1282–1290. doi:10.1016/j.wasman.2008.10.018

Wijkman A, Skånberg K (2015) The circular economy and benefits for society jobs and climate clear winners in an economy based on renewable energy and resource efficiency. Club of Rome. Available: https://www.clubofrome.org/wp-content/uploads/2016/03/The-Circular-Economy-and-Benefits-for-Society.pdf. Accessed 8 Feb 2017

Yuan Z, Bi J, Moriguichi Y (2006) The circular economy: a new development strategy in china. J Ind Ecol 10(1–2):4–8. doi:10.1162/108819806775545321

Chapter 8
Financial System, and Energy and Resource Husbandry

R. Andreas Kraemer

Abstract This chapter explores aspects of the relationship between the financial system and resource industries, starting with general criteria for sound investment and an overview of the various materials and resources that need to be distinguished. To this end, the focus is first placed on fossil energy commodities that do not lend themselves to management in a circular economy, before the metals and mining sector and its regulation are presented. The global transformation of energy systems presents an opportunity to phase out a non-circular industry and replace it with one that is characterised less by commodities for consumption and more by commodities for the manufacture of energy conversion equipment and durable investment goods. Combining the energy and mineral resource industries, the impact of the decline of fossil energy industries is discussed, including the implications for international trade, economic activity, public finance and the financial sector. The chapter concludes with the general argument that the financial system is affected by changes in the resource industries and their shift to a circular economy, and that it can facilitate that shift if the political, legal and regulatory framework is right. Finally, a suite of criteria for investment in support of resource sector transformation and the circular economy is proposed.

Keywords Resources • Raw materials • Fossil • Mineral • Metals • Mining • Rare earths • Circular economy • Recycling • Energy transformation • Renewable energy • Investment • Investment risk • Stranded assets • Trade • Economic policy • Growth • Taxation • Fiscal policy

R.A. Kraemer (✉)
Institute for Advanced Sustainability Studies (IASS),
Berliner Str. 130, 14467 Potsdam, Germany
e-mail: kraemer@ecologic.eu

H. Lehmann (ed.), *Factor X*, Eco-Efficiency in Industry and Science 32,
https://doi.org/10.1007/978-3-319-50079-9_8

8.1 Introduction: Investment Criteria and the Variety of Resource Plays

> Invest in businesses that "*offer products and services for reducing the use and consumption of non-renewable resources, or help in substituting non-renewable with renewable resources from sustainable production*".

> Do not invest in businesses that "*extract natural resources or enable such extraction*".

These two rather simple, complementary statements are starting points for defining what financial investors can and, in their self-interest, should do in the field of *resource husbandry*—in the mining industry also referred to as materials stewardship (ICMM 2007)—from extraction or mining through the life cycle to recycling and waste management. They are drawn from the criteria that define the investment philosophy of Ökoworld Lux SA (2016, p. 8–9), a financial asset manager specialising in sustainable investment products. However, for such criteria to work, they must not be treated as absolutes but interpreted, weighed and applied in context.

The Ökoworld philosophy is based on environmental, social and ethical criteria that guide the search for investment opportunities in companies that contribute to building a sustainable economy and society. Such companies can be expected to thrive as global changes force transformations in the dominant patterns of extraction, production, infrastructure, consumption and post-consumption resource management. The criteria are meant to provide a viable system for identifying businesses that will be "winners of the future", i.e., companies that will likely benefit from the changes that are necessary for humanity to live within its planetary boundaries. Likewise, the criteria also help identify businesses that must close—or undergo costly restructuring—as the world undergoes the necessary transformations. Most financial investors will want to avoid such companies and the losses they are doomed to generate.

Resources are extracted, processed, traded and used in value chains that link producers and consumers across continents, countries and business sectors. These value chains are highly complex, and no two criteria can do justice to the practical challenges of assessing resource management. This becomes obvious when considering this typology of resources:

- *Fossil resources*, such as coal, lignite, oil, gas and, some say, peat, that are extracted mostly for one-off consumption, although a small part may be converted into durable products and wastes;
- *Mineral resources*, other than fossil resources, such as ferrous and non-ferrous metals, metalloids and non-metals, which are extracted for both one-off consumption and subsequent dissipation and for repeated use in systems that allow for recycling. Some such resources, like gold or gemstones, enter relatively stable stocks that are managed by humans or as raw materials, before becoming part of the approximately 30 trillion tonne technosphere comprised of buildings and products (Zalasiewicz et al. 2016);
- *Biological resources* resulting from growth in natural, managed or artificial ecosystems, such as fish (and marine mammals), kelp, bushmeat, timber, agricultural

crops, farm animals and their products (e.g. chicken eggs), kitchen garden produce etc., all the way to pharmaceuticals grown in biological reactors. Most biological resources are for one-off consumption, but some products may be durable;
- *Environmental flow resources*, such as wind, sunlight, rain and water flow, ocean currents, tides and waves or geothermal heat. By their nature, these resources are permanently renewed in many locations but vary in their availability over time. They must be used when they exist or will be lost (to humans).

All these resources have different industrial and other uses, investment cycles and capital needs; they provide a multitude of financial opportunities. The number and variety of businesses in the resource sector is large; there are many "resource plays"—large and small, specialised or diversified—for investors with different appetites for risk, time horizons, geographic preferences etc. There are financial analysts and reporters working on technologies and companies for every relevant element in the Mendeleev periodic table, from Lithium (Li) to Uranium (U), even if there are no good uses for the latter. Carbon (C) in solid, liquid and gaseous form is the basis of an industry with a perhaps inflated $33 trillion in asset value to lose as the world shifts to low-carbon technologies (Ryan 2016).

More criteria than the two highlighted at the beginning of this chapter are obviously needed to do justice to the variety of resources; more differentiation and more nuances. Some additional ones are proposed at the end of this chapter. The practice of investment decisions also requires priorities for the short, medium and long term, especially in a sector where the "long term" is really long. Many mining activities are more durable than the states that are supposed to govern and regulate them (Kraemer 2016). In some cases "mining and mining law" are—proverbially—"above the constitution" in the sense that pre-existing rights, the economic might and political influence of mining companies, their links to foreign powers, as well as international agreements (notably on investor protection) limit the ability of governments to regulate the sector, especially in areas of limited statehood suffering from the "resource curse" (Auty 1993, 1994; Ross 1999; Global Witness 2017).

In financial markets, priorities are expressed in prices and price changes over time. The value or price of the various resources, and therefore their attractiveness for investors, depends on their scarcity, the benefits generated by their use, the cost of extraction, and other characteristics, such as the inert nature of some metals like gold that make them useful as near-permanent stores of value.

In society and policy, priority can also be the result of the consequences of resource use that are not captured by financial markets. The chief example is the "price of carbon", which is meant to reflect the damage that carbon emissions do to the Earth's atmosphere, hydrosphere and biosphere, damage which is material if not financial, for current and future generations. There are only a few examples, all of them recent, of collective (non-financial) preferences being expressed through policy, law and regulation to affect prices on financial markets and thus influence the allocation of capital. The political creation of price signals that reflect environmental or social externalities is strongly resisted by many in the financial markets, not only because they are designed to disrupt incumbent businesses but also as a matter of principle. Policy must not, so the credo, be allowed to interfere with financial markets, which are, in the self-interested eyes of many in the sector, best left unregulated. Still, others point to the discrepancy between humanity's need for changes in production and consumption patterns and the direction in which the financial markets push business behaviour and demand a political reconfiguration in the way financial markets work.

As markets evolve, linkages between regions, sectors or products can change in surprising ways. This was illustrated by the spike in energy prices and the heightened interest in bioenergy, notably in the United States, roughly in the period 2005–2008. High energy prices resulted in increased demand for land, soil and water, and the prices for food and feed crops rose accordingly to match the revenue and profit expectations of farmers and land owners. The knock-on effect of energy policy favouring biofuels—prominently in the form of corn-derived ethanol as a high-octane additive to car fuel—resulted in a direct price link between food and fuel on world markets, to the detriment of the poor.

Although all types of resources, raw materials and commodities are captured by the concept of a circular economy, the focus in this section is on fossil energy resources and minerals. This is because the discussion on the economic and financial consequences of changes in the energy economy on the financial system are a topical concern, and financial markets have already learned to deal with a number of specific challenges emanating from the metals and mining sector.

8.2 Fossil Energy Resources with Few Options for Circular Management

The decline of the fossil energy industry has become a global concern and regular item on the agenda of G20 Heads of State and Government as well as G20 Finance Ministers and Central Bank Governors. The issue rose to prominence and could no longer be ignored after being the focus of a speech by the Governor of the Bank of England, Mark Carney (2015), before insurers at Lloyds of London. Subsequently, the leaders of the G20 countries asked the Financial Stability Board (FSB) to establish a Task Force on Climate-Related Financial Disclosures (TFCD) to provide analysis on the issue and advice on how to deal with it (TFCD 2016). The main concern is that a decline in asset values may trigger crises that would then spread through contagion in the wider financial system; crises that need to be guarded against and prepared for.

Important as that may be, the decline of the fossil industry is both inevitable and desirable. There are attempts to "think circular" in relation to fossil carbon and to develop technologies for carbon capture and storage (or sequestration), commonly referred to as CCS. The primary aim here is to avoid the release of carbon dioxide (CO_2) into the atmosphere or to strip CO_2 out of the atmosphere and dispose of it in such a way that it no longer contributes to the heating of the planet. The advantages from the point of view of the incumbent fossil industry and the financial sector are clear. CCS is an add-on technology that protects existing plants, businesses and investments. It prolongs the path on which the industry depends. It is also capital intensive, which is attractive for rent-seeking investors, especially if policy-makers can be persuaded to grant subsidies to the industry. It is furthermore a technology that some hope can be appropriated through patents (and other intellectual property rights) so that it may earn high long-term rents. However, CCS closes no cycle.

There are technological options for managing carbon in cycles, both in biological and in technical systems. The carbon that is taken out of the air (or water) and fixed through photosynthesis in biomass can be used as a fuel. Bioenergy can be stored and transported in solid, liquid and gaseous form. When burned for heat (and power, as the case may be), the carbon is released back into the atmosphere.

Technical processes that work in a similar way are being developed for the conversion of renewable "power to gas" and "power to liquid", using renewable electricity in times of abundance of supply. These processes would provide new options for energy storage, transport and conveyance of renewable energy using existing storage and distribution infrastructure of the oil and gas industry. Also, the various approaches in the conversion of power-to-gas or power-to-liquid can provide fuels or feed-stock for the chemical industry. The central concept is simple: when there is insufficient demand, renewable wind or solar power, which fluctuates in availability irrespective of demand, is used to split water ($2H_2O$) into hydrogen ($2H_2$) and oxygen (O_2). The H_2 is then combined with carbon from CO_2, which is taken from the atmosphere or the off-gas from combustion processes, to form methane (CH_4) first and then longer chains of alkanes and their derivatives.

The products are gaseous and liquid fuels, which can be integrated into existing distribution or "down-stream" infrastructure that is a legacy of the fossil oil and gas industries. Power-to-gas and power-to-liquid technologies are expected to be cost-competitive as soon as the penetration of renewable energies is such that there is "surplus" renewable power often and long enough to operate the conversion plants 1000s rather than 100s of hours in a year.

The implications for strategic energy security are clear: synthetic gas, transport and heating fuels as well as feed-stock for industry derived from domestic renewable energies can substitute for fossil energy that otherwise has to be imported. Each step in the conversion also improves the storability of the product. The technology facilitates the creation of reserve stocks of derived gaseous and liquid fuels, which can be used in times of no wind during a dark and cold winter period to power efficient heat-and-power plants (Kraemer 2016).

8.3 Circular Management of Mineral Resources: Metals, Metalloids, Non-metals

The metals and mining industry provides many opportunities for circular resource management as many, but not all, elements occur in aggregate, solid form and can be used, re-used and recycled many times. Some elements tend to dissipate more easily, because of their chemical and physical properties or because how they are used. For instance, while platinum in jewellery is rarely lost, platinum in the catalytic converters of cars is rarely fully retrieved.

The mining industry presents a number of challenges with respect to environmental protection, human rights—in regards to the rights of indigenous peoples, working conditions and worker rights—and the industry's influence in government. The emergence of repressive, autocratic regimes that seek to capture the rent from resource extraction, tend to be or become corrupt, suppress opposition and eventually need to maintain control through violent means.

Those companies that are listed on stock exchanges and are therefore subject to the scrutiny of financial analysis have an interest in understanding the possible reputational or financial implications of such behaviour (whether it comes from the companies or from government officials acting in the interest of the mining sector). Activists around the world also highlight abuses by smaller companies and other operations, including artisanal mining under repressive or dangerous conditions and illegal operations. Prominent issues are "conflict minerals", "conflict diamonds" or "blood diamonds", but the list also includes the impact of mining operations on water supply and water quality, general levels of corruption and money laundering. Some groups call for divestment from the sector; they demand that institutional investors shun projects and companies that engage in inhumane mining practices.

The financial and the mining sectors have developed responses, such as a collaboration on materials stewardship, eco-efficiency and product policy (ICMM 2007). Furthermore, environmental, social and ethical rating agencies have developed detailed approaches on how to analyse and judge companies in the sector (e.g. Oekom Research 2013). These go beyond the assessment of financial risk and include broader aspects such as reputational risk, regulatory risk or a potential loss of social licence to operate. In effect, these agencies translate concerns from activist organisations, such as Global Witness, into metrics and valuations that can be integrated into financial analysis. At the 2002 World Summit on Sustainable Development, the United Nations (UN) initiated the establishment of the Intergovernmental Forum on Mining, Minerals, Metals and Sustainable Development (IGF). The forum is hosted by the International Institute for Sustainable Development (see IISD 2016) and holds general meetings annually to discuss the main issues affecting the industry, including the internationally agreed Mining Policy Framework (MPF).

An alternative to (some) mining is recycling, which is also fraught with challenges, especially when it comes to recycling those materials and elements that have low value. The recycling industry grew out of the waste management industry and is often still based not on market values of the recycled materials but on regulatory requirements. The goods the industry handles often have a negative economic value, which results in what economists term "moral hazards": When dealing with substances of positive value, all involved in trading it can be expected to stop release into the environment because that would cause an economic loss, but no such precautions can be expected if the value of substances or materials is negative. More regulation and more enforcement are needed to counter illegal release or dumping. The intensity of regulation in the sector also creates risks of collusion between businesses and regulators, and many in the law enforcement agencies regard the waste management and recycling industries as vulnerable to organised crime (Klima et al. 2009).

The recycling industry, like the metals and mining industry, is also internationally connected. Again, there are problems, such as the illegal export of electronic and other hazardous waste from the EU and other developed countries for "recycling" in China and other emerging economies and developing countries. Some such wastes are not processed for material recovery but simply dumped. Others are processed in ways that would not be legal in the country of origin and disregard health, worker safety and environmental protection. Again, the animus of such activity is criminal energy and intent, and there is relatively little the financial sector can to.

One approach to deal with the challenge is the Dodd-Frank Wall Street Reform and Consumer Protection Act of 2010 in the United States (which US President Donald Trump has pledged to repeal). The core of the Act deals with changes to the regulation of financial services and institutions in the US after the recession starting in 2008. However, the Act also includes provisions that might be considered of marginal relevance to avoiding another financial crisis in the US. Some of these address the metals and mining and fossil energy industries and notably establish requirements for the:

- Disclosure of conflict materials in or near the Democratic Republic of the Congo;
- Reporting on mine safety; and
- Reporting on payments by the oil & gas, the coal and minerals industries for the acquisition of extraction or mining licenses.

By virtue of the Act, it is now commonplace for affected industries to work with their suppliers and trace the origin or provenience of minerals and materials that could be tainted with blood as a consequence of the situation in Central Africa. Cooperation across sectors has resulted in a monitoring system that captures most of the relevant smelters and approximately 90% of the materials traded. Company reports usually contain a specific number of smelters that they may be exposed to but cannot verify, and a percentage of materials procured from places they treat as "grey zones". The evolution of the coverage and accuracy of these reports is watched not only by (US) government regulators but also by financial analysts, because failure to keep up with the evolving good practice in the industry might expose a company to regulatory and reputational risk, but also because it would serve as an indicator of general management failure that may affect the governance of a company as a whole. Conversely, continuous improvement in coverage and accuracy can be regarded as a sign of good management.

This example shows that the financial system can act in the interest of resource as well as development policy, but that in order to do so effectively, it requires a policy-induced regulatory framework and an innovation dynamic that aligns short-term micro-economic decisions at firm level with longer-term macro-economic welfare for society. The intended result is a technological and economic transformation of the resource industry, including the waste management and recycling industry, towards an ecologically sustainable circular resource economy. The transformation of the energy sector holds some lessons and has implications for the wider resource sector. These will be explored in the next section.

8.4 Implications of the Unstoppable Shift Out of Fossil Energy Resources

The current shift from fossil energy resources to "green" energy—renewable energy plus storage in many smart grids—is now a global phenomenon (IEA 2016). For economic reasons, this energy transformation (or *Energiewende*) has become self-sustaining and self-accelerating where it has already started, and self-replicating in an increasing number of countries, regions and locations.

The main reason for this boom in green energy is decreasing costs for key energy technologies and equipment, especially wind turbines, solar panels, storage systems and smart energy management systems. The cost reductions are consequences of technology learning that is projected to continue for years to come. The costs of fossil energy, by contrast, are rising. The currently low-to-medium world market prices for fossil energy commodities cannot hide the fact that these resources are getting ever more difficult and expensive to extract and bring to market.

The urgency of climate change and its impacts—global overheating, extreme weather events, desertification, ocean acidification and sea-level rise leading to flooding of coastal areas and low-land river plains—is increasing. The environmental and social costs of fossil energy use have moved from economics textbooks onto front pages and news channels. Financial analysts at Trucost conclude that the highest external damages (so-called externalities) are caused by coal-fired power in Eastern Asia and Northern America and are estimated at US$ 453 billion per annum and US$ 317 billion, respectively. These consist of the damage impacts of GHG emissions, health costs and other damage due to air pollution. "In both instances, these social costs exceeded the production value of the sector" (Trucost 2013). Policy-makers respond by removing some of the subsidies and privileges for the fossil industries and by placing "a price on carbon" through taxes or emission trading systems.

The fossil energy industries are reaching the end of their historical cycle and losing their social license to operate. Bankruptcies of major coal companies signal a trend that is spreading to the oil and gas industry, a trend that is increasing investor risk in an industry with dim prospects of long-term recovery. The result may be a denial of capital as institutional investors pull out of the industries, as they are being asked to do by activist investors and the growing divestment movement, and as they are being nudged to do by ever more stringent risk-disclosure requirements.

The demise of the fossil energy industry overshadows the parallel decline of the much smaller nuclear power industry. Devoid of economic justification, it cannot compete without massive subsidies and privileges, and these are increasingly difficult to hide. This industry will probably not die, but shrink to what it can be without economic rationale: a military technology with marginal relevance to the energy economy.

The contrasting evolutions of the major energy technologies (green energy systems on one side and fossil and nuclear on the other) result in a historic cross-over. The cross-over point, called (cost) "parity" by economists, is usually represented by the intersect of the two curves in their diagrams. In the real world, parity is both a range and an episode. Renewable smart energy systems are generally getting cheaper, but have different costs in different locations. The same is true for fossil (and nuclear) energy systems. The competitive advantage becomes relevant primarily when investment decisions are made. In other words, there are discrete points in time when parity becomes relevant.

The overall effect, however, is simple: in ever larger parts of the world, all new investment goes into renewable smart energy, and progressively, capital is being withheld from the fossil and nuclear industries. Because it is based on sound economics, this shift is not only unstoppable but also transformational. At the end, the winner takes all; essentially, the whole market will shift to renewable energy, similar to the way motor cars replaced horses in transport 100 years ago.

This energy transformation will have significant and potentially disruptive impacts in other areas, notably trade, investment, growth and tax revenue and the ability of companies and countries to service debt. This disruption should be anticipated and prepared for, so that negative effects can be mitigated and risks of contagion can be contained. The disruptions are balanced by a range of obvious benefits of the energy transformation, which are summarized in the next section.

8.5 Financial, Economic and Trade Impact of Phasing Out Fossil Resources

The ongoing shift from fossil energy resources towards renewable energy has implications for capital formation, finance and investment, the wider economy and trade. These are worth exploring because their significant impact and the resulting financial risks and consequences are not yet generally understood.

8.5.1 Impact on Export, Trade and Import of Various Resources and Materials

Simply put, there is an ongoing shift from trading chemical energy commodities for consumption to trading durable equipment for the conversion of (kinetic) energy in environmental flows into electricity as a first product.

The old fossil energy systems mine coal and extract oil and gas, which are stocks of preserved energy, from sites where they occur and uses an extensive—and expensive—global processing and long-distance transport infrastructure to bring derivative products to market for consumption. The energy itself is traded in its chemical form as a commodity. Once a site is exhausted of its resources, the industry moves on, progressing from locations that are easy and cheap to work in to those that are more difficult and expensive. Although technical innovation may obscure it at times, the operational costs of the fossil energy system rise over time (Lovins et al. 2005; summarized and updated in Lovins 2012).

In contrast, the new renewable energy systems harvest environmental flow resources, which are near ubiquitous so that the energy, often in the form of electricity, does not have to be transported over long distances. Many technologies are suitable for deployment at small-scale, which creates the option of self-supply for many users, which then produce the energy they consume and become "prosumers". In such cases, the energy itself is no longer traded. Once a site has been developed for harvesting renewable energy flows, it can be used theoretically in perpetuity, even if physical structures and equipment must be replaced from time to time in practice. In consequence, the new energy systems tend to get cheaper over time, especially as prices for the initial equipment fall rapidly.

Overall, this results in lower trade volumes, both in total and in terms of international trade. The main fossil energy producing countries will have declining exports, and the importing countries will save on their fossil energy import bill. This overall phenomenon can be broken down into three components:

1. *Substitution effect*: whether resulting from fracking for oil and gas (in North America) or the growth of renewables (everywhere), renewable energy reduces the market share of fossil energies;
2. *Quantity effect*: in consequence, the volume (or mass) of internationally traded energy commodities declines;
3. *Price effect*: with lower demand and thus increased competition in the fossil energy industries, the revenue per unit of volume or mass declines; the value of trade declines even faster than the volume of trade.

These three components are likely to persist because they are underpinned by the changing economics of the energy systems, which have resulted in renewable energy being cheaper than fossil (or nuclear) energy (Randall 2016a, b; Sussams and Leaton 2017).

The old and the new energy systems still require large capital investment in infrastructure and equipment, but the raw materials used are likely to change over time (cf. Angerer et al. 2016). Even if technologies evolve and the long-term evolution of the energy system is uncertain, one can surmise that there may be a shift from ferrous metals that dominate the manufacture of equipment for the coal, oil and fossil methane industries to more non-ferrous metals and other elements used in to manufacture renewable energy, storage and ICT equipment for smart energy systems.

The decline of the fossil energy industry will leave behind "stranded assets" in the form of redundant equipment that can be recycled in the building of the new energy industry, which will also require new resources to enter the technosphere. The current overcapacity in the steel sector is already a concern for investors and even governments, to the point that the G20 leaders meeting at the summit in Hangzhou, China, in September 2016 addressed the issue in their final declaration (G20 2016, para. 31).

These developments might result in sectoral and geographic shifts in the trade of raw materials with new opportunities for some mineral-exporting countries. It is possible that the shift toward the new energy systems might be slowed in the short and medium term by bottlenecks in mining for elements that will be in higher demand. The expectation of higher demand in volume and value is already attracting new investors into the field, as the results of a web search for "lithium plays" will demonstrate.

The processing of materials and manufacture of equipment (for investment in the energy sector) and goods, such as batteries for electric cars (for medium-term use followed by recycling), is currently finding new locations around the world. There is a trend to build very large manufacturing plants, so-called giga-factories, to reap economics of scale and accelerate the system transformation. It is unclear if this trend will persist, but the emerging industry structure might facilitate the circular economy of the resources employed through take-back and collection systems with quality control and high expertise in the processing of the materials involved.

There may be also significant growth in down-stream trade in renewable energy, including fuels from (renewable) power-to-gas and power-to-liquid conversion processes. There is furthermore a potential for (small or incremental) increases in international trade in electricity and renewable-power-derived fuels, where interconnections exist and temporal, geographic differences in the available of natural environmental energy flows make such trade advantageous. However, the overall impact of the shift from the old fossil energy to smart renewable systems will likely lead to the decline and cessation of (bulk) fossil energy commodity trade, resulting in a loss of trade in value and volume. These losses will likely not be compensated by net increases in the trade in other raw materials and manufactured goods (Kraemer 2017).

8.5.2 Impact on Investment, the Economy (GDP), Tax Revenue and Subsidies

Furthermore, the total capital needs of the new energy system may well be significantly lower than those of the old energy system. "Every time the world's solar power doubles, the cost of panels falls by 26%" (Randall 2016a), a far above average effect of technology learning. The corresponding effect for on-shore wind is not as high, but at 19% is still above average. So far, technology learning in storage technology seems to replicate the downward cost trajectory of solar power (AECOM Australia 2015; Lazard 2016). Judging from the reports about new findings in material research laboratories, it can be assumed that the trend will continue for at least another 5–7 years before technology learning may settle at a more average pace.

New configurations of equipment using low-voltage, direct-current (LVDC) technology are not only cheaper to build but also much more energy efficient to run. They can provide low-cost modern and smart energy in areas that are not currently served by the power grid, at a technical complexity that is similar to motorcycle maintenance and smartphone applications. With such installations mushrooming in rural areas with no grid or unreliable grids, large scale investment may no longer be required for central power stations and regional or national grids. The investment needs of the new energy system can be shouldered by individual households (Vinci et al. 2017).

In consequence, there is a greatly reduced need for central coordination of the electricity system and accordingly a much lower need for the deployment of large (aggregated) capital. Indeed, capital formation may shift, at least partly, from large aggregators (e.g. stock markets, funds, governments) to individuals, households, microcredit institutions or mutual saving institutions serving local communities. Lower overall capital needs of the new energy industry imply lower opportunities for capital deployment, and a downward pressure on interest levels.

It follows that not only international trade, overall trade, value of the energy system and capital needs decline, but that business volume may also be greatly reduced, due to both a rising share of self-supply that is neither business nor taxable and the reduced cost of the industry and its products and services. Economic activity (as measured by GDP) would be smaller compared to the business-as-usual scenario.

Tax bases are also likely to shrink, because of the lower capital values employed and the lower volumes and values of energy bought and sold. This effect should be roughly in line with the decline in GDP, except that the impact on public finance would be mitigated by phasing out subsidies for the fossil energy industries and a lower overall need for public funding to deal with the external environmental and social costs, the "externalities" imposed by the fossil energy industries on the public. The value of subsidies (in 2012) was estimated to be Euro 57 billion per annum for Germany, of which 90% is linked to the energy system and climate damage (Köder and Burger 2017). The OECD (2015) estimates that direct subsidies to the energy sector account for hundreds of billions per year, and the total value of global subsidies, including external costs, has been estimated by the International Monetary Fund (IMF) to be $5.3 trillion (Coady et al. 2015).

8.5.3 Value Creation from Electrification and an Economic Paradox

Despite the reductions in cost, value, trade, economic activity and tax revenue from the energy sector that can be expected, the energy transformation is likely to be beneficial. This is not only because the money saved on energy supply can go to other and potentially better uses but also for the following two reasons:

1. The total value of damages caused by the industry will diminish; the external cost, for example, in the form of an overheated planet and a legacy of radioactive waste that needs to be kept safe and managed for thousands of future generations, will stop rising;
2. The shift towards electricity as the main energy carrier will help in the creation of additional value beyond what was possible with the chemical and thermal energy from fossil resources.

Electricity is a noble form of energy, more physically valuable than the equivalent chemical energy contained in fossil fuels. The cost of extracting, processing and bringing fossil energies to market, plus profits, determine the price consumers pay. Those prices tend to rise as fossil fuels become more difficult to extract. Renewable energies, especially solar and wind energy, rely on the harvest of free environmental flows; their costs are determined largely by the capital expenditure for photovoltaic equipment and wind turbines, divided by the respective life-time output of electricity (in kWh). These costs tend to come down over time. Paradoxically, the energy with lower physical value has higher costs and prices, and the physically superior electricity is getting cheaper.

The physical value of electricity translates into properties that increase the economic value of electricity for end users. It can be transformed quite easily into other forms of energy—movement, light and heat. Electricity also enables modern information and communication technology (ICT), which is at the heart of the transformations or "digital disruptions" (Khare et al. 2017) that improve efficiencies at many levels.

The demise of fossil fuels and the rise of renewable electricity together produce an apparent paradox in economic development, to the benefit of end users. In return for lower total energy costs they obtain a more valuable form of energy that allows them to create additional value in manifold ways. There is a very large additional "consumer rent" they can enjoy. The value of this rent is difficult to estimate, and it is likely that a large part of it will be enjoyed in ways that are not captured by economic statistics or subject to taxation. Well-being (and perhaps happiness) may rise but not be reflected in GDP growth or an increase in tax revenue.

8.6 Criteria for Resource Husbandry and Circular Economy in Finance

The challenges for moving towards resource and material management guided by the logic of a circular economy must be met by businesses of many sectors. This is obviously the case for the fossil energy and the metals and mining sectors, but is also true for the waste management and recycling industries. It even applies to all companies that design, produce, service or maintain, or repair goods of many kinds. Ways to improve the "eco-efficiency" of business activities, products, services, and consumption patterns include the "cradle-to-cradle" approach (Braungart and McDonough 2002) or (total) life cycle assessment (LCA) among many others.

8.6.1 Criteria for Resource and Material Flow

Investors could look for companies or use their influence as shareholders to ensure that the companies they are invested in have resource and material management systems in place (Jasch 2009) not only as part of their own environmental management systems but also in their relations with and influence over suppliers and (business) customers. Such management systems within companies and along value chains are useful for understanding resource flows, (potential) bottlenecks and other vulnerabilities to disruption in supply quantities or quality, potential alternative sources and substitutes.

Assessing the impact of resource and material use on natural capital and ensuring extraction and mining operations or production and processing systems have the least negative impact on ecosystems and social conditions can be important for customer relations, access to markets and minimizing the impact of future legislation adopted to protect social and environmental interests. The example of conflict minerals and the legislation forcing businesses to cooperate along value chains to trace the sources of their raw materials is an illustration.

Linking resource and material management systems to operational decision-making as well as product development and innovation can be done through internal "shadow pricing". The direction of research and innovation should be to seek out opportunities to improve resource and material efficiency, substitution of scarce, dangerous or environmentally harmful resources and material with more benign alternatives.

Substitution can be through "drop-in substitutes" that leave the production system and product design essentially unchanged. They can be through changes in production processes that require reconfigurations in factories but leave product definitions largely unchanged, or they can completely new products or services. In any case, substitution is likely to affect suppliers and customers in some way, and it is best to involve them as member of a co-development community of new approaches to resource and material management. Financial investors will know that the market embeddedness and thus future business of their investee companies is thus protected.

Including such aspects in the investor analysis of businesses (and even whole sectors) will help identify resources, commodities, value chains and businesses that are at risk of scarcity and supply disruption, the impact of regulation. Investors might avoid investing in—or choose to divest from—companies that fail to track sourcing and use of materials, value their social and environmental impact and investigate substitution options. Such companies are vulnerable to disruption of their business model.

Investors would be wise to anticipate government action in the form of policy, law and regulation wherever the social and environmental cost of resource extraction, processing, use and waste are particularly high in general or at the global level, or where consequences are concentrated and lead to localized extremes that may trigger consequences. The underlying thinking is that governments will be forced to address the social and environmental consequences sooner or later, and that disruption will be harder for those companies that fail to anticipate and prepare.

8.6.2 *Criteria for Product Stewardship, Reuse and Recycling*

Companies in the fossil energy and metals and mining sectors, from extraction or mining to processing of raw materials and trading of commodities, as varied as they are, are still not as diverse as the companies in the many sectors that produce goods and services. And yet, most or all such companies have an impact on resource and material management through their, choice of raw materials, product design, durability and ease of repair, after sales services including maintenance and repair, take-back policies, refurbishing of their own goods to give them a "second life" and the reuse and recycling possibilities they offer. Financial analysts and investors should therefore develop and apply appropriate sets of criteria for the various sectors and types of business.

8.7 Conclusion

This chapter began with an example of relatively simple criteria for (portfolio) investment guided by environmental, social and ethical concerns. Such non-financial considerations are becoming more accepted as the financial services industry learns that they can help improve performance, help ensure due diligence in the monitoring of investee company governance and ward against losses from bad management and cyclical downturns. The financial services industry is also developing its capacities for managing investments with the help of non-financial criteria and supports divestment from companies with business models or practices that do more harm than good (Lewis and Pinchot 2016). In essence, ever larger parts of the financial services industry have now moved to include non-financial criteria as proxies for long-term financial parameters.

The example of the fossil energy industry—a sector that is subject to divestment campaigns as well as being denied capital by some institutional investors because it produces unacceptable external environmental and social cost—shows that non-financial criteria can help anticipate the decline and eventual demise of an industry. Thinking through the implications of that decline on the wider economy, and the metals and mining sector in particular, reveals opportunities for strengthening the resource and materials economy starting from the extractive primary stage. The example of the Dodd-Frank Act demonstrates the importance of policy, law and regulation in guiding the industrial sector and the financial services industry to address concerns that are not financial in the short run but are nevertheless important and potentially material in their longer-term financial implications.

The conclusion from this reflection is that the financial system cannot be expected to take on non-financial arguments, unless they are proxies for financial criteria. Where economic and financial considerations steer investment in an undesirable direction, policy-makers must respond with clear signals and economic incentives (or disincentives). That will be no different when it comes to promoting the circular economy; governments must get the political, legal and regulatory framework right.

References

AECOM Australia (ed) (2015) Energy Storage Study. A storage market review and recommendations for funding and knowledge sharing priorities. AECOM Australia, Sydney. Retrieved from http://arena.gov.au/files/2015/07/AECOM-Energy-Storage-Study.pdf (7 Jan 2017)

Angerer G, Buchholz P, Gutzmer J, Hagelüken C, Herzig P, Littke R, Thauer RK, Wellmer F-W (2016) Rohstoffe für die Energieversorgung der Zukunft: Geologie—Märkte—Umwelteinflüsse. München: acatech. Retrieved from http://www.acatech.de/fileadmin/user_upload/Baumstruktur_nach_Website/Acatech/root/de/Publikationen/Kooperationspublikationen/ESYS_Analyse_Rohstoffe_fuer_die_Energieversorgung.pdf (7 Jan 2017)

Auty RM (1993) Sustaining development in mineral economies: the resource curse thesis. Routledge, London/New York

Auty RM (1994) Industrial policy reform in six large newly industrializing countries: the resource curse thesis. World Dev 22(1):11–26. doi:10.1016/0305-750X(94)90165-1

Braungart M, McDonough W (2002) Cradle to cradle: remaking the way we make things. North Point Press, Berkeley. ISBN 978-0865475878

Carney M (2015) Breaking the tragedy of the horizon—climate change and financial stability. Speech at Lloyds of London on 29 September 2015. Bank of England, London. Retrieved from www.bankofengland.co.uk/publications/Pages/speeches/2015/844.aspx (7 Jan 2017)

Coady D, Parry I, Sears L, Shang B (2015) How large are global energy subsidies. IMF Working Paper WP/15/105. International Monetary Fund (IMF), Washington, DC. Retrieved from www.imf.org/external/pubs/ft/wp/2015/wp15105.pdf (7 Jan 2017)

G20 (2016) G20 leaders' communiqué: Hangzhou summit, 5 September 2016. Retrieved from http://www.g20.utoronto.ca/2016/160905-communique.html (7 Jan 2017)

Global Witness (2017) Oil, gas and mining: corruption and fraud in the oil, gas and mining industries keeps poor countries poor and props up brutal regimes. Retrieved from https://www.globalwitness.org/en/campaigns/oil-gas-and-mining/

ICMM (ed) (2007) Materials stewardship. Eco-efficiency and product policy. International Council on Mining & Metals (ICMM), London. Retrieved from hub.icmm.com/document/19 (7 Jan 2017)

IEA (ed) (2016) World energy investment 2016. IEA—International Energy Agency, Paris. Retrieved via http://www.iea.org/bookshop/731-World_Energy_Investment_2016 (7 Jan 2017)

IISD Reporting Service (2016) 2016 Annual general meeting of the Intergovernmental Forum on Mining, Minerals, Metals and Sustainable Development (IGF)—the Sustainable Development Goals (SDGs) and Mining. 24–28 Oct 2016. Retrieved via http://www.iisd.ca/igf/agm/2016/

Jasch CM (2009) Environmental and material flow cost accounting. Springer, Dordrecht. ISBN 978-1-4020-9028-8

Khare A, Schatz R, Steward B (eds) (2017) Phantom ex machina: digital disruption's role in business model transformation. Springer, Cham/New York/Heidelberg/Berlin. ISBN 978-3-319-44467-3

Klima NJ, Vander Beken T, Van Daele S (2009) Vulnerabilitat der Wirtschaft ffir Organisierte Kriminalitat. Vulnerabilitatsstudien zur Pravention am Beispiel der Abfallentsorgungsindustrie. Kriminalistik 1:28–35

Köder L, Burger A (2017) Umweltschädliche Subventionen in Deutschland. Aktualisierte Ausgabe 2016. Umweltbundesamt (UBA), Berlin. Retrieved from https://www.umweltbundesamt.de/sites/default/files/medien/2546/publikationen/umweltschaedliche_subventionen_2016_.pdf (7 Jan 2017)

Kraemer RA (2016) Twins of 1713: energy security and sustainability in Germany. In: Looney R (ed) Handbook of transitions to energy and climate security. Routledge, London/New York, pp 413–429. ISBN 978-1-857-43745-4

Kraemer RA (2017) Green shift to sustainability: co-benefits & impacts of energy transformation, vol 109. CIGI Policy Brief 109, pp 10. Retrieved from https://www.cigionline.org/publications/green-shift-sustainability-co-benefits-and-impacts-energy-transformation

Lazard (ed) (2016) Lazard's levelized cost of storage—version 2.0. Lazard, New York. Retrieved from https://www.lazard.com/media/438042/lazard-levelized-cost-of-storage-v20.pdf (7 Jan 2017)

Lewis E, Pinchot AC (2016) Debunking 4 myths about sustainable investing. Retrieved from http://www.wri.org/blog/2016/12/truth-about-sustainable-investing-4-myths-debunked (7 Jan 2017)

Lovins AB (2012) A farewell to fossil fuels. Answering the energy challenge. Foreign Aff 91(2):134–146. Retrieved via http://www.rmi.org/Knowledge-Center/Library/2012-01_FarewellToFossilFuels (22 Jan 2017)

Lovins AB, Datta EK, Bustnes O-E, Koomey JG, Glasgow NJ (2005) Winning the oil endgame. Innovation for profits, jobs, and security. Rocky Mountain Institute, Snowmass. Retrieved via http://www.rmi.org/Winning%20the%20Oil%20Endgame (22 Jan 2017)

OECD (ed) (2015) OECD companion to the inventory of support measures for fossil fuels 2015. OECD Publishing, Paris. Retrieved via http://www.oecd.org/environment/oecd-companion-to-the-inventory-of-support-measures-for-fossil-fuels-2015-9789264239616-en.htm (7 Jan 2017)

Oekom Research (2013) Oekom industry focus: metals & mining. Oekom Research, München. Available via http://www.oekom-research.com/index_en.php?content=industry-focus

Ökoworld (ed) (2016). SRI- und Jahresbericht zum 31. Oktober 2015. Ökoworld Lux SA, Luxemburg. Retrieved from https://www.oekoworld.com/fileadmin/downloads_oekoworld/classic/jahresbericht_ovc.pdf (7 Jan 2017)

Randall T (2016a) Wind and solar are crushing fossil fuels. Record clean energy investment outpaces gas and coal 2 to 1. (6 April 2016). Retrieved from https://www.bloomberg.com/news/articles/2016-04-06/wind-and-solar-are-crushing-fossil-fuels (22 Jan 2017)

Randall T (2016b) The world nears peak fossil fuels for electricity. Coal and gas will begin their terminal decline in less than a decade, according to a new BNEF analysis. (13 June 2016). Retrieved from https://www.bloomberg.com/news/articles/2016-06-13/we-ve-almost-reached-peak-fossil-fuels-for-electricity (22 Jan 2017)

Ross ML (1999) The political economy of the resource curse. World Politics 51(2):297–322. doi:10.1017/S0043887100008200
Ryan J (2016, July 11) Fossil fuel industry risks losing $33 trillion to climate change. Bloomberg. Retrieved from https://www.bloomberg.com/news/articles/2016-07-11/fossil-fuel-industry-risks-losing-33-trillion-to-climate-change (7 Jan 2017)
Sussams L, Leaton J (2017) Expect the unexpected: the disruptive power of low-carbon technology. Grantham Institute at Imperial College London, London. Retrieved via http://www.carbontracker.org/report/expect-the-unexpected-disruptive-power-low-carbon-technology-solar-electric-vehicles-grantham-imperial (5 Feb 2017)
Task Force on Climate-Related Financial Disclosures (TFCD) (ed) (2016) Recommendations of the task force on climate-related financial disclosures. Financial Stability Board (FSB), Basel. Retrieved from https://www.fsb-tcfd.org/wp-content/uploads/2016/12/16_1221_TCFD_Report_Letter.pdf (7 Jan 2017)
Trucost (ed) (2013) Natural capital at risk: the top 100 externalities of business. Trucost, London. Retrieved from https://www.trucost.com/publication/natural-capital-risk-top-100-externalities-business (7 Jan 2017
Vinci S, Nagpal D, Parajuli B (2017) Accelerating off-grid renewable energy—key findings and recommendations from IOREC 2016. International Renewable Energy Agency (IRENA), Abu Dhabi. Retrieved from http://www.irena.org/DocumentDownloads/Publications/IRENA_Accelerating_off-grid_renewables_2017.pdf (5 Feb 2017)
Zalasiewicz J, Williams M, Waters CN, Barnosky AD, Palmesino J, Rönnskog A-S, Edgeworth M, Neal C, Cearreta A, Ellis EC, Grinevald J, Haff P, Ivar do Sul JA, Jeandel C, Leinfelder R, McNeill JR, Odada E, Oreskes N, Price SJ, Revkin A, Steffen W, Summerhayes C, Vidas D, Wing S, Wolfe AP (2016) Scale and diversity of the physical technosphere: a geological perspective. Anthropocene Rev 4:1–14. doi:10.1177/2053019616677431. Retrieved from http://anr.sagepub.com/content/early/2016/11/25/2053019616677743.full.pdf+html (7 Jan 2017)

Chapter 9
Developing Resource Competence – Anchoring Resource Conservation and Efficiency in the German Education System

Carolin Baedeker, Holger Rohn, Michael Scharp, and Jaya Bowry

Abstract Commissioned by the Federal Environmental Agency (Umweltbundesamt), the project "Identifying and Developing Opportunities for All Areas of Education in the Fields of Resource Conservation and Resource Efficiency" (BilRess) contributed to raising awareness for resource conservation and resource efficiency by further incorporating these subjects in all areas of education in Germany. After an inventory analysis of existing offerings on resource conservation and resource efficiency, the areas in need for action have been successfully identified and possible approaches for different educational contexts were suggested. Based on this, the "Road Map Resource Education" was developed. This lays the foundation for the integration of the subjects resource conservation and resource efficiency. Concurrently, the BilRess website (www.bilress.de) and the network "Education for Resource Conservation and Resource Efficiency" was created.

The project was funded through the German Federal Ministry for the Environment, Nature Conservation and Nuclear Safety (BMU) within the context of the Ufoplan 2012 (FKZ 371293103). After the closure of the project, the BilRess-Network continues to be carried out in the framework of the "Kompetenzzentrum Ressourceneffizienz 2015–2019" (Competence Centre for Resource Efficiency 2015–2019), which is run by the VDI Centre for Resource Efficiency (VDI ZRE). This paper outlines the course of the project and presents its results. It focuses on developing recommendations, strategies and best practices for further anchoring the

C. Baedeker (✉)
Wuppertal Institute for Climate, Environment and Energy. [Wuppertal Institut für Klima, Umwelt, Energie], Döppersberg 19, 42103, Wuppertal, Germany
e-mail: carolin.baedeker@wupperinst.org

H. Rohn • J. Bowry
Faktor 10 – Institut für nachhaltiges Wirtschaften gemeinnützige GmbH, Alte Bahnhofstrasse 13, 61169, Friedberg, Germany

M. Scharp
IZT – Institute for Future Studies and Technology Assessment [Institut für Zukunftsstudien und Technologiebewertung], Schopenhauerstr. 26 14129, Berlin, Germany

H. Lehmann (ed.), *Factor X*, Eco-Efficiency in Industry and Science 32,
https://doi.org/10.1007/978-3-319-50079-9_9

subjects resource conservation and efficiency in the German educational system. Additionally, the development of the BilRess-Network is outlined and a forecast on the topic of resource education is provided.

Keywords Resource efficiency • Resource conservation • Resource education • Competencies • Roadmap for resource education • Network

9.1 Resource Education: Initial Situation and Problem Context

The natural resources available on Earth are limited. Their exploitation goes hand in hand with considerable negative effects for humans and the environment, In light of these and other challenges linked to climate change, the conservative and efficient use of natural resources are important goals we should be pursuing with all the means available to us. Economic costs linked to the exploitation of resources are also considerable. Material costs represent a hefty cost factor for the processing industry in Germany, amounting to 57.4% (Destatis 2015) of the total turnover (approximately 1976 billion euros, as of 2013). Whilst there still is no (global) decoupling of economic growth from the exploitation of resources, we remain on a non-sustainable path.

The resource transition, like the energy transition, is an unavoidable step towards sustainable development (Liedtke et al. 2014, 2015). It is essential that we drastically curb our consumption of resources (metals, minerals, etc.) and the pollution or consumption of the environmental media (earth, air, water) if we are to reach a more sustainable level (German Federal Government 2012; BMU 2012). The exploitation of natural resources due to human activity has continually risen over the past few decades, and global resource extraction is now more than twice as high as it was thirty years ago. Whilst in 1980 the figure was still around 36 billion tons a year, by 2011 it had risen to 78 billion tons (SERI/WU Vienna 2014; BMUB 2016). By 2050, the global population will have reached around ten billion people, and if our consumption habits remain the same, then "more than 140 billion tons of minerals, ores, fossil fuels and biomass" (BMUB 2016: 11) will be required (UNEP 2011). Economic studies show that the total material consumption (TMC)[1] in most industrialised countries amounts to an average of 40 to 50 tons per capita each year. It has therefore exceeded a manageable level four or fivefold (Dittrich et al. 2012; Bringezu et al. 2009; Lettenmeier et al. 2014). According to calculations made in Germany, the total consumption of raw materials in the area of production and consumption amounts to 73.3 tons per capita (Wang et al. 2013; Bringezu et al. 2009). In contrast, for an average citizen – based on the area of consumption – sustainable resource consumption is estimated at around 8 tons per capita, per year. Corresponding measures and altered consumer behaviour could enable us to reach this level by 2050

[1] http://www.un.org/sustainabledevelopment/sustainable-development-goals/

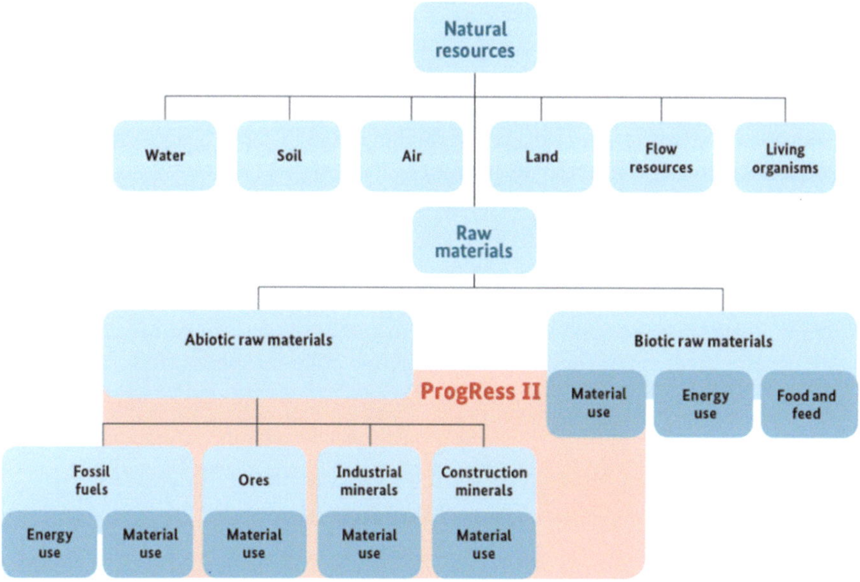

Fig. 9.1 Focus of ProgRess (BMUB 2016: 38)

(Lettenmeier et al. 2014). Here, a fundamental shift in the approach to resources must be based on two pillars: resource conservation and resource efficiency (Schmidt-Bleek 2007). Given its significance, the topic of resource consumption has found its way onto the agendas of Europe's environmental and economic policy makers (e.g. "Thematic Strategy on the Sustainable Use of Natural Resources", 2005; the flagship initiative "Resource-efficient Europe", 2011; "Roadmap for a Resource-Efficient Europe", 2011). At the national level, the German Federal Government had already formulated the goal of doubling raw-material productivity by 2020 (compared to 1994) back in 2002 in the German Sustainability Strategy. Continuing on from this, in 2007 it became a primary goal to develop the German economy as a resource-efficient economy by 2020 (BMU 2007). Adopted on 29 February 2012, the "Programme for the Sustainable Use and Conservation of Natural Resources for Germany", German Resource Efficiency Programme (ProgRess), has been put in motion (BMU 2012), and was further developed with ProgRess II in 2016. ProgRess II also aims to increase resource conservation and efficiency along the entire value chain. So far, however, there has been a focus on the material use of natural resources, but not on their use for energy. In order to better dovetail the efforts towards energy and material savings so that they can be mutually beneficial, the energy and material flows should be incorporated more effectively into a joint approach, where this makes sense (BMUB 2016: 9) (Fig. 9.1).

Sparing and efficient use of natural resources requires an awareness of the problem and knowledge among the actors involved, as well as corresponding competence in design. It is therefore important to raise awareness of resource conservation and efficiency and to promote the development of a culture of sustainable use of resources.

That is where the "BilRess" project "Education for Resource Conservation and Resource Efficiency" comes in. The overarching goal of the project was to develop an education strategy in order to make a contribution to current resource policy (including the German Resource Efficiency Programme – ProgRess I and II). An active effort toward anchoring the topics of resource conservation and resource efficiency in the educational areas of school, vocational training, university and further education must be promoted. In order to curb the use and consumption of resources through greater resource efficiency, there also needs to be qualification and competence development among actors in the economy, political sphere, and society.

9.2 Education in Resource Conservation and Efficiency: Where Do We Stand Now? An Inventory

The German Resource Efficiency Programme (ProgRess) and education in resource conservation and efficiency are both at an equally early stage – in fact, they are only just beginning. There are various approaches in all areas of education in the form of pilot projects, occasional teaching and study materials or even websites that pick up on the topic or present subject-specific examples. In order to push forward with education in resource conservation and efficiency in a systematic manner, there needs to be a clear expansion and continuation of previous activities (Baedeker et al. 2014).

In the course of project implementation, initially a comprehensive analysis of the status quo was undertaken in relation to resource education and the obstacles in the four educational areas of school, vocational training, university, and professional development. It has been shown that in the different educational areas there has so far not only been inadequate teaching and study materials and training for teachers, but there has also been a lack of structural anchoring in teaching plans and curricula. When considering areas of education individually, some specifics become apparent. In teaching and learning materials of primary and secondary schools as well as of vocational training, the concept of resources is often understood in terms of energy (fossil resources) and frequently also includes the environmental mediums (air, earth, water). Resource conservation and resource efficiency as actual concepts are rarely featured in curricula. In non-formal professional development, the demand for resource-related education has long been very insignificant. Formal professional development suffers from the fact that in the underlying curricula, the topic of resources – if featured at all – is allocated to other concepts, such as "materials management". In the university context, it becomes clear that the diversity of universities on the one hand, and their autonomy on the other, make it difficult to anchor resource conservation and efficiency at a central point. So far, there have been hardly any study programs that include resource competencies.

9.3 The Status Quo in the Four Educational Areas of School, Vocational Training, Professional Development, and University

As part of the BilRess analysis, 48 interviews with key actors from the various areas of education and ten focus groups specific to the educational areas were held, alongside a comprehensive study. The interviews and focus groups served to provide better access to the ways of understanding resource education in each educational area and, in doing so, to reveal the status quo in each field and to explore obstacles and possibilities. These are summarised below for each of the individual educational areas.

Primary and Secondary Education
The topic of resource conservation and resource efficiency in school focuses on the area of energy. Almost all curricula elements relating to resources are framed around energy use – especially at the higher levels. In primary schools, resources are examined in connection with waste separation and collection. Teaching on these themes is also impoverished by the fact that only a few references are made to school life (sparing use of paper, writing materials, waste separation) or the lives of pupils (mobile phone recycling).

Vocational Training
The topic of resource conservation and resource efficiency is dealt with primarily in the sense of sparing use of resources used as materials and also in relation to reducing energy consumption. The key argument in favour of addressing resource efficiency is reduction of material and operational costs. Conservation of resources for the reduction of environmental harm is not taken into account.

University Education
The topics of resource conservation and resource efficiency remain abstract terms that need to be made more specific. According to various experts, raising awareness of this topic is essential. A further problem in the German university landscape is institutional teaching through the Bologna Process and the tightly packed curricula, leaving little room for knowledge about the basics. There is also a lack of time and resources for the creation of teaching and study materials. Experts in the field suggest the development of trans-/interdisciplinary learning and teaching material as one possible approach for integrating resource education in a wider array of subject areas.

Further Education
Many actors categorise the topic of resource conservation and efficiency under the broader subject area of "energy and material efficiency". In the trade sector in particular, it is crucially important to adapt educational offers to potential participants in terms of their language and form. Anchoring these in practice, for example through in-house seminars and long-term guidance of participants, is a similarly important aspect. An approach that focuses on offerings for specific key figures in companies and organisations (e.g. works councils, procurement officers, etc.) is considered particularly effective. The key argument for addressing resource

efficiency at all is the reduction of material and operational costs. Statutory incentives can foster the demand for existing educational offerings. Increased networking of the actors within the area of further education with actors in other areas of education is considered important in order for the topic of resource conservation and efficiency to become lastingly established in the educational area.

9.4 BilRess-Roadmap: Perspectives Spanning the Educational Areas

Even in its first version back in 2012, ProgRess called for an increase in public awareness of resource conservation and efficiency. The eighth strategic approach highlights the necessity for sparing resource-use approaches and for the promotion of resource-efficient technologies, as well as for information, motivation and qualification (BMU 2012: 44). Thus, it is essential that educational measures are implemented at all levels.

One means of doing this is activating relevant actors who can further raise awareness. In addition, various activities must be undertaken in order to anchor resource efficiency in the different areas of education (BMUB 2012). In order to support this strategic approach, the BilRess project was conceived as having two pillars: the construction of a network and the development of a roadmap.

The development of the roadmap led to various challenges, some specific to a certain educational area and others relevant for all areas. These were structured in four areas of action, which address a variety of activities. Generally, the actions called for are political (campaigns), legislative (anchoring of the resources in the educational plans) or involve the specific commissioning or testing of measures (e.g. development of teaching and study materials or educational measures). In addition to this, the study also revealed needs for action in research (e.g. examination of the elementary level). From this, various recommendations can be derived for different target groups or individuals. For the establishment and dissemination of resource education to be successful, other efforts must be undertaken on the one hand, and further studies carried out on the other.

In the creation of the roadmap there were also suggestions as to which actors might play a supporting or coordinating role for the individual steps of work (see roadmap graphics). In these suggestions, actors who are in charge of certain budgets or capital and could therefore potentially realise the suggested steps are specifically highlighted. In the course of the BilRess project, it became clear that education for resource conservation and efficiency is still in its early stages, and that companies rarely promote resource education as an end in and of itself. In order to expand further, as has partly already been done in the area of energy education, the following strategic approaches are essential.

9.4.1 Information, Sensitisation, and Activation

There is a lack of awareness and knowledge about the significance of the topics of resource conservation and resource efficiency in the education system. Information, sensitisation, and activation are essential elements in anchoring resource conservation and resource efficiency across the educational landscape and within society. A public campaign that comprises diverse target-group-oriented information materials and event formats is a crucial key to raising awareness of the topic. Only through a broad public discourse and by addressing multipliers, the topic will be included in long-term agendas of educational actors, institutions, social partners, and policy-makers and create incentives to anchor resource education extensively.

9.4.2 Supporting Teachers and Students

The prerequisites for resource education are, most importantly, easily usable and accessible teaching and learning materials, as well as target-group-specific professional development offerings for trainers. As part of the BilRess project, a comprehensive status quo analysis of existing projects, media, and materials, as well as of existing internet information was carried out. These were made accessible via the BilRess-Wiki[2] (www.bilress.de). The result showed, however, that the available pool of material is still very small, very specific, and, in parts, outdated. There is a lack of breadth and depth in the offering of good, current teaching and learning materials.

9.4.3 Creating Incentives for Projects and Research

Incentives are the most important intervention measures for fostering sustainable behaviour. Incentives may be monetary, but may also stem from specific support offerings. In many areas of education, there is room for individual development, e.g., for independent project work. Resource projects can be implemented in schools and vocational colleges as project weeks or through excursions, for example. Universities can carry out resource projects as part of internships, seminars, or dissertations. Project support in the form of educational programmes, research programmes, or support for practical projects (e.g. within companies), can aim at promoting resource education to an even greater extent.

[2] www.bilress.de

9.4.4 Encouraging Formal Anchoring in the Education System

One important key for comprehensive resource education is the application of regulations, teaching plans, and curricula. If the topic of natural resources were to be anchored within these or even simply mentioned explicitly as part of Education for Sustainable Development (ESD), then teaching and learning materials would be developed and demanded in classes. However, this change process is very labour intensive and aimed at the long term. Ideally, therefore, the topic of resources is incorporated into ongoing reform processes. For schools, this can be linked to ESD, and it is particularly important to highlight, that natural resources form a fundamental part of ESD. In vocational training, resource education can be integrated explicitly via the topic areas of "environmental protection within the firm".

9.5 Recommendations for Action

In summary, the following ten key recommendations for action can be identified:

1. **Establishment of a BilRess-Competence Centre in existing structures**: Without coordination and without ongoing impetus, resource education will not gain the same status as energy education. A competence centre can be responsible for the overarching coordination and development of various activities, such as the organisation of information campaigns, networking of actors, or the initiation of projects for resource building.
2. **Encouraging network development**: When it comes to resource education, it is crucial that key actors are able to connect. Networking should be encouraged at various levels, both local and national as well as industry-specific and cross-sector. The actors need a forum where they can exchange ideas and thoughts on their activities, experiences and content.
3. **Interface management**: A "strategy" for resource education has yet to be developed, since the topic is barely anchored in the educational areas. However, knowledge builds on knowledge, so the interfaces between the educational institutions need to be outlined in order to generate target-oriented offerings. When it comes to interface management, it is important to set out which learning objectives are pursued in the educational areas and how these are coordinated between the different areas.
4. **BilRess-Platform 2.0**: Resource education is a specific topic, thus, it should ideally be offered on a purposefully designed BilRess 2.0 network platform. In this case, "2.0" means the possibility of an enhanced form of communication among users in order to facilitate joint work, for example. Offers on the platform may take the form of professional development search engines, links to teaching and learning material platforms, the BilRess materials, and an integrated appointment calendar for resource education.

5. **Developing teaching and learning materials**: In order to ensure that educational actors can take on resource topics, easily usable materials and lessons or seminar units that can be integrated into current educational work are needed. These materials must be broadly applicable in order to foster awareness of resources and possibilities for action (basic material). Subject-specific materials should be developed to complement this so that the specific references to the various professions can be highlighted in a more sophisticated way.
6. **Offering qualifications**: Teachers should develop their own resource competence and incorporate this into their knowledge transfer within their educational area. The range of offers should be tailored to the knowledge requirements and should span from easily accessible topics for general educational teaching to subject-specific, in-depth topics. One important element is on-site consultations in schools and universities and among vocational training and further education providers, national educational institutions, companies, and committees. They stimulate processes, because they also influence the multipliers' immediate environment and hence lay the foundations for resource projects.
7. **Supporting and implementing resource projects**: There is room for individual development in the form of independent project work in many areas of education. Independent, specific educational work can be encouraged by means of incentives. Resource projects can be implemented in schools and vocational colleges as project weeks or through excursions, for example. Universities can implement resource projects in the form of seminars or dissertations. Project support in the form of educational programmes, research programmes, or support for practical projects (e.g. within companies), can aim at promoting resource education to an even greater extent.
8. **Images and exhibits**: Resource conservation and resource efficiency are abstract terms that need to be clarified. For primary and secondary education in particular, but also for further education, images and tangible exhibits can help make these concepts comprehensible.
9. **Competitions and prizes**: Both incentive formats have proven effective in education. Competitions can be implemented in the form of a prize for the "best project" or as honorary recognition for a successful or planned achievement. Prizes and awards are particularly well received by institutions. In the educational area of "schools" there are many examples, which are nevertheless more energy- or waste-oriented.
10. **Integration of resource conservation and efficiency into the formal education regulations**: The key to comprehensive resource education is the application of regulations, teaching plans, and curricula. The goal is to anchor natural resources or cite them explicitly as part of education in sustainability. As a consequence, teaching and learning materials would be developed and used in classes.

9.6 BilRess-Network

The "Education for resource conservation and resource efficiency" network was established on 22 September, 2014, under the patronage of Environment Minister Dr. Barbara Hendricks and with the support of 50 founding members. In its foundation phase, four biannual network conferences were held, with venues alternating between Berlin and Frankfurt, and by May 2016 a total of around 160 members were part of the BilRess-Network.

Members and target groups of the BilRess-Network are actors with an interest in the implementation and anchoring of resource education in the educational areas – primary and secondary education, university, vocational training and further education. The BilRess-Network is coming to life through events, the exchange of knowledge and information among its members and all those interested in education, and the continuous strengthening of ties between them. In the process, the network brings together representatives of all educational areas with actors from politics, business and academia.

The BilRess-Network:

- aims to improve education in resource conservation and efficiency,
- fosters an exchange of experiences among the actors in resource education,
- develops proposals for the formation of framework conditions that offer incentives and eliminate obstacles,
- facilitates active participation and the opportunity for the exchange of teaching and study materials and
- contributes to the implementation of the German resource efficiency programme ProgRess and makes proposals for its further development.

Against the background of the very positive resonance the BilRess-Network has achieved, one important success has been the progression of networking which is an initial important building block for the implementation of the BilRess-Roadmap. Since March 2016 the BilRess-Network is carried out in the framework of the "Kompetenzzentrum Ressourceneffizienz 2015–2019" (Competence Centre for Resource Efficiency 2015–2019) which is run by the VDI Centre for Resource Efficiency (VDI-ZRE). The implementation of the network is being carried out by the Faktor 10 – Institut für nachhaltiges Wirtschaften and the IZT – Institute for Future Studies and Technology Assessment.

The key tasks of the BilRess-Network still include the hosting of biannual networking events as a central platform for exchange and dialogue among the network members and any interested parties. The fifth BilRess-Network conference was held successfully in cooperation with the Technical University of Berlin, with a total of approximately 100 participants, 19 exhibitors and nine learning stations. The new event concept with learning stations for practical experience of resource competence at the network conferences was very warmly received.

In addition to this, the activities are oriented towards continuous networking, consolidation and expansion of the network and greater public awareness. This includes not only the maintenance of a website and the publication of a newsletter, but also the network PR and the inclusion of new target groups and partners. In 2016 the BilRess-Network was represented nationwide at 23 events, disseminating the existing and newly created BilRess products and carrying out varied PR work. Furthermore, numerous on-site appointments were arranged for the acquisition of new network members and the further development of the BilRess-Network. The number of network members developed positively thanks to these activities, rising from 160 (05/2016) to 254 (12/2016).

9.7 Outlook

The aim of the BilRess project was to make a contribution to resource education and, thereby support the push to anchor the topic in the German education system. The roadmap that has been developed and the BilRess-Network set up have thereby made a substantial contribution to mobilising, involving and integrating a significant number of actors.

Through the BilRess project and its activities, the topic of education in resource conservation and efficiency has considerably gained in prominence and significance since 2012. The importance of resource education was clearly emphasized on the political side through the involvement of the Federal Environment Minister Dr. Barbara Hendricks as patron of the BilRess-Network and through the Minister's participation in the BilRess conference on 22 September 2015 in Berlin.

This is also reflected in the continuation of the German resource efficiency programme (ProgRess II). The findings resulting from the BilRess project should be integrated into the implementation of the resource efficiency programme (ProgRess II), in which BilRess is already prominently represented. Alongside the resource efficiency network (NeRess), the BilRess-Network is proving to be an important nationwide network and its continuation and promotion is being encouraged. The anchoring of the topic of resources in the education system is also being made an unequivocal item on the agenda.

Further specific steps towards the implementation of education for resource conservation and efficiency should and must take place within the implementation of ProgRess II. To this end, the BilRess project has developed and outlined key recommendations for action in the available roadmap.

The recommendations for action that have been formulated now need to be picked up on by the different actors and incorporated into the individual educational areas. In order to achieve the results stated in the roadmap, which spans the different educational areas, it is not only important to have the indispensable commitment of the educational actors, but also financial fostering of this commitment and an (educational-) political framework.

So far, the BilRess project has looked at improving resource education in Germany exclusively. However, just as resource exploitation and resource policy are not simply national, but rather European and global topics, resource education should also be considered in the international context in future.

At a global level, a conserving and efficient treatment of natural resources is seen as a key competence of future societies. The United Nations claims within the Sustainable Development Goals (SDG) that sustainable growth and development require minimizing the natural resources and toxic materials used, and the waste and pollutants generated, throughout the entire production and consumption processes. Therefore sustainable practices should be adopted in production and consumption (see SDG-Goal 12: Responsible Production and Consumption). For the implementation, the education for resource preservation and efficiency plays a major role. Within SDG-Goal 4: "Ensure inclusive and quality education for all and promote lifelong learning" it is claimed that by 2030 it should be ensured that all learners acquire the knowledge and skills needed to promote sustainable development.

This also corresponds to the requirements of the European Raw Materials Initiative[3] and the Strategic Implementation Plan (SIP)[4] of the European Innovation Partnership (EIP) on Raw Materials. These European strategies highlight the fact that there is a shortage of specialists in Europe in some of the areas related to raw materials production and processing (e.g. mineral processing), and call for the overall knowledge and skills of people working in the raw material sector to be increased. A crucial element of this is the development of a European network for resource education, which primarily relates to professionals. In this endeavour, it should link up actors from professional training centres, companies, and research and innovation sites along the entire raw materials value chain, i.e., exploration/mining – processing/metallurgy – circular design – production – use – recycling. Comparable to the BilRess-Network on the German level, the European network should be used to develop a roadmap for improving skills and knowledge in the raw materials sector in Europe. The roadmap should describe steps to increase competence and expertise in the field of primary and secondary raw materials in all important professional raw material educational contexts in the future, and will show how the availability of a qualified and skilled workforce will be increased. To this end, it is also important to find out what education initiatives and training programmes exist in the raw materials sector in Europe, as well as what skills and expertise are available and what gaps and emerging needs for skills and expertise are likely to arise in future.

Since Germany and Europe import large quantities of their raw materials from other continents, resource education is crucially important on an international level as well. Essential raw materials and in some cases also their initial processing steps are located in the BRIC countries, as well as the Middle East and Southeast Asia. Another key starting point lies in the increased integration of resource education into international cooperation, including for the purposes of development.

[3] https://ec.europa.eu/growth/sectors/raw-materials/policy-strategy_de

[4] https://ec.europa.eu/growth/tools-databases/eip-raw-materials/en/content/strategic-implementation-plan-sip-0

This may take place, for example, within the context of existing activities by the Federal Ministry for Economic Cooperation (BMZ) and the Society for International Cooperation (GIZ). Correspondingly, the European and international implementation of resource education could support European and international resource strategies.

References

Baedeker C, Bowry J, Rohn H, Scharp M (2014) Bildung für Ressourcenschonung und – effizienz. In: Umweltwirtschaftsforum, vol 22/2–3, pp 169–175

Baedeker C, Rohn H, Scharp M, Schmitt M, Fesenfeld L, Bowry J, Bielke J (2016) BilRess-Roadmap. Wuppertal Institut für Klima, Umwelt, Energie GmbH; Institut für Zukunftsstudien und Technologiebewertung gGmbH, Faktor 10-Institut gGmbH

BMU – Bundesministerium für Umwelt, Nachturschutz und Reaktorsicherheit (2007) Strategie ressourceneffizienz. Impulse für den Ökologischen und Ökonomischen Umbau der Industriegesellschaft, Berlin

BMU – Federal Ministry for the Environment, Nature Conservation, Building and Nuclear Safety (2012) German resource efficiency programme (ProgRess). Programme for the sustainable use and conservation of natural resources, Berlin. http://www.bmub.bund.de/fileadmin/Daten_BMU/Pools/Broschueren/progress_broschuere_en_bf.pdf. Accessed Mar 2016

BMUB – Federal Ministry for the Environment, Nature Conservation, Building and Nuclear Safety (2016) German resource efficiency programme II. Programme for the sustainable use and conservation of natural resources, Berlin. http://www.bmub.bund.de/fileadmin/Daten_BMU/Pools/Broschueren/german_resource_efficiency_programme_ii_bf.pdf. Accessed Mar 2016

Bringezu S, Schütz H, Saurat M, Moll S, Acosta-Fernández J, Steger S (2009) Europe's Resource Use. Basic Trends, Global and Sectoral Patterns and Environmental and Socioeconomic Impacts. In: Bringezu S, Bleischwitz R (eds) Sustainable resource management. Global trends, visions and policies. Greenleaf, Sheffield, pp 52–154

Bundesregierung – German Federal Government (2002) Perspectives for Germany. Our Strategy for Sustainable Development. https://www.bundesregierung.de/Content/. Accessed Jun 2015

Bundesregierung – German Federal Government (2012) National Sustainable Development Strategy. 2012 Progress Report, Berlin https://www.bundesregierung.de/Content/. Accessed Mar 2016

Destatis – Statistisches Bundesamt (2015) Industrie, verarbeitendes Gewerbe. Auf einen Blick – Kennzahlen 2013. https://www.destatis.de/DE/ZahlenFakten/Wirtschaftsbereiche/IndustrieVerarbeitendesGewerbe/ IndustrieVerarbeitendesGewerbe.html. Accessed Aug 2015

Dittrich M, Giljum S, Lutter S, Polzin C (2012) Green economies around the world? implications of resource use for development and the environment. Sustainable Europe Research Institute, Wien

Lettenmeier M, Liedtke C, Rohn H (2014) Eight tons of material foot-print – suggestion for a resource cap for household consumption in Finland. Resources 3(3):488–515

Liedtke C, Bienge K, Wiesen K, Teubler J, Greiff K, Lettenmeier M, Rohn H (2014) Resource use in the production and consumption system – the MIPS approach. Resources 3(3):544–574

Liedtke C, Baedeker C, Borrelli LM (2015) Stellschrauben für Nachhaltigkeit. Trends und Verantwortungen in Produktion und Konsum. In: Roth M, Ulbert C, Debiel T (eds) Stiftung Entwicklung und Frieden, Institut für Entwicklung und Frieden, Käte Hamburger Kolleg/Centre for Global Cooperation Research: Globale Trends 2015. S. Fischer Verlag GmbH, Frankfurt, pp 299–314

Schmidt-Bleek F (2007) Nutzen wir die Erde richtig? Die Leistungen der Natur und die Arbeit des Menschen: Von der Notwendigkeit einer neuen industriellen Revolution. Fischer, Frankfurt a. M

SERI, WU Vienna (2014) Global resource extraction by material category 1980–2011. http://www.materialflows.net/trends/analyses-1980–2011/global-resource-extraction-by-material-category-1980–2011. Accessed Aug 2015

UNEP – United Nations Environment Programme (2011) Decoupling natural resource use and environmental impacts from economic growth. Report of the Working Group on Decoupling to the International Resource Panel. Fischer-Kowalski M, Swilling M, von Weizsäcker E U, Ren Y, Moriguchi Y. Crane W, Krausmann F, Eisenmenger N, Giljum S, Hennicke P, Romero Lankao P, Siriban Manalang A, Sewerin S. Switzerland

Wang H, Yue Q, Lu Z, Schuetz H, Bringezu S (2013) Total material requirement of growing China: 1995–2008. Resources 2:270–285

Chapter 10
The Way from Problem Scope Towards the Vision of a Low Resource Society – The First Working Period of the Resources Commission at the German Environment Agency (KRU)

The Resources Commission at the German Environment Agency (KRU)

Sascha Hermann and Christa Liedtke

Abstract In Chap. 10, the Resources Commission at the German Environment Agency (KRU) introduces itself and its tasks within the policy frame of resource efficiency topics in Germany.

In its first working period, the KRU has oriented itself outlining first a framework paper addressing the change towards a "low resource society" and has started a vision process focusing urgent action fields for an effective resource policy. The results underline and confirm the need to apply an integrated policy mix in Germany, in Europe or internationally. Chapter 10 focuses on the core visions and ideas of the KRU and their current initial starting points as a contribution to the discussion in search for the best solutions. To achieve the goal of low resource products, processes and services along the supply chain, the KRU suggests a product labelling

---

Faulstich, Martin; Hinterberger, Friedrich; Lutzenberger, Alexa; Meyer, Bernd; Oberle, Bruno; Reller, Armin; Tischner, Ursula; Tschesche, Julia; Wilken, Hildegard; Wilms, Herwart with friendly support from Anke Niebaum.

The Resources Commission thanks the German Environment Agency for support, especially Alexandra Lindenthal, who leads the commission's office.

S. Hermann (✉)
VDI Technologiezentrum GmbH (VDI Technology Centre),
VDI-Platz 1, D-40468 Duesseldorf, Germany
e-mail: hermann@vdi.de

C. Liedtke
Wuppertal Institute for Climate, Environment and Energy Research,
Doeppersberg 19, D-42103 Wuppertal, Germany
e-mail: christa.liedtke@wupperinst.org

H. Lehmann (ed.), *Factor X*, Eco-Efficiency in Industry and Science 32,
https://doi.org/10.1007/978-3-319-50079-9_10

approach in the form of a state-run product labelling office which supervises the mandatory labelling of products with regard to resource efficiency and recyclability. Additionally, the KRU outlines scenarios of an ambitious policy mix at least at the European level that leads to market success and a low resource society and economy. Furthermore; the KRU examined the legal feasibility for a law governing resource conservation ('Resource Conservation Act'). Synthesizing the societal and scientific discourse on climate and resource policy this targeted approach seems to be a manageable undertaking. Finally, the KRU provides an outlook on future issues on its agenda in a further working period.

Keywords Natural resources • Low resource society • Low resource economy • Resource efficiency • Resource consumption • Vision process • Scenario analysis • Integrated policy mix • Product labelling • Resource Conservation Act

10.1 Political Background and Aims of the Resources Commission at the German Environment Agency (KRU)

The use of natural resources[1] (VDI 2016a) is the basis for human life and economic activity. However, many of these resources are already overused today so that the pressure on global ecosystems is constantly increasing, with as yet incalculable consequences, both economic (e.g. land no longer usable for agriculture, scarcity of raw materials) and social (e.g. increasing migration due to social and ecological crises and uninhabitable areas).

In its review of 2014, the KRU described the situation as follows:

"Global extraction of resources is expected to double by 2030. This will be due to a – fundamental and intended – rise in enabling welfare for a wider part of the world population.[2] As a result, the number of people having resource-intensive lifestyles will increase disproportionately not only in today's developed countries.

The current situation is highly problematic with regard to the following aspects in particular:

- Even the moderate goals that have currently been set to limit resource and land use are not being achieved so far.
- Resource justice does not currently exist.

[1] Natural resources are defined to include renewable and non-renewable primary raw materials, physical space (such as land), flow resources (such as geothermal, wind, tidal and solar energy), the environmental media water, soil and air, and ecosystems.

[2] Extension of the global middle-class to 25% of humankind until 2025 (Meadows et al. 2004) or, according to McKinsey Global Institute (2012): 1 billion people in 2025, of which 600 millions will be living in about 440 cities in emerging markets.

- Value chain networks and infrastructures maintain and catalyse each other as regards resource consumption. The spread of resource-intensive lifestyles is accelerating.
- Our economy and its value chain networks are not resilient enough to work in the interests of national and global resource justice and sustainability (Umweltbundesamt 2014)

These problems, and the urgency of conserving resources as a global and cross-cutting issue of providing ecosystem services for the future in the interests of sustainable development, have been worked on by policymakers for quite some time (see Fig. 10.1).

In Germany, this is reflected by the adoption of the German Resource Efficiency Program I (ProgRess I) on 29 February 2012 (Federal Ministry for the Environment, Nature Conservation, Building and Nuclear safety (Bundesministerium für Umwelt, Naturschutz, Bau und Reaktorsicherheit (BMUB) 2012)), which systematically continued on the path mapped out by the EU flagship initiative "A resource-efficient Europe".

The adoption of the programme made Germany one of the first countries to define principles and strategies for action for the conservation of natural resources. In the programme, the German federal government committed to reporting every 4 years on the development of resource efficiency in Germany, assessing the progress of the Resource Efficiency Program and updating it accordingly. The first update was adopted by the German cabinet on March 2, 2016 (ProgRess II, Federal Ministry for the Environment, Nature Conservation, Building and Nuclear safety (Bundesministerium für Umwelt, Naturschutz, Bau und Reaktorsicherheit (BMUB) 2016)). Like ProgRess I, ProgRess II focuses, in particular, on market

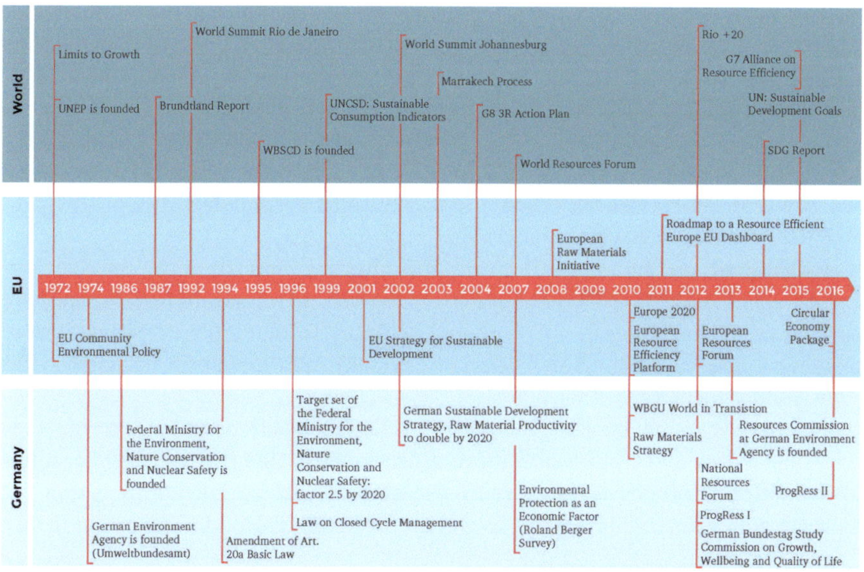

Fig. 10.1 Milestones of national, European and international resource policy (Adapted and translated from Berg et al. 2017)

incentives and on strengthening voluntary measures and initiatives in industry and society. Information and advisory services for companies are a key module, aimed at identifying and tapping resource efficiency potential in the value chain.

With its programme, Germany points the way to a "resource-efficient Europe" and steadily consolidates its leading role on this issue. The KRU believes that ProgRess II and subsequent updates should go further, aiming to establish a policy mix that is integrated, cuts across all levels and adequately addresses all relevant target groups in order to ensure that possible potentials can be increased. In the following sections, this is described in more specific terms.

In 2015, Germany took advantage of its G7 presidency to put resource efficiency on the agenda of a G7 summit for the first time. This led to the establishment of the G7 Alliance for Resource Efficiency as a new forum of cooperation within the G7. In addition, the Heads of State or Government requested the United Nations' International Resource Panel (IRP) and the OECD to present solutions and approaches for resource efficiency. The reports by the OECD and the IRP were presented on the occasion of the G7 Environment Ministers' Meeting in Toyama in May 2016. In their reports, the OECD and the IRP warn that, if the current trend ("business as usual") continues, global resource use is expected to at least double by 2050. The reports highlight the advantages which a more efficient and effective use of raw materials and other natural resources would have for the economy and the environment. Analysis in the IRP report shows that effective policies to improve resource efficiency and to tackle climate change can reduce global resource extraction by some 28% and cut global greenhouse gas emissions by around 60% by 2050 while at the same time boosting the value of world economic activity by 1% (UNEP 2016).

The OECD additionally analyzed existing policy instruments and found that these primarily address the end of the value chain. It therefore called for the consideration of the entire value chain, from raw material extraction and processing, manufacturing, distribution and use of the product, to its reuse, recovery or disposal. The OECD finds it necessary to address more upstream parts of the value chain, such as product design, production and consumption. Overall, it identifies a great need to apply an integrated policy mix (OECD 2016). In a position paper on elements of a successful resource policy, published in October 2015, the German Environment Agency (Umweltbundesamt (UBA)) likewise called for a policy mix of this kind (Umweltbundesamt 2015a).

Established on July 3, 2013, the Resources Commission at the German Environment Agency (KRU) has set itself the task of supporting the Agency by making concrete proposals for further developing resource policy, so as to give resource conservation a greater voice in Germany and implement the vision of a 'low resource society'.

Since its inception, the KRU has published several position papers (see Fig. 10.3) which underline and confirm the above assessments by the IRP and OECD. It has also developed proposals for specific measures in selected fields of action for German policymakers and society as well as for possible ways to implement them.[3]

[3]https://www.umweltbundesamt.de/themen/abfall-ressourcen/ressourcenkommission-am-umweltbundesamt-kru

The contents of the position papers, identified urgencies, developed measures and obtained results by the KRU are summarized in the following.

10.2 Low Resource Lifestyles and Economy

Reducing resource use in absolute terms means for Germany that it must define concrete targets to meet these challenges. "Targets for reductions by factors ranging from 4 to 20 by 2030 or 2050 are currently being discussed. The challenge lies in preserving prosperity while at the same time reducing the consumption of resources. This requires changes in both, production and consumption as well as new policy frameworks. It requires resource-friendly lifestyles, new economic models and radical technological and social innovations for more resource efficiency and the circular economy" (Umweltbundesamt 2014, p. 3).

Change can be induced by creating incentives, by voluntary measures and through initiatives by business and society (for details, see ProgRess I and ProgRess II). However, this is far from sufficient: policy options for shaping regulatory and economic instruments must be utilized wisely and in a way geared to mutual learning. The necessary transformation can only be achieved through an integrated fiscal, industrial, economic, research, education, consumer and environment policy that is innovation-oriented and aligned with the German and international sustainability strategies.

In its position paper (Umweltbundesamt 2016b) for the EU's POLFREE[4] (policy options for a resource efficient economy) project, the KRU therefore recommends that "a differentiated policy mix and, based on that, a master plan or a roadmap with short, medium and long term goals and steps should be developed." The assumption "that resource policy only means costs and burdens the economy is radically called into question by these results [of the project, see the following gray box]. They show that it is economically and socially worthwhile to be offensive and formative here. [...] A proactive German resource strategy and its implementation, with integrated consideration of all input-oriented resources, AND actively calling for such an economically so relevant strategy at European level is fundamental for the European economy and its competitiveness. Especially the contribution that socio-technical potentials could make, as well as integrating and researching and testing them, would be extremely important for reducing absolute resource consumption, i.e. it is the key to combine the more technical optimisation via economic/industrial policy with a societal perspective (social innovation processes, integration of education, research, production and consumption, stronger actor-integrated use of the potentials of industry and digitalisation 4.0 for a resource strategy and sustainable development)"(Umweltbundesamt 2016b).

[4] http://www.polfree.eu/polfree

A Resource-Efficient Europe – Results of the POLFREE Project

In the POLFREE project, modelling was conducted to determine the effects of different and separate policy mixes on resource consumption, climate targets (2 degrees goal), economic growth and employment (Meyer et al. 2015).

In the **EU Goes Ahead** scenario, the goal of an ambitious climate and resource policy not broadly supported by non-EU countries is to achieve the following sustainability targets in the EU by 2050:

- Per-capita consumption of abiotic materials: 5 t
- Cropland footprint 30% less than in 2005
- Reduction of the water exploitation index to 20%
- CO_2 emissions 80% of 1990 levels.

The results show that if the EU alone pursues an ambitious policy (non-EU countries having weaker climate policies which only allow achievement of a 4 degrees goal) with predominantly market-based instruments:

- prices for resources remain high because EU countries only account for a small share of global resource consumption, but
- the "first mover advantage" effect is larger than in the global scenario (exports are higher and imports lower), and
- real GDP in the EU is 12% higher and employment increases by 3.4 million compared to a "business as usual" scenario.

The "EU Goes Ahead" scenario is based on the following policy mix:

Climate policy:

- Reform of the existing European emissions trading system: elastic supply,
- Directly compensated taxes on coal, gas and oil for all other industries,
- Quota for renewable energies in power generation (globally, lower in non-EU countries),
- Support for electric mobility through regulations and economic instruments (globally),
- Tax on air transport,
- Subsidisation of land-based public transport,
- Subsidisation of investments in the energy efficiency of buildings.

Abiotic resources:

- Quotas for the recycling of ores and non-metallic materials,
- Tax on the use of non-metallic minerals,
- Taxes on final demand, not including exports, according to the goods' raw material content (RMC)
- Subsidisation of goods with a low RMC,
- Tax on water from public supplies,
- Support for resource efficiency in manufacturing industries.

Food, agriculture and forestry:

- Information programme for consumers and producers on how to reduce food waste,
- Information programme on how to reduce the “crop yield gap”,
- Tax on meat, dairy products and eggs,
- Autonomous reduction of the demand for meat,
- Limitation of agricultural land use,
- Limitation of water abstraction in agriculture.

Environmental tax reform:

- Compensation of environmental taxes through reduction of general corporate taxes
- Reduction of taxes on goods and services with low emissions of harmful substances and low resource consumption.

Based on a scenario process (like the one carried out in the POLFREE project) and the development of a roadmap, the following fundamental societal questions have to be answered:

- “How do we want to live together?
- What resources can and do we want to use to sustain our lives?
- How do we want to jointly shape the needed transformation towards a resource-just and resource-friendly society?
- How can a sustainable post-industrial society be forged?” (Umweltbundesamt 2014).

The following sections describe eight key approaches (cf. Fig. 10.2) which the KRU has been considered in a first prioritization for a vision process (see Sect. 10.4 and Umweltbundesamt 2014):

Mental models or, put simply, ideas shape the actions of entire societies over an extended period. They sometimes come in the form of megatrends and ultimately manifest themselves in our actions in research and development, in realization and implementation, hence in our infrastructures, products and services.

Actions and corresponding decisions in production and consumption are always culturally determined (production patterns, lifestyles). The underlying resource culture **(8.)** and the associated routines and practices are the actual drivers of global resource consumption and, hence, of climate change. They form the basic structural and mental framework – the non-physical infrastructure – within which high or low levels of resource consumption for our lives and economies are embedded and up-scaled (= physical infrastructure).

That is why the core areas focus on the future **(1.)**, or future visions, and the existing (unsustainable) resource culture **(8.)** encompass the behaviour and perspectives within the system – the perspective on ecosystems and our dealings with eco-

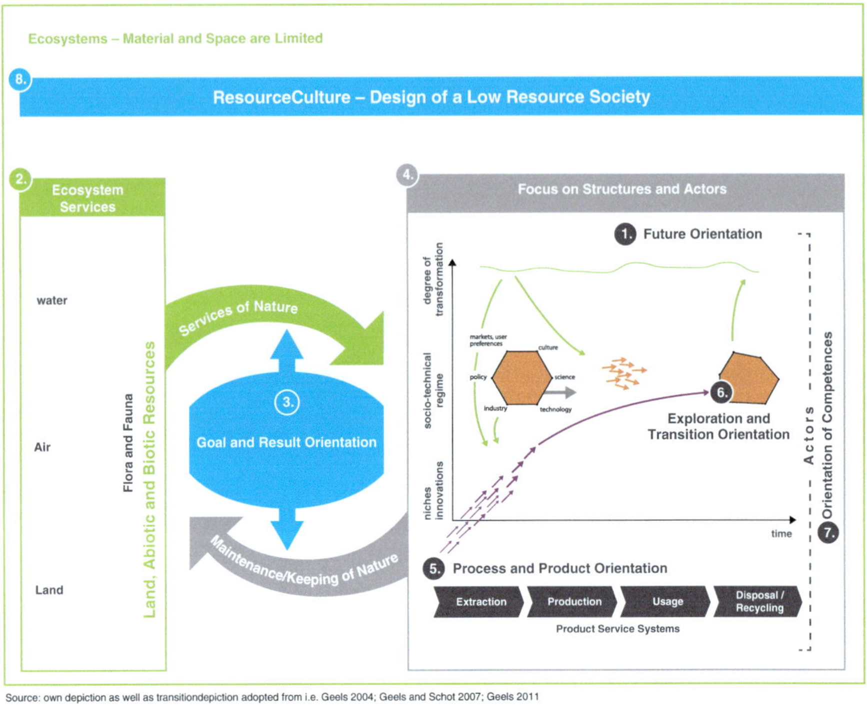

Fig. 10.2 Perspectives and actors for shaping a low resource society (Translated from Umweltbundesamt (2014) and adapted by Geels (2004), Geels and Schot (2007), Geels (2011), Liedtke et al. (2015))

systems, their services, the perception of effects on the various media **(2.)** and the shifting of problems between these and between continents or regions. The level of commitment with regard to the goals set **(3.)**, and the focus on delivering results crucially depend on how ecosystem services are valued **(3.)**. Starting out from these future visions and goals which may or may not be widely acceptable, the social or societal system with the economic functions adjusts – with decades- or centuries-long implications (e.g. transport or energy systems, World Wide Web). But even these can be changed, as Germany's *Energiewende* or the ongoing industry/digitalisation 4.0 processes show.

Change, diversity and dynamism are the principles of a functioning social or ecological system. Pure mono-cultures, in contrast, pose a threat to the system. The diversity and dynamics of existing structures, and actor competences and constellations **(4.)** shape the world of infrastructure, technology, processes, products and services, user behaviour, and economic trends (e.g. development of hybrid value chains, sharing, the green, circular and bio-economies **(5.)**). All of these are ultimately materializations of actor groups' ideas and innovativeness – inducing or reducing resource consumption, and with it, climate change.

Change – a transition **(6.)** – can happen only through the actors within the system themselves, through their determination, value judgements, competence and their curiosity in making society and the economy sustainable. Through explorations **(6.)** they must empower **(7.)** themselves and develop resilient patterns of behaviour and innovation systems and structures **(7.)**. Such resilience- and design-oriented transformation structures for research, development, testing and implementation which are based on the integration of all system actors concerned and appropriate sustainability monitoring are essential for achieving the international Sustainable Development Goals (SDGs).[5] Flagship projects and suitable innovation infrastructures can offer key actors the chance to test new service systems and to develop and phase in business models based on the learning. Failures will be possible (like they are today) and must be considered in the learning process.

Strategic concerted actions which address desired scenarios and render transformation tangible and malleable should be jointly developed and implemented by policymakers, society, business and science. Spaces of experience that help develop, test and implement both horizontal and vertical solutions should be created to that end. Analyzing and creating conditions for success and structural factors, and evaluating their potential contribution to sustainability and resource conservation are essential to a successful system of learning and innovation.

10.3 Developing Visions – Making Ideas Transparent[6]

The wide-ranging discussion on resource policy's and the KRU's role brought up the central question of what main goal should be pursued. This led to the KRU deriving the task of jointly developing a vision of a low resource society in order to make the various futures easier to grasp both for the KRU members and for policymakers and society.

Focusing on the issues raised in Sect. 10.2, the following overarching questions were formulated to develop the vision:

- How does a low resource society look like?
- What does this mean for day-to-day life and economic activity and the requisite structures?

At a two-day future workshop in March 2015, these questions were intensively dealt with.[7] The main results in the areas of an ethical framework; research and development (R&D); the role of the state; and economy and finance are summarized below (Umweltbundesamt 2016a). For each of the four areas, a focused vision of a successful low resource society is formulated. Moreover, concrete measures are proposed in order to reach the visions.

---

[5] www.sustainabledevelopment.un.org

[6] Based on and citing from: Umweltbundesamt (2016a).

[7] A two-day future workshop and its results cannot claim to be exhaustive. However, they do give an insight into the main aspects considered relevant by members of the KRU and focus on key elements of transformation dynamics towards a low resource society.

While a vision that looks into a future 30 or more years away from now may seem provocative, it does illustrate concrete ideas that the authors find possible, useful or desirable. These are now presented for discussion.

10.3.1 Ethical Framework

Vision
A wide range of successful, resource-friendly living and work models exist. Their innovativeness and creativity make it possible to achieve compatibility of time patterns within and between the natural environment, society and the economy. Time frames and structures for exchange processes and regenerations within and between geological and ecological systems, the economy, civil society, education, politics etc. vary widely. Their needs and requirements are recognized and integrated.

Communication on sustainability has become effective enough to emotionally reach different target groups and cultures of thinking, encouraging them to try out new, resource-saving models of living and working.

Specific Measures
The aim is to design a "low resource society". Diversity in business and living ideas and models that do their part to conserve resources is deemed desirable. Barriers such as "lock-ins" in existing systems or counterproductive subsidies and legislation are removed by the public sector while respecting fundamental societal values. Innovations in the organization of life and work are supported by the following measures, among others:

- Social security systems are integrated and give citizens more flexibility in managing their time.
- More risk capital for innovations for resource-friendly living and work patterns is available, e.g. through crowd-funding or support programmes.
- Innovation advisory centres for citizens are set up at regional level.
- Education systems for all age groups integrate information about these measures in their curricula and impart helpful skills and experience (with a focus on learning by experience).
- Communication measures in all media inform about resource conservation approaches in a transparent, target group-specific and authentic manner and invite addressees to actively participate.

10.3.2 Research and Development

Vision
The research and development sector integrates socially relevant and practical subjects and issues and cooperates with relevant actors and stakeholders while

maintaining a high level of scientific quality. Citizens and small- and medium-sized companies are involved in R&D processes and participate in shaping social and technical innovations. Through intelligent knowledge and information management, research contributes to increasing resource productivity by orders of magnitude. Fundamental research remains independent. Researchers and developers are encouraged to take into account resource conservation and the common good.

Specific Measures

- Mono-, inter- and/or trans-disciplinary "future radar" projects are carried out within the multi-level system from the local to the international level – to provide a reference framework (vision, foresight and scenario processes).
- Each publicly funded R&D project must prove what contribution it will make to conserving resources. In the area of fundamental research, that contribution will likely be prospective, and in the application domain it must be goal-oriented and focused on delivering results.
- Support is provided for research and development activities in schools and neighbourhoods (researching society, society that shapes transformation).

10.3.3 Role of the State[8]

Vision

Resource conservation, as an element of the German sustainability strategy, has constitutional status. Resources must be safeguarded and ecosystem services preserved in the long term.

Germany has contributed to the global stabilization of resource use. Resource justice has been achieved.

Specific Measures

- The state acts as a moderator and driver of innovation in environmental policy and resource conservation by:
  - creating innovation infrastructure including experimental settings,
  - adjusting innovation policy so as to promote resource-saving innovations,
  - integrating resource efficiency in education policy and programmes.
- Resource policy sets resource goals on the basis of the precautionary principle and pursues the internalization of external costs based on the principles of origin and precaution through taxes on resources and other instruments.
- Resource policy instruments are designed so as to be socially fair.
- A resource conservation law is established, which:

[8]The section on the role of the state is still being finalised. Changes to the vision and measures may be made after preparation of a more detailed paper on the subject.

- formulates resource conservation principles,
- establishes an independent institution that regulates resources:
  - This institution decides which resources require regulation (monitoring) and reviews them regularly ("red list").
  - In the case of resources from the red list, the resource quantity used is controlled by that institution through long-term targets (Bundesbank model).[9] The institution also makes policy suggestions, e.g. regarding taxes, producers' responsibility. Subsidiarity is maintained.

10.3.4 Economy and Finance

Vision

There is a wide variety of public and private-sector initiatives (in economy and finance) by actors ranging from individuals to the international community. Public and private-sector action complements each other in the spirit of a transparent principle of subsidiarity.

Acting sustainably, as individual, community or company, in the interests of conserving resources and the common good pays off – not only in monetary terms – both for individual businesses and society.

Specific Measures

- Germany campaigns for the global harmonization of high environmental and occupational health and safety standards as well as for tariff preferences being granted where these standards are met.
- To improve and control product development and recycling, users and other stakeholders from the lifecycle phases are legally required to be represented on boards of directors or similar company bodies.
- Resource balances open for inspection become mandatory for companies and public institutions.
- A product licensing authority[10] that evaluates the appropriate length of product lifetimes/useful lives and subsequent recycling is established.
- Financial market institutions are required to report in a transparent manner on the resource use caused by their products and on associated consequences and risks.
- Long-term impacts are considered in the assessment of financial products.

---

[9] A national bank inter alia regulates the money supply so as to ensure currency stability. A regulatory authority for resource use should, in the same way, ensure long-term stability of the quality and quantity of natural resources by controlling resource supply (e.g. by granting licences for production, use or consumption). The proposed "Regulatory Authority for Natural Resources" should be able to assess and control resource markets independently of legislative and executive authorities.

[10] In the further development and elaboration of the KRU's internal discussion regarding the product-relevant measures, the term "product licensing authority" developed in the direction of "product labeling office". Compare the current state of the KRU's recommendations section 10.4.2 and the upcoming KRU's Paper „Produktkennzeichnungsstelle zur Förderung der Ressourceneffizienz und Kreislauffähigkeit von Produkten" (in prep.).

- Incentives are created for investments in projects that conserve resources in the long term.

10.4 Initial Explorations and Starting Points – Contribution to the Discussion in Search for the Best Solutions

Based on the analysis of the current situation and the vision, the KRU agreed its further work on the following topics:

- Requirements for a law governing resource conservation – focus on a 'Resource Conservation Act'
- Production, processes, services – focus on product labelling
- Scenarios for a low resource society and economy
- Transferring R&D into economic practice.

10.4.1 Requirements for a Law Governing Resource Conservation – Focus on a 'Resource Conservation Act'

The measures considered by the KRU and described in Sect. 10.3 were reviewed regarding their legal feasibility. Based on the recorded discussion, the measures formulated in very general terms were specified and interpreted as far as possible and their legal feasibility was examined.

The vast majority of the measures can be assessed as legally feasible.

It would, however, be necessary to integrate corresponding provisions in the respective sectoral laws. For example, more flexible time management could be achieved through e.g. more flexible working arrangements (like home office), variation of school-starting hours and more flexible shop opening hours. This would require changes to labour legislation, education laws and ordinances, and the legislation on shop opening hours, for example.

Some of the measures proposed, namely the adaptation of education programmes and the strengthening of information and awareness-raising activities, could be integrated in existing or new programmes under the government's responsibility.

Complementing today's environmental protection legislation, horizontal provisions should be laid down in a future "Basic Act on Resources". Such an act might introduce basic resource protection principles (precautionary and polluter-pays principles), define guiding values for resource use (particularly aggregated indicators) or provide for the establishment of specialized bodies, institutions or institutes entrusted with resource protection functions.

Some of the measures discussed were assessed as being politically – and therefore legally – critical, because they affect important distributional aspects (e.g. further internalization of costs that are currently external) or strongly interfere with the balance between different legal interests. Examples of the latter are the establishment of a product labelling office that evaluates the appropriate length of product lifetimes and useful lives and subsequent recycling (see Sect. 10.4.2), and the formation of an institution regulating resources which operates independently of the legislative and executive branches (conflict with freedom of enterprise and the fundamental right to property).

Finally, some measures are related to international treaties and regulations. This is the case for e.g. the introduction of border tax adjustments and the extension of national targets.

Synthesizing the societal and scientific discourse on resource policy and devising targeted approaches seems to be a manageable undertaking. Translating the insights gained by this into specific legal action requires a deeper understanding of interest and stakeholder networks within the existing structures and of the relevant legal principles. Doing so was possible only in regard to some specific points.

10.4.2 Production, Processes, Services – Focus on Product Labelling

A low resource society offers interesting opportunities and challenges for companies. The primary aim is to gain optimum benefit from a minimal input of resources. A range of dimensions play a role, from a resource-efficient design of goods (products and services) which takes into account their entire lifecycle, to the planning and realization of resource cycles; from the question of what offers support to the transition towards a low resource society and low resource consumption, to vetting the supply chain for wastage of resources and redesigning products accordingly. Eliminating material consumption can often be achieved by avoiding the use of products (e.g. packaging), and service offers can replace product sales, thus leading to greater resource efficiency in the production/consumption system. It is also crucial that companies provide their customers with information that allows them to get an idea of the amounts of resource used for their products or services. In that regard, the KRU suggests a product labelling approach in the form of a state-run product labelling office which supervises the mandatory labelling of products with regard to resource efficiency and recyclability. This office would have the task to collect, examine and monitor specific information to be delivered by companies on products placed on the market in Germany or Europe and would assess whether mandatory labelling of the products by their manufacturers corresponds to the office's labelling requirements. The product labelling office should first deal with goods with a fairly long lifetime (e.g. consumer goods which unlike non-durables can be used for some time and only wear out through repeated use). The labelling mechanism should be extended to non-durable goods at a later date. The overarching goal is for this type of labelling to become mandatory for all products placed on the market in the European Union once the mechanism has been fully established.

The following key figures should be reported to the labelling office by companies and included on labels:

- Lifetime/functional/useful life, expressed in units of time: Companies declare the guaranteed lifetime of their products when used as intended, expressed e.g. in years.
- Resource inputs over the entire lifecycle: resources (used and unused, primary and secondary), energy, water, land and their availabilities (criticality). The indicators here include embodied energy (VDI 2012) and embodied raw materials (VDI 2016b), for example.
- Specific consumption levels for a typical and real use situation (cooperation with actors/use of values from the Eco-design and Energy Consumption Labelling directives could be appropriate here).
- Recyclability: information on reusability/recoverability, possibilities for dismantling and disassembling the product, components and materials under plausible end-of-life scenarios and given proper use.

Large companies should determine these values in specific terms. SMEs can apply average values (determined e.g. using software tools for lifecycle analysis which would have to be made available to them through statistical offices, relevant consulting organisations such as efficiency/consulting agencies or other sources).

Competition between companies in these areas can evolve through the labelling process and the integration of the collected data on resource consumption and recyclability in product communication activities. Consumers can compare products directly using these data and take the findings into account in their purchasing decisions. Overall, a positive effect towards greater resource efficiency and recyclability is expected – not only for products of suppliers in Europe but also for imported goods. The mandatory generation, collection and transfer of data in the value chain would also bring a significant gain in knowledge and thus enhance the companies' innovative capacity.

10.4.3 Scenarios for a Low Resource Society and Economy

The results of the simulations done for the POLFREE project's "business as usual" scenario using the GINFORS model (cf. excursus above) show that planetary boundaries will be clearly transgressed if we do not succeed in implementing an ambitious climate and resource policy on a global scale. Global development under business-as-usual conditions involves great risks and will be associated with a slowdown in economic growth, a deterioration of social conditions and increasing risks on financial markets in the EU.

Alternatively, a global agreement to apply a mix of resource and climate policies could be put in place. Its initial intensity would be so low as to be hardly noticeable, gradually increasing over a period of 30–35 years. Such a "Global Cooperation" scenario was simulated using the GINFORS model. The results are extremely positive. They show that the ambitious emission and resource consumption targets needed to prevent environmental disaster can be achieved. At the same time, lower prices for raw materials and investments in new technologies would induce clearly positive effects on income, employment and national debt.

To make this happen, a comprehensive package of measures, comprising economic, regulatory and informational instruments, must be agreed internationally. These measures must be aimed directly at keeping the consequences of economic and population growth for greenhouse gas emissions, the extraction of abiotic raw materials and land and water use within planetary boundaries. It might be questioned whether an international agreement of this kind could be reached. If this is not possible, the remaining option for the EU is to go this path alone (see gray box in Sect. 10.2). The design of the economic instruments would, however, have to be modified so as to prevent direct competitive disadvantages for EU countries. The simulations done for the "EU Goes Ahead" scenario using the GINFORS model show that the EU can achieve its environmental objectives, while the global targets would be missed for the simple fact that the EU only plays a minor role on a global scale. This also means, however, that, prices for raw materials would remain high, which would diminish the competitiveness of non-EU countries and lead to a marked improvement of the EU's balance of payments vis-à-vis non-EU countries. This "first mover advantage" and the persistent investment boom triggered by the policy mix would lead to the EU enjoying major economic successes and rising employment.

It can therefore be argued that these economic advantages would gradually prompt other countries to follow the example set by the EU. If no international agreement can be reached, it would thus still be possible to achieve global environmental goals, with a delay, by exerting mild competitive pressure. The only conclusion that can be drawn from the results of the GINFORS simulations is that the EU should urgently implement an ambitious climate and resources policy.

10.4.4 Transferring R&D into Economic Practice[11]

The structure of research support and the value creation culture makes it difficult for research results to reach entrepreneurial practice, i.e. they rarely prove to be of practical relevance and feasibility for companies. One of the main reasons for this is that science and business pursue different goals; also, their performance criteria and incentives are often contradictory (in this regard, see the call for new visions in Sect. 10.2). Whilst in research the focus is on publishing as a "competitive advantage", companies are focused on gaining an economic competitive advantage through patents.

[11] Researchers and representatives of companies (SMEs and large companies), institutions and politics explored the question of why it is so difficult to transfer results from research of resource conservation into practice. Some results of this expert discussion are presented here (Umweltbundesamt 2015b).

Research programmes are often designed without the participation of companies, particularly SMEs, and are often less relevant to the goals of business, market development or society. Furthermore, research proposals are often optimized along the lines of existing research programmes, which in turn focus on megatrends which might not necessarily be relevant for most companies.

The time frames of application processes for funding do not meet the needs of companies, who require a quick decision as to whether or not their projects will receive funding. It would be of urgent necessity here to respond to companies' different time dynamics, system logics and innovation cycles.

Another important aspect is the need to be aware of companies' system of action (structural logic and business DNA[12]) and develop solutions which reach key company decision-makers in a targeted way (i.e. a way that meets practical needs). This requires that solutions developed in research must relate to reality and also to the final customer.

Building on this knowledge, tailored interfaces between actors (researchers, translators, users) and needed transfer times must be taken into account. It is important here to understand the transfer of technologies and results not as an act, but as a process. For SMEs, providing funding and accompanying the process is important. Also, innovations in companies must be flanked by appropriate framework conditions for resource efficiency because, for example, energy prices for industrial uses of energy are too low to support the market diffusion of energy- or resource-efficient applications/solutions (this calls for regulatory intervention by government, e.g. via the instrument of taxation). As a result, even measures not requiring investment fail to be carried out although some could bring energy and/or resource savings of ten to 20%.

In addition, the KRU considers that the process of designing products, services or product-service systems should be more strongly addressed as a starting point of a comprehensive resource efficiency strategy. Depending on the way a product or service and its use options are developed, resource efficiency can be increased two-fold, three or more fold. So designers fill the role of "hidden agents" for a change towards sustainability.

10.4.5 Research and Development

In the research and development sector, the KRU sees the need for research to collect data on ecosystem services and map these data to serve as a basis for decisions on a policy process. This will be highly relevant for an integrated formulation of goals, for related assessment approaches, for the development of visions and scenarios, and for the integration of various modelling approaches (including resource calculators) which currently often address different questions. The integration of data and information on resources, energy, climate, land, and biodiversity would be a crucial step.

[12] The "business DNA" consist corporate values, corporate objectives and all elements that create identification and loyalty.

In addition to providing such information and data, there is a need for R&D in areas such as alternative materials, functional integration and optimized production, disposal and recycling processes. Designing resource-efficient products and services is particularly important here. The R&D process should be designed to include the development of business models and market introduction – however, measures to ensure this have yet to be developed and implemented.

10.5 Outlook

In its previous working period, the KRU identified a large number of challenges and put the focus on what is feasible and necessary. The scenarios, the discussion on translating research into company practice, and analyses regarding familiarity with the term "resource conservation and efficiency" show that

- Consumers are aware of the term and can provide a competent description of what it means. However, very few resource-efficient products or companies are identified in the market,
- Research results on the subject are available in the production sector, but only rarely find their way into practice.

This provides different starting points which the KRU wishes to work on in greater depth in its second working period (see Fig. 10.3).

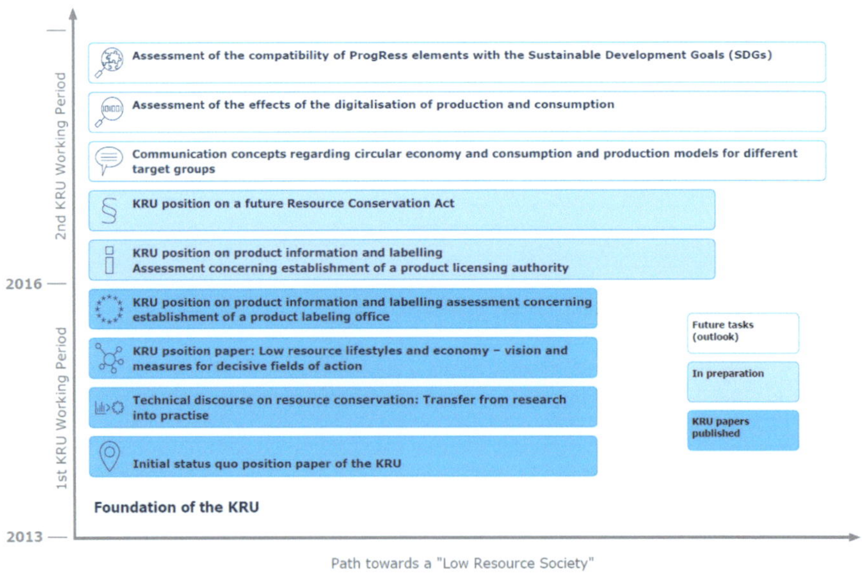

Fig. 10.3 Position papers already published by and future topics of the Resources Commission at the German Environment Agency (KRU)

The KRU is already working on the topic of product information and necessary elements thereof – with orientation towards consumer information in particular – as well as the topic of circular economy, particularly on recycling, dissipation and substitution quotas. The aim is to ensure already at the development and design stage that products are sustainable according to the standards set by the Sustainable Development Goals (SDGs) and the German sustainability strategy, taking into account characteristics such as choice of materials, recyclability, longevity, modular design, user- and service-friendliness, etc. It must be possible to assess the contribution of products to reducing absolute resource consumption across their entire lifecycle. The implementation of measurements for assessing and managing these contributions would be a promising step in this direction. That is why the KRU will assess ProgRess II and ProgRess III, which is currently being prepared, for their compatibility with the SDGs and work out corresponding recommendations with regard to what contribution ProgRess III and a fundamental resource management could make and what instruments, both hard and soft, could help to develop an implementation of a low resource society.

An important area which has received little attention so far in the debate on resource efficiency is the digitalization of production and consumption – this is an area which will define energy and resource consumption's infrastructure for the coming decades. A comprehensive assessment of its effects on resource consumption or a corresponding methodology does not yet exist. The KRU cannot develop such an assessment or methodology but it can raise the questions needed to help create awareness in this area of rapidly developing new production and consumption structures.

Floating above all this, ultimately, are the mental models that influence production and consumption patterns, including the discussion and development of a circular economy and the idea of an absolute goal for dematerialization. In regard to this, the KRU will develop a concept for communicating its position and proposals to the public, which will enable active interaction and engagement with different target, multiplier and user groups such as companies and consumers. The scenarios modelled in the POLFREE project show that it is possible to achieve concrete, ambitious goals. People are open to being reached by this positive message – as awareness studies and representative surveys show – but it must become much clearer to them what can be done and how relevant a change in behaviour is in any concrete case – both in production and consumption. Hence, we fundamentally need formats for implementing and communicating knowledge that already exists as well as further innovation and development approaches for low resource lifestyles and a low resource economy.

Disclaimer This paper does not necessarily reflect the opinion or the policies of the German Federal Environment Agency.

References

Berg H, Liedtke C, Welfens J (2017) Hintergrundrecherche und Definition einer ressourcenleichten Gesellschaft (Background research and definition of low resource society). Working paper in the project "Erfolgsbedingungen für Systemsprünge und Leitbilder einer ressourcenleichten Gesellschaft" (Success factors for system leaps towards and guiding principles of a low resource society) (in press)

Bundesministerium für Umwelt, Naturschutz, Bau und Reaktorsicherheit (BMUB) (2012) Deutsches Ressourceneffizienzprogramm (ProgRess) – Programm zur nachhaltigen Nutzung und zum Schutz der natürlichen Ressourcen, Bundesministerium für Umwelt, Naturschutz, Bau und Reaktorsicherheit (BMUB) (Eds), Berlin

Bundesministerium für Umwelt, Naturschutz, Bau und Reaktorsicherheit (BMUB) (2016) Deutsches Ressourceneffizienzprogramm II – Programm zur nachhaltigen Nutzung und zum Schutz der natürlichen Ressourcen, Bundesministerium für Umwelt, Naturschutz, Bau und Reaktorsicherheit (BMUB) (Eds), Berlin

Geels FW (2004) From sectoral systems of innovation to socio-technical systems: insights about dynamics and change from sociology and institutional theory. Research policy No 33. pp 897–920

Geels FW (2011) The multi-level perspective on sustainability transitions: responses to seven criticisms. Environ. Innov Soc Trans 1:24–40, Elsevier B.V., doi:10.1016/j. eist.2011.02.002

Geels FW, Schot JW (2007) Typology of socio-technical transition pathways. Res Policy 36:399–417

Liedtke C, Baedeker C, Borrelli L-M (2015) Transformation towards a sustainable society – key intervention areas. Innov Energy Policies 4(2)

McKinsey Global Institute (2012) Urban world: cities and the rise of consuming class. McKinsey & Company, New York

Meadows D, Randers J, Meadows D (2004) Limits to growth – the 30-year update. Earthscan. Chelsea Green Publishing, London/New York

Meyer B, Distelkamp M, Beringer T (2015) Report about integrated scenario interpretation. Deliverable 3.7a of the POLFREE project. http://www.polfree.eu/polfree. Accessed 12 Dec 2016

OECD (2016) Policy guidance on resource efficiency. OECD Publishing, Paris

Umweltbundesamt (2014) Position Juni 2014 – Ressourcenleicht leben und wirtschaften – Standortbestimmung der Ressourcenkommission am Umweltbundesamt (KRU). Ressourcenkommission am Umweltbundesamt (KRU) (Eds.), Dessau-Roßlau

Umweltbundesamt (2015a) Position Oktober 2015 – Elemente einer erfolgreichen Ressourcenschonungspolitik. Umweltbundesamt (Ed.). Dessau-Roßlau

Umweltbundesamt (2015b) Position September 2015 – Fachgespräch Ressourcenschonung – Transfer aus der Forschung in die Praxis. Ressourcenkommission am Umweltbundesamt (KRU) (Eds.). Dessau-Roßlau

Umweltbundesamt (2016a) Position der Ressourcenkommission am Umweltbundesamt Februar 2016 – Ressourcenleicht leben und wirtschaften – Vision und Maßnahmen in zentralen Aktionsfeldern. Ressourcenkommission am Umweltbundesamt (KRU) (Eds.). Dessau-Roßlau

Umweltbundesamt (2016b) Position of the resources commission of the German environment agency April 2016 – a resource-efficient Europe – a program for climate, competitiveness and employment. Ressourcenkommission am Umweltbundesamt (KRU) (Eds.). Dessau-Roßlau

UNEP (2016) Resource efficiency: potential and economic implications. A report of the International Resource Panel. Ekins P, Hughes N et al

VDI (2012) VDI 4600:2012-01 Kumulierter Energieaufwand (KEA) – Begriffe, Berechnungsmethoden (Cumulative energy demand (KEA) – terms, definitions, methods of calculation). Beuth Publishers, Berlin

VDI (2016a) VDI 4800 Blatt 1 (Part 1):2016-02 Ressourceneffizienz – Methodische Grundlagen, Prinzipien und Strategien (Resource efficiency – methodological principles and strategies). Beuth Publishers, Berlin

VDI (2016b) VDI 4800 Blatt 2 (Part 2):2016-03 Ressourceneffizienz – Bewertung des Rohstoffaufwands (Resource efficiency – evaluation of the use of raw materials). Beuth Publishers, Berlin

Chapter 11
Implementing Resource Efficiency in Europe – Overview of Policies, Instruments and Targets in 32 European Countries

Paweł Kaźmierczyk

Abstract This chapter presents an overview of national approaches to material resource efficiency and the circular economy in 32 European countries. It is based on a detailed survey carried out by the European Environment Agency (EEA) and its Eionet network in the second half of 2015.

The chapter explores similarities and differences in policies and strategies adopted by the countries, policy drivers, priority resources and sectors, policy objectives as well as targets, indicators and institutional set-up. To provide the context for country-level information, the chapter includes a short overview of the evolution of the EU regulatory framework in this policy area.

The analysis covers material resources (rather than the more broadly defined natural resources) as well as those initiatives which countries take to close material loops in the circular economy.

A selection of specific country examples is presented to illustrate the wide variety of national approaches. Many more examples are included in the June 2016 EEA report "More from less – material resource efficiency in Europe", which serves as the basis for this chapter. All national examples are described in more detail in the 32 detailed country profiles published alongside the main report.

Keywords Environmental Policy • European Union • European Environment Agency • Eionet Network • Resource Efficiency • Circular Economy • Raw Material Supply • Environmental Legislation • Sustainable Development • Green Economy • National Policies • Policy Driver • Driving Forces • Raw Materials • Natural Resources • Priority Resources • Food Waste • Waste • Waste Policy • Energy Policy • Energy Efficiency • Secondary Resources • Recycling • Sustainable Consumption • Policy Objective • Policy Target • Indicators • Material Flow Analysis MFA • Domestic Material Consumption DMC • Raw Material Consumption RMC •

P. Kaźmierczyk (✉)
European Environment Agency, Kongens Nytorv 6, 1050, Copenhagen, Denmark
e-mail: pawel.kazmierczyk@eea.europa.eu; https://www.eea.europa.eu/resource-efficiency

H. Lehmann (ed.), *Factor X*, Eco-Efficiency in Industry and Science 32,
https://doi.org/10.1007/978-3-319-50079-9_11

Resource Productivity • Monitoring Framework • Closing Material Loops • Critical Materials • Institutional set-Up • Institutional Arrangements • Economic Considerations • Environmental Concerns • Regulatory Requirements • Europe 2020 Strategy for Smart, Sustainable and Inclusive Growth • Flagship Initiative for a Resource-Efficient Europe • Roadmap to a Resource Efficient Europe • Seventh Environment Action Programme (7EAP) • EU Action Plan for the Circular Economy

11.1 Introduction

The European Union is estimated to have adopted more than 200 pieces of environmental legislation since the 1970s. Throughout the last decade, the sustainable use of natural resources and the refocusing of waste management towards prevention and recycling have steadily moved up the EU environmental policy agenda. They have also gained prominence on the economic agenda under the heading of resource efficiency. Table 11.1 presents examples of specific policy initiatives related to material use and resource efficiency, grouped by theme in a loosely chronological order.

National implementation of policies for resource efficiency—in some cases driven by the EU requirements and in others motivated by domestic priorities—is a topic of increasing interest to policymakers and practitioners across Europe.

In 2015, the European Environment Agency (EEA) and its Eionet network of 39 member and cooperating countries set out to review national approaches to material resource efficiency and to explore similarities and differences in policies, strategies, indicators and targets, policy drivers and institutional set-up.

The work resulted in the publication of the EEA report "More from less – material resource efficiency in Europe. 2015 overview of policies, instruments and targets in 32 countries" (European Environment Agency 2016). The analysis in the report, published in June 2016, is illustrated with short examples of countries' policy initiatives, which are then described in more detail in the 32 country profiles, i.e., self-assessments prepared by the countries and published alongside the main report.

The analysis covers material resources (rather than the more broadly defined natural resources) and initiatives which countries take to close material loops in the circular economy. Beyond the analysis of information reported by the participating countries, the report offers several considerations for future policies on material resource efficiency and the circular economy.

This work followed an earlier EEA survey, carried out in 2011 in support of the work on the 2011 Resource Efficiency Roadmap, that collected, analysed and disseminated information about national experiences in developing and implementing resource efficiency policies.

Table 11.1 Examples of EU policies related to material use and resource efficiency

Theme	Examples
Energy	Energy 2020: A strategy for competitive, secure and sustainable energy
	A policy framework for climate and energy for 2020–2030
	Energy roadmap 2050
	European energy security strategy
Waste and recycling	Waste framework directive
	Landfill directive
	Packaging and packaging waste directive
	Thematic strategy on the prevention and recycling of waste
Sustainable management of natural resources	Sixth Environment Action Programme (6EAP)
	Thematic strategy on the sustainable use of natural resources
	EU Forest Strategy
Sustainable consumption and production, and business-oriented initiatives	Sustainable consumption and production and sustainable industrial policy (SCP/SIP) action plan
	Eco-innovation action plan
	Industrial policy for the globalisation era and innovation union
	Single market for green products
	The green action plan for small and medium enterprises (SMEs)
Raw materials	Raw materials initiative
	Strategy on commodity markets and raw materials
	European innovation partnership on raw materials
	EU list of critical raw materials
Resource efficiency	Europe 2020 strategy for smart, sustainable and inclusive growth
	Flagship initiative for a resource-efficient Europe
	Roadmap to a resource efficient Europe
	Seventh environment action Programme (7EAP)
Circular economy	Towards a circular economy: A zero waste programme for Europe (2014)
	Flanking communications on sustainable buildings, green employment, SMEs
	Closing the loop: An EU action plan for the circular economy (2015)

11.2 Driving Forces for Material Resource Efficiency

As reported by the countries, the factors and concerns which drive work on material resource efficiency policy roughly fall into three groups: economic interests, environmental concerns and regulatory requirements.

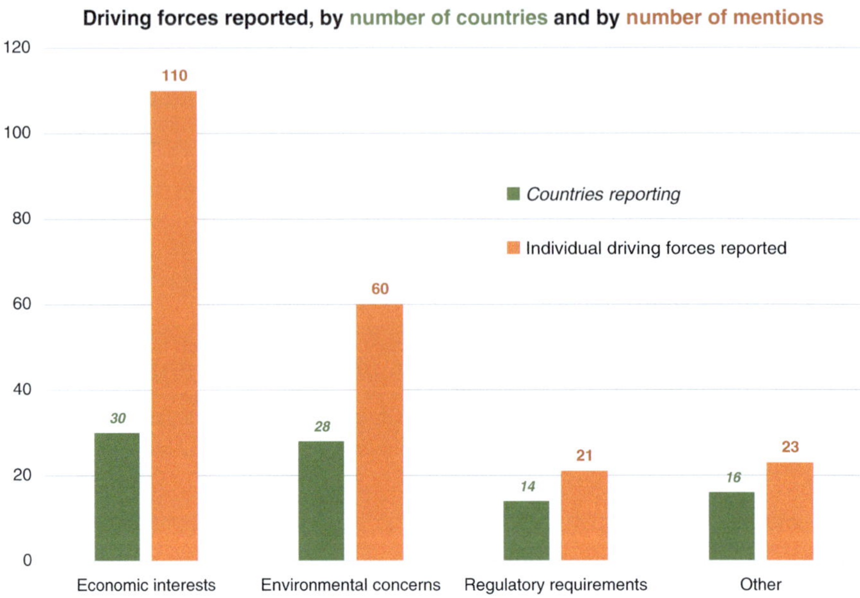

Fig. 11.1 Driving forces for material resource efficiency, by number of reporting countries and by number of individual driving forces reported

Although most countries reported a combination of all three categories, economic considerations seemed the most important in 2015, with economic drivers outnumbering those related to environmental concerns in most countries (Fig. 11.1). This contrasts with the situation in 2011, when most countries listed environmental concerns as the most common driving forces.

This finding seems to indicate that material use and resource efficiency are now core economic and strategic issues, and that the logic of 'doing more with less' has been widely embraced.

The most recurrent drivers were the desire to increase competitiveness and to secure the supply of raw materials and energy as well as to reduce dependence on imports on the one hand (economic interests), and the need to reduce pressures on the environment on the other (environmental concerns), as shown in Fig. 11.2.

Some countries emphasised the potential contribution of material resource efficiency initiatives to economic competitiveness and (green) job creation. This dimension may deserve stronger emphasis in future policies on material resource efficiency in light of growth and jobs being high on the EU policy agenda.

Programmes for reducing import dependency mainly referred to fossil fuels and critical minerals. Only nine countries specifically pointed to the need to reduce greenhouse gas (GHG) emissions as a driver of material resource efficiency, which indicates that the two policy areas are seen as separate topics.

Somewhat surprisingly, only eight EU Member States identified compliance with EU requirements as a driver for national policies on material resource efficiency. This may reflect the fact that the current regulatory framework for material resource efficiency is still limited, with few hard targets. At the same time, the

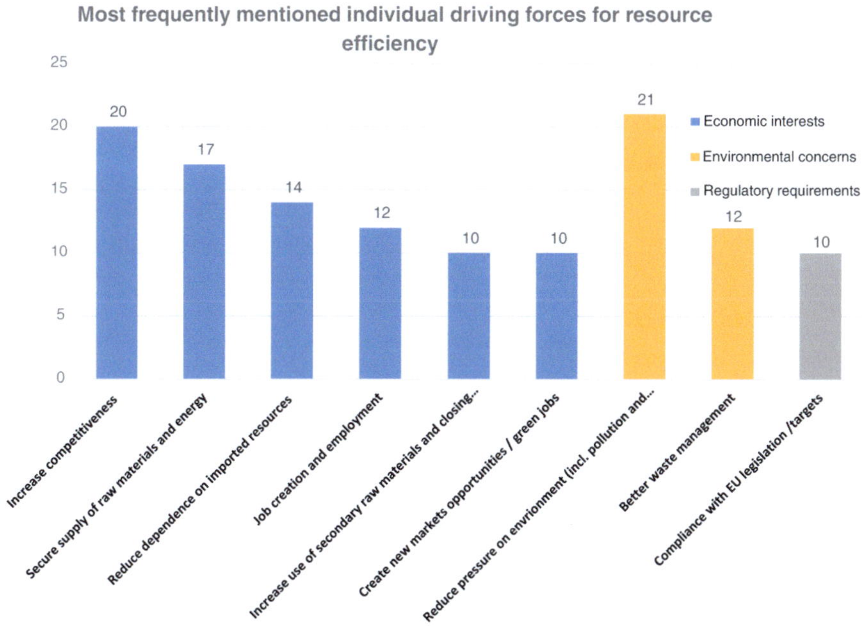

Fig. 11.2 Most frequently mentioned individual driving forces for resource efficiency policies

significance to countries in terms of energy and waste policy shows that EU policies can—and indeed do—provide a powerful impetus where the regulatory regime is strong and/or includes targets.

11.3 Priority Resources and Sectors

A majority of countries (27) identified a number of waste streams and secondary materials as the most common group of priority materials. This is an interesting development, illustrating the ongoing shift of emphasis in waste policies from end-of-pipe approaches to a focus on reuse and recovery. Key individual waste streams are plastic and packaging (17), construction and demolition waste (16) and food waste (15).

The prevention of food waste, identified by about half the countries as a priority, is an interesting example of how improvements in material resource efficiency can also result in a reduction in GHG emissions—with related climate benefits. It may be worth identifying and highlighting other cases with synergistic co-benefits.

Waste was followed by different types of biomass, which were identified as a priority by 18 countries. Wood was the most common material in this category (mentioned 15 times), primarily as timber and only in a few cases as a renewable energy carrier.

In the non-metallic minerals group (reports by 19 countries), construction minerals were most commonly mentioned (by 10 countries).

Metals and metal ores were mentioned as priority materials by 18 countries (24 reports).

Energy carriers, including renewables, were mentioned by 18 countries as priority resources. Both energy and waste were also top priorities in the 2011 EEA review.

Manufacturing was singled out most frequently as the key economic sector for material resource efficiency policies, followed by agriculture and forestry, construction and waste management.

There were very few initiatives focusing on consumption patterns, and only three countries identified the service sector, which accounts for two thirds of most European economies, as a priority, indicating that the potential role of services in improving material resource efficiency could be explored further.

11.4 National Strategies and Policies Related to Material use and Resource Efficiency

The number of national programmes and policy initiatives incorporating material resource efficiency as a core component has increased in recent years. This can be at least in part attributed to the adoption of the EU Flagship initiative for a resource-efficient Europe and the Roadmap to a Resource Efficient Europe.

However, despite the call in the 2011 Resource Efficiency Roadmap for Member States "to develop or strengthen existing national resource efficiency strategies, and mainstream these into national policies for growth and jobs by 2013", only three countries—Austria, Finland and Germany—had dedicated national strategies for material resource efficiency as of late 2015. Two further countries had dedicated strategies at a regional (i.e., subnational) level—in Flanders (Belgium) and Scotland (United Kingdom).

Instead, most countries reported that they incorporate material use and resource efficiency in a wide variety of other strategies and policies, including those on waste and energy, industrial development and reform programmes, or in national environmental or sustainable development strategies. This may in part result from the lack of a clear definition of key terms, as discussed in Box 11.1.

A mosaic of diverse policies—in total 240 examples were reported—can be roughly grouped into the following categories:

- national plans and programmes concerning **waste management and recycling**, for example, the Czech Republic's Secondary Raw Materials Policy (2014), Denmark's resource strategy, Denmark Without Waste, and the Dutch circular economy programme, From Waste to Resource (2014);
- national strategies and action plans addressing **energy and energy efficiency**, including the French Energy Transition for Green Growth Act (2015), and Poland's National Energy Efficiency Action Plan;
- general national **sustainable development and environmental strategies** addressing material resource efficiency, such as Hungary's National Framework Strategy on Sustainable Development (2012–2024), Ireland's Our Sustainable

Box 11.1: The Scope of Materials, Resources and Resource Efficiency

It is worth noting that there is no uniform definition or even implicit understanding of key terms such as *materials*, *raw materials* or *resources* in EU and national policy documents that deal with resource efficiency and raw materials.

The term 'resource efficiency' denotes the political goal of 'allowing the economy to create more with less, delivering greater value with less input, using resources in a sustainable way and minimising their impacts on the environment'. The December 2015 EU Action Plan aims to support a circular economy in which 'the value of products, materials and resources is maintained in the economy for as long as possible, and the generation of waste is minimised'.

Similarly, in most countries key concepts remain undefined. The EEA analysis of country policies on resource efficiency carried out in 2011 showed that there was neither a clear definition nor a common understanding of key terminology. Terms such as *resource efficiency*, *decoupling* or *sustainable use of resources* often seemed to be used as synonyms. Very few countries formally defined 'resources' in their policies, and some used the narrower 'raw materials' when addressing resource efficiency. The responses provided by the countries in 2015 indicate that key terms and concepts continue to defy definition. Two thirds of the countries reported that neither the terms nor the scope of 'resources', 'materials' or 'resource efficiency' are explicitly defined in their policies relating to material resource efficiency.

While the intuitive shorthand of 'doing more with less' seems sufficient for current policy needs, such a vague scope will make it difficult to carry out an insightful assessment of progress towards resource efficiency objectives.

Future, Latvia's Sustainable Development Strategy (2010), and the Swedish Generational Goal and Environmental Quality Objectives;

- national plans and programmes for **waste prevention** or the **prevention of specific wastes**, including the national waste prevention programme Towards a Resource Efficient Ireland (2014), the Italian National Plan to Prevent Food Waste, and Towards Zero Waste in Wales (United Kingdom);
- strategies for **innovation and industrial development**, including Hungary's National Environmental Technology Plan (2011–2020), and Turkey's 10th Development Plan (2013) with its Priority Transformation Programmes;
- strategies for a **green economy**, including Promoting the Green Economy in Iceland, the Green Growth Policy in the Netherlands (2013), the Green Growth Commitment in Portugal (2015) and the Green Economy Action Plan in Switzerland (2013);
- strategies to **reduce import dependency and secure raw materials**, including the Czech Republic's Action Plan for Self-Sufficiency in Raw Materials, and Turkey's Programme for Reducing Import Dependency;

- thematic strategies or plans addressing **specific resources or sectors**, such as Liechtenstein's Action Plan on the use of recycled material for public buildings (2010) or the Lithuanian National Forestry Sector Development Programme for 2012–2020;
- programmes for **agricultural and rural development**, such as the Estonian Rural Development Plan for 2014–2020, Latvia's Rural Development Programme 2014–2020 or the Polish Strategy for Sustainable Development of Rural Areas, Agriculture and Fishing.

A majority of countries (28 out of 32) presented waste management and recycling initiatives as the core of their national approach to material resource efficiency. Examples of good practice reported by countries were also dominated by waste prevention and/or recycling measures. This indicates an opportunity to address both themes together, through for example, the circular economy, the recovery of secondary materials or industrial symbiosis.

Almost all countries (29) reported various initiatives related to energy use, energy efficiency and the use of renewables as part of national policies on material resource efficiency. In most countries, however, energy policy is a separate long-standing policy field, and energy and resource efficiency are still largely disconnected from a programmatic point of view. This might warrant more attention in future, as there are many potential synergies between the two, in line with the Seventh Environment Action Plan (7EAP) objective to 'turn the Union into a resource-efficient … low-carbon economy'.

11.5 Policy Objectives

In contrast to targets, which are measurable, concrete and have a deadline (and are discussed in section 11.8), general policy objectives are a statement of intent and an indication of goals to be achieved with respect to material use and resource efficiency.

Countries reported almost 400 examples of general and specific policy objectives directed at all kinds of natural resources, sectors and topics. Table 11.2 presents the most frequently reported (seven or more entries) policy objectives for material resource efficiency, grouped by theme

11.6 Closing Material Loops in a Circular Economy

Flanders (Belgium), Germany and the Netherlands reported having a dedicated strategy for closing material loops in the circular economy. Several more countries acknowledged the need to move away from the current linear economic model, stating that the circular economy and closing material loops are already policy priorities.

Table 11.2 The most frequently reported policy objectives for material resource efficiency, grouped by theme

Waste and recycling (144 in total)	Increase recycling rates and recovery (24)
	Waste prevention (17)
	Promotion of reuse and secondary resources (17)
	More efficient waste management (16)
	Efficient construction and alternative building materials (11)
	Reduce environmental impacts of waste (9)
	Apply waste hierarchy (8)
	Reduce food losses and waste (8)
Material resource use objectives (64 in total)	Sustainable use of natural resources (14)
	Increase (material) resource efficiency (13)
	Reduce environmental impacts associated with material use (10)
	Reduce use of minerals (7)
Managing energy more efficiently and increased share of renewables (54 in total)	Increase share of renewables/energy transition (14)
	Improve energy efficiency in (certain) sectors (12)
	Reduce greenhouse gas emissions (8)
Economic considerations – Competitiveness and security of supply (54 in total)	Green/sustainable economic growth (13)
	Innovation/resource-efficient production/clean technology (12)
	Secure supply of energy and raw materials (10)
	Market introduction of innovative, resource-efficient technologies and services/eco-design (9)
	Increased competitiveness (8)
Conservation of natural resources (47 in total)	Protect biodiversity (14)
	More sustainable forest management (12)
	Reduce waste of water/increase water-use efficiency (10)
	More sustainable utilisation of land and soil (7)
Societal interests – Education and consumption patterns (20 in total)	Improve education/knowledge (7)
	Decrease unsustainable consumption patterns (7)

Note: Table 11.2 lists only those policy objectives that were mentioned by at least seven countries. However, these account for almost 80% of the total reported and as such are considered representative of the most common priorities.

Complying with existing waste legislation and targets—as well as the Circular Economy Action Plan—appear to be the most important drivers for initiatives to close material loops. This is a clear illustration of how initiatives at the EU level stimulate national action.

As shown in Fig. 11.3, the majority of reported initiatives on the circular economy are targeted at waste and secondary raw materials (a downstream policy option) and at the abiotic part of the economy. Only two countries explicitly commented that the circular economy requires going beyond increasing recycling rates and a

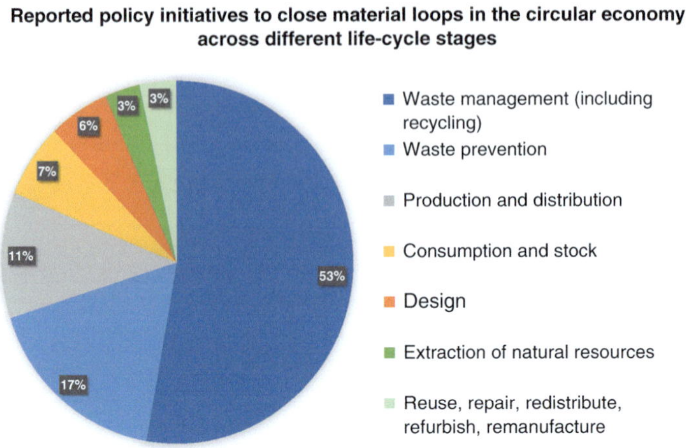

Fig. 11.3 Reported policy initiatives to close material loops in the circular economy across different life-cycle stages

higher use of secondary raw materials. It might therefore be worth reflecting on how policies on the transition to a circular economy could encourage initiatives beyond waste and recycling.

It is worth noting that the situation with respect to policies for the circular economy is changing dynamically. Figure 11.3 is based on information provided by participating countries in late 2015. Since then, a variety of national initiatives have been taken, including the adoption of national circular economy plans in, among others, Belgium (at the federal level), Finland, the Netherlands and the United Kingdom (Scotland).

11.7 Indicators

The indicators most commonly reported as being used to monitor material resource efficiency are Eurostat-produced ones based on material flow accounting (MFA). The most frequently reported indicators in this group were, in descending order: resource productivity (the ratio of gross domestic product to domestic material consumption, GDP/DMC – 17 reports); Domestic Materials Consumption, DMC – 16 reports); DMC per person – 9 reports; and direct material input (DMI).

Interestingly, MFA-based indicators are in use both within and outside the EU, as demonstrated by reports from Serbia, Switzerland and Turkey.

In addition, eight countries reported various initiatives to estimate their global material footprints, typically using raw materials consumption (RMC) as the indicator.

Countries also tend to use indicators on waste generation and management as a measure of material resource efficiency. For this group, the high number of reporting countries results from the requirements laid down in EU waste-related directives.

The EU Resource Efficiency Scoreboard was frequently mentioned as a common source of indicators. This suggests that a model which brings together several relevant indicators in one place has been well received. Finally, very few countries have developed their own indicators on material resource efficiency and closing material loops.

One of the challenges for both material resource efficiency and the circular economy is to develop adequate indicators to monitor trends, measure progress and set targets where appropriate. No country reported having a coherent set of indicators in place for the circular economy and closing material loops. Measuring the degree of circularity is quite challenging within the established statistical system in Europe, so it may be useful to monitor the progress of ongoing experiments in this field, for example in Belgium, Germany or the Netherlands.

11.8 Targets

For the majority of countries, compliance with existing EU legislation is a strong driver of policy action taken at the national level. The two areas for which targets are common are waste and energy. Action here is clearly driven by EU regulations, though some countries have adopted targets that are more ambitious than those required by current EU legislation. Some non-EU participating countries also reported having targets for waste and energy in line with EU directives.

All in all, the formulation of resource efficiency objectives and targets is clearly a challenge, at both the EU and national levels. The EU itself does not have a target for material resource efficiency. However, nine EU member states have adopted targets for national material resource efficiency: Austria, Estonia, France, Germany, Hungary, Latvia, Poland, Portugal and Slovenia. In most cases, these targets are based on gross domestic product relative to domestic material consumption (GDP/DMC), the EU's lead resource productivity indicator:

- Austria is striving for a 50% increase in resource efficiency (GDP/DMC) by 2020 relative to 2008, and aspires to a four- to ten-fold improvement by 2050, as presented in its national Resource Efficiency Action Plan.
- Estonia reports aiming for a 10% increase in resource efficiency to EUR 0.46/kg (GDP/DMC) as part of a Coalition Agreement of the Estonian Government for the period 2015–2019.
- France aims for a 30% increase in resource productivity (GDP/DMC) between 2010 and 2030 as well as a decrease in per person DMC over the same period.
- Germany has a target of doubling abiotic material productivity within the period 1994–2020, included in its 2002 National Sustainable Development Strategy.
- Hungary stipulates reducing its material intensity (DMC/GDP) to 80% of the 2007 level, by 2020, in the 2011 National Environmental Technology Innovation Strategy, which is part of the Hungarian National Reform Programme.

- Latvia adopted a target for resource productivity (GDP/DMC) to reach EUR 710/tonne in 2030, with intermediate targets of EUR 540/tonne in 2017 and EUR 600/tonne in 2020.
- Poland, in its Strategy for Innovation and Efficiency of the Economy, adopted a target of increasing resource productivity to EUR 0.45/kg by 2015 and EUR 0.5/kg by 2020 (GDP/DMC).
- Portugal stipulates an increase in national resource productivity from EUR 1.14/kg in 2013 to EUR 1.17/kg in 2020 and EUR 1.72/kg in 2030 in its Green Growth Commitment.
- Slovenia's target for resource productivity anticipates that overall resource productivity by 2023 should increase to EUR 1.5/kg DMC, from 1.07 in 2011.

No countries reported having targets for reducing the use of primary materials (metals, minerals or biomass) or for specific materials, including those on the EU list of critical raw materials. Very few targets have been adopted at the level of individual economic sectors.

One interesting finding emerging from the 2016 EEA report is an increasing number of initiatives on material resource efficiency taken by public authorities and local governments within their areas of competence. The reported targets go beyond the most obvious topic of sustainable public procurement, and include reducing energy consumption in public administration, reducing the use of paper, or increasing the use of sustainable transport. Examples of specific targets set for the public sector are presented in Box 11.2.

11.9 Institutional Set-up

Material resource efficiency is a cross-cutting issue involving several domains and policy levels. The institutional set-up reflects this, having evolved in different ways depending on national governance structure, and ranging from fairly centralised approaches, such as in France or Poland, to more decentralised ones, as in Belgium and the United Kingdom.

Almost all countries reported having an institutional structure to develop material resource efficiency policies. The most frequently occurring model is a shared ministerial responsibility, typically involving ministries of the environment, economy, energy and agriculture. However, this frequently leads to overlapping responsibilities and competencies—twenty countries reported having four or more ministries or agencies with responsibility for material resource efficiency. Further streamlining of institutional arrangements in which several ministries are involved could ensure the more effective use of capacities and help improve policy coherence.

Box 11.2: Examples of Targets set for the Public Sector and Government
Finland

- All new public buildings should be near zero energy by 2017.
- Finnish public-sector employees are to strive to rearrange their working routines so that by 2015 they will travel 10% less than in 2010.
- By 2015 all vehicles purchased for mass transport should have emissions lower than 100 g/km; or at least 30% of vehicle fleets should use electric, ethanol, gas or hybrid solutions.
- Purchased electricity for public buildings must come 100% from renewable sources.
- Catering in central government organisations should increase the share of organic food to 10% by 2015 and 20% by 2020.

France

- For public authorities, reduce office paper consumption by 30% by 2020.
- Achieve a 25% share of recycled paper in all paper use by 2017, and a 40% share by 2017 for public authorities.
- Achieve a 50% share of reused or recycled building waste materials in road construction materials purchased by national and local authorities in 2017, rising to 60% by 2020.

Portugal

- Reduce energy consumption in public administration by 30% by 2020 and 35% by 2030.

Turkey

- Reduce annual energy consumption in public enterprise buildings and facilities by 10% by 2015 and 20% by 2023.

11.10 In Conclusion

The overall picture that emerges from the EEA survey of 32 countries carried out in late 2015 is that the economic benefits of improved efficiency and circularity of resource use are increasingly being recognised and acted upon.

However, despite a growing number of national initiatives, the wide scope and conceptual complexity of the issues around materials resource efficiency and circular economy leave much room for improvement in policy initiatives and their implementation. Further integration of policies regarding energy, material resources and waste would appear particularly beneficial.

There is also scope for an increased focus on upstream measures to close material loops (such as eco-design, business models, consumer behaviour and corresponding incentives) to complement the well-established downstream policy measures for waste management and prevention laid down in the EU environmental acquis.

Various targets have been adopted and corresponding monitoring mechanisms are in development in many countries, but major gaps still exist regarding compatible waste and material flow statistics and accounts, sectoral performance indicators, enablers of progress and environmental and socio-economic co-benefits.

The need for capacity building is widely recognised, with the exchange of national experiences and propagation of effective practices seen as central to the harmonisation of key concepts and methods as well as to increased policy coherence and impact.

Reference

European Environment Agency (2016) More from less — material resource efficiency in Europe. 2015 overview of policies, instruments and targets in 32 countries. European Environment Agency, Copenhagen, June 2016. Available online: www.eea.europa.eu/resource-efficiency

Chapter 12
The Resource Nexus and Resource Efficiency: What a Nexus Perspective Adds to the Story

Raimund Bleischwitz and Michal Miedzinski

Abstract This chapter addresses the resource nexus, a concept that has recently become quite popular. It is often framed as a water-energy-food nexus and discussed at international organisations, within infrastructure planning units and among actors on the ground concerned with the Sustainable Development Goals (SDGs) of the United Nations. The chapter seeks to clarify a common understanding of what the concept entails and proposes an augmented five-node nexus, which includes materials and land. We define the nexus as a set of context-specific interlinkages between critical natural resources used as inputs into socio-economic systems of provision. Furthermore, we discuss how the nexus fits into the narratives of resource efficiency and eco-innovation. The chapter argues in favour of the complementary strengths of both and gives an outlook on governance options.

Keywords Sustainability • Eco-innovation • Resource nexus • SDGs • Transitions

12.1 The Challenge of the Resource Nexus

The narrative of the water-energy-food nexus is nowadays often used to portray the complexity of nature and its interactions with societies.[1] This approach refers to the numerous interlinkages and competing demands for the use of natural resources, perhaps best illustrated by water needed for both energy and food production. Research conceptualizes the nexus as a set of interactions, comprising important drivers for the use of resources. Natural resources serve as direct inputs in the production processes of another resource or they can substitute the use of another resource. Indirect effects related to specific uses of resources also have to be taken

[1] See also the Future Earth Knowledge Action Network on the nexus at: http://futureearth.org/future-earth-water-energy-food-nexus; the work of the UK nexus network at: http://www.thenexusnetwork.org; or one of the origins: http://www.water-energy-food.org

R. Bleischwitz (✉) • M. Miedzinski
UCL Institute for Sustainable Resources (ISR), Central House, 14 Upper Woburn Place, London WC1H 0NN, UK
e-mail: r.bleischwitz@ucl.ac.uk

H. Lehmann (ed.), *Factor X*, Eco-Efficiency in Industry and Science 32,
https://doi.org/10.1007/978-3-319-50079-9_12

into account because claims for a particular use of one resource can compete with other demands, as in the case of land use for either food or bioenergy production.

Such a systems approach has a long tradition in sustainability research (Wichelns 2017). The nexus debate has emerged only at the beginning of the 2010s (Hoff 2011; Andrews-Speed et al. 2012, 2014) to offer a more integrated approach to studying resource use and management. In fact, nexus terminology is increasingly popular, possibly at risks of becoming a 'buzzword,' as a recent editorial in *Nature* (2016) suggests, and being subject to quite different perceptions from stakeholders (Cairns and Krzywoszynska 2016; Green et al. 2016).

Our contribution takes those uncertainties about what the nexus means as a starting point to discuss a perspective. What can be considered new is, first, **a systemic resource-based approach to environmental challenges** taking into account issues emerging across all resource use patterns rather than originating from one or few selected flows and, second, **a distinct attempt to grapple with social issues on the ground**, such as access to resources and security.

Against the background of research, planning and management often being organized along single 'silos' of water, energy, etc., the aim of the resource nexus approach is to look at the connections between the resources in a more integrated manner. These interlinkages are manifold and complex, as all resources need others as inputs for their production and along value chains to the delivery of goods and services for final consumers. The nexus attracts attention because it provides a holistic and systemic view enabling fresh thinking on emblematic sustainability issues. We shall discuss below how such concept could facilitate adequate innovative solutions, engaging researchers, entrepreneurs and innovators on the ground, dubbed here 'nexus innovations'.

12.1.1 Definition and Scope of the Nexus

Our contribution defines the resource nexus as the set of context-specific critical interlinkages between critical natural resources used as inputs into socio-economic systems of provision. The nexus can be conceptualized as a set of critical interlinkages between the different natural resources, with human activities shaping the drivers, intensity and efficiency of resource use, and humans and the environment either benefiting or being impacted by the outcomes of resource use. Criticality in such interlinkages may result from different drivers, such as overusing minimum supply conditions, passing critical threshold values, and indeed numerous tradeoffs. Hydroelectricity may serve as a case where an expected provision of electricity won't occur under conditions of an extended drought, creating knock-out impacts for customers.

Recent nexus scholars (Bleischwitz et al. 2017) propose a scope for the nexus that comprises all direct and indirect resource inputs into socio-economic processes at appropriate scales, taking into account:

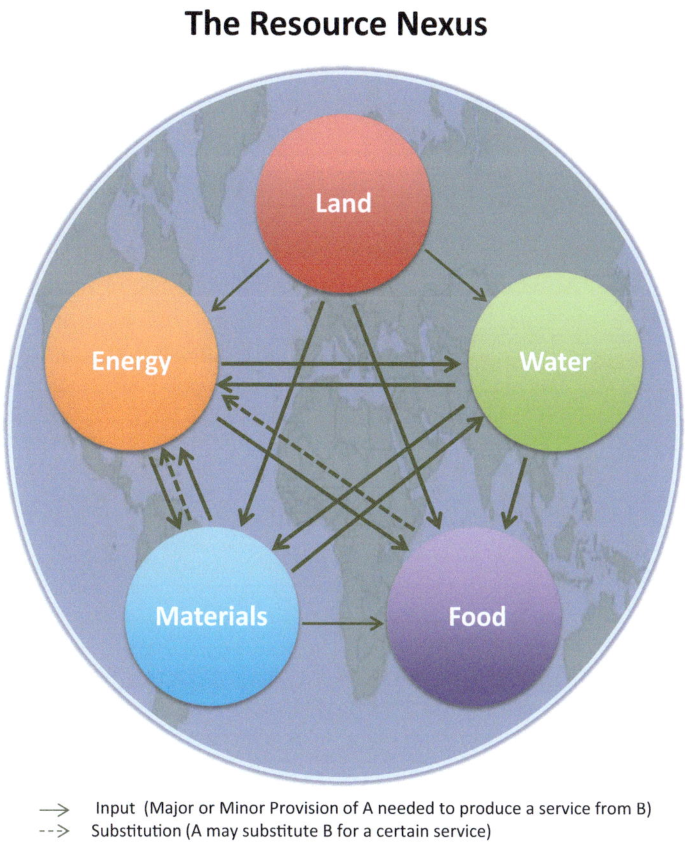

Fig. 12.1 The resource nexus (Source: Own illustration after Andrews-Speed et al. 2014)

- Water
- Energy: Fossil fuels and other fuels such as nuclear, REN
- Food: food is seen as a system of provision, as there is usually a series of processing steps between biomass production and consumable food, all of which depend on inputs of energy, water and other resources. In Fig. 12.1, biomass is part of food, energy, and land, illustrating the interlinkages and putting more emphasis on the systems of provision rather than resources itself.
- Land: Land is an ultimately limiting factor of production and serves all environmental functions of support, regulation, provisioning, and cultural services.
- Materials are relevant because:
  - they make up for ~50% of resource use in most industrialized countries (measured in physical units as used in Material Flow Analysis methodology);
  - The costs for manufacturing industry are significant (Wilting and Hanemaaijer 2014) and the potential for eco-innovation enormous;

- Base metals, critical materials and construction minerals are relevant for the SDGs related to water, energy, and urbanisation; mineral fertilizers are relevant for food production;
- Materials have been assessed as being important intermediaries of environmental impacts (UNEP 2010: 81).

We propose to have subcategories for metals and critical materials, construction and industrial minerals and a separate account for mineral fertilizers.

Having such **a five-node nexus** of water-energy-food-land-materials leads to more complexity compared to the majority of previous studies that analyse a two-node or a three-node nexus. In line with recent sustainability research (Liu et al. 2015: 3), it also captures 'a bigger picture' and facilitates bringing in the social dimension. We consider this approach as flexible and open: case studies may focus on a few core critical interlinkages, and may also analyse interlinkages within some of these dimensions, such as biomass, land use and food.

Box 1: The Many Faces of Nexus

The contemporary debate features a number of approaches to the resource nexus with different views on which resources ought to be considered as part of the concept. The most widely acknowledged scope covers *water – energy – food* (Hoff 2011; Bazilian et al. 2011; Lawford et al. 2013). Other studies focus on:

- The *water – energy* nexus (Ackerman und Fisher 2013; Glassman 2011; Howells and Rogner 2014), inspired by the huge amount of energy needed for water pumping and by the impact a drought might have on electricity production;
- Water – energy – food – *land* (European Commission 2012; Ringler et al. 2013; Sharmina et al. 2016) as main biotic resources originate from land use patterns;
- Water – energy – food and *mineral fertilizer* (Mo and Zhang 2013), pointing at the potential depletion of such resources, their relevance for food security, and their complex supply chain with recovery opportunities from, e.g., waste water;
- Water – energy – *minerals* (Giurco et al. 2014) illustrated by declining ore grades and the high intensity of using water and energy during extraction processes.

The studies published by Chatham House (Lee et al. 2012) and by the Transatlantic Academy (Andrews-Speed et al. 2012 and 2014) share a wider recognition of resources as manifold inputs into economic processes in line with the approach proposed in this Outlook; so has McKinsey Global Institute (Dobbs et al. 2012) with a focus on opportunities for some industrial sectors.

There is also a large number of regional case studies, e.g., on India (Rasul 2014), South Asia (Mukherji 2007; Rasul 2014), and the MENA region (Siddiqi and Anadon 2011), which assess those resource interlinkages that are most relevant in the region. Without being exhaustive here it can be said that the resource nexus concept is fairly often applied on the ground at different scales.

Figure 12.1 shows the many ways in which the use of key resources interacts. Some nexus issues may be more obvious than others, such as the connection between food and water suggests. Others have become more pressing recently, such as the water inputs needed for energy production when droughts occur.

12.1.2 The Relevance of the Nexus

The nexus challenges are pertinent for many actors, involving private companies, development agencies, infrastructure planning units for water and energy, and international organizations. The challenge for decision-makers is that all activities that are intended to use a specific resource should be based on the knowledge about the following factors: the estimated inputs needed from other resources in the future, how those may compete with other demands, and what critical events might arise that may put constraints on such supply in the future.

The extended novel narrative of the resource nexus should, therefore, also address:

- The resource interlinkages across use patterns, especially along what is known from consumption research as 'systems of provision' (Ben Fine and Ellen Leopold), i.e., the essential services of public importance;
- Human security, a 'nexus on the ground', and livelihoods of the one billion + people living below the poverty line (Biggs et al. 2015);
- Political and economic security, partly as a tool for analysing conflicts related to natural resources within regions or across borders and partly as a tool to assess supply chain securities.

The resource nexus concept becomes relevant for risk assessments especially in water and energy planning, but also for land use planning and for strategic investments. Furthermore, it can be seen as systemic in addressing all relevant issues that can be related to the use of natural resources in societies and across many scales.

12.2 Eco-Innovation to Address the Nexus Challenges

12.2.1 Nexus and Eco-Innovation Opportunities

Much of the nexus narrative has been built-up to understand risks stemming from critical interlinkages and how they would affect people. Yet, it is also a compelling narrative for opportunities – and here the narratives of resource efficiency and eco-innovation kick in. Minimizing tradeoffs and exploiting synergies across the use of resources is a common understanding of both nexus approaches and those opportunities. It should also be common ground to avoid waste and making resource use more circular, from business operations onto supply chain management. What needs to be recognized is a wider opportunity of both approaches coming together. In comparison to the nexus with its focus on security especially in fragile regions, the current understanding of eco-innovation has strong bearings in pioneering manufacturing industries and policy actors across the environment and economy – quite often in mature or emerging regions with import dependencies on commodities.[2] The overarching challenge of bringing those two narratives together thus shouldn't be underestimated. Yet it seems promising to try, and it may help turning the bias on risks and threats inherent to the nexus concept into a joint narrative of delivering SDGs and opportunities to eco-innovate. Needless to say any understanding of innovation should include both technological and non-technological innovations introduced by various actors (including public sector and partnerships).

Nexus innovation needs to be a change that addresses critical interlinkages of at least two resources, while not posing additional risks on others. Therefore, it would need to go beyond measures to focus on single materials. One should also strongly consider equity issues and a focus on improving access for deprived actors – giving it a boost in developing and emerging societies. Accordingly, it could be small such as an application of drip irrigation in farming and land use and should typically be related to an SDG. By design it should correct systemic deficiencies and reconfigure system-level structures and dynamics by introducing mutually reinforcing innovations to respond to a nexus challenge. These changes will often include new products or services but, in order to ensure systemic impacts, they have to come with an enabling mechanism. The latter may be co-created by actors on the ground such as farmers or SMEs, by introducing collaborative business models, innovation alliances, and go ahead via ambitious standards and norms, as well as conducive policy and regulatory frameworks. The boundary of nexus innovation should encompass elements in the system to create Schumpeterian dynamics. Innovating together and enabling people may influence the entire system dynamics towards a more sustainable mode of production and consumption. Such eco-innovation needs to internalize changes within innovation systems that usually reside outside innovators' strategy. It will often require new collaborations, often

[2] See e.g. the excellent work done by the EU's eco-innovation observatory at: www.eco-innovation.eu

involving public-private partnerships, within and across sectors, value chains and supply chains, and may require adapting regulatory frameworks.

Nexus innovation has to differentiate and compare key resource interlinkages, and likely environmental pressures and impacts associated with alternative eco-innovative solutions. Energy transitions towards a low carbon system have partly underestimated the nexus challenges of dealing with water, land use and the need for materials. The concepts of a circular economy and resource efficiency are very much in line with main aims of a nexus approach. Some related indicator systems (DMC/RMC) are more narrow and should be widened to capture nexus dimensions. Material flows analysis (MFA), for instance, does neither address water nor land.

Furthermore, due to its broad boundaries, it needs to include both economic and social rationale behind the choices to support one or another innovation pathways. This requires a reflection, collective deliberations and new forms of evidence on how trade-offs between different uses of different resources can be addressed.

12.2.2 Energy Transitions

The Paris agreement of 2015 has paved the way for a deep decarbonisation of our economies, i.e., a major transformation of the energy system towards low carbon societies. A nexus innovation approach would underpin such new direction by adding insights on how all alternative energy sources – unconventional fuels, bio-energy, wind and solar – come with additional demand for water and materials and, partly, for land. It is thus of utmost importance to go beyond a 'carbon' indicator and add such nexus dimensions in scenarios, integrated assessment modelling as well as other modelling tools. Nexus modelling tools are emerging (WEAP-LEAP, CLEW, ENV-Linkages, ENGAGE) and should be used for the development of energy transitions, especially if regions are exposed to nexus risks of water stress and food security issues. However, issues of critical materials needed for low carbon technologies point straight forward at the international nexus dimension of any national lead market for clean energy – such national forerunners usually import critical materials and should trace the sustainability of supply conditions. Similar nexus dimensions arise with using bio-energy that may have been produced under conditions elsewhere that are unsustainable for water, land, and food dimensions of bio-energy. It is also worth noting coal production as a source for nitrogen, and thus there is a need to innovate for alternative supply paths for fertilizers.

12.2.3 Implementing the SDGs

It will be decisive to bring both a nexus and an eco-innovation perspective into the implementation of the sustainable development goals (SDGs) launched in 2015. The SDGs are likely to have major implications for future resource markets.

However, those implications are mixed. On the one hand, many of the new SDGs will lead to an increase in demand for a number of materials:

- Goal 2: "End hunger, achieve food security and improved nutrition and promote sustainable agriculture" – implies increasing demand for land, mineral fertilisers, water, biomass and food.
- Goal 6: "Ensure access to water and sanitation for all" – implies investments in water supply and a water distribution infrastructure, i.e., increasing demand for materials.
- Goal 7: "Ensure access to affordable, reliable, sustainable and modern energy for all" – is likely to imply increasing demand for bio-energy and renewable energy, plus more traditional energy sources, which again implies more demand for land, biomass, water and materials.
- Goal 9: "Build resilient infrastructure, promote inclusive and sustainable industrialization and foster innovation" – will require more construction materials, metals and other materials.

Adding the promotion of economic growth to it, as well as efforts to eradicate hunger and enhance health, the signals for future demand for resources stemming from the SDGs are clearly upwards. At least for key metals (aluminium, iron ore, copper and nickel, which altogether make up for more than 80% of world production of metals), for construction minerals, for biomass and food, for water, and for arable land, the SDGs are very likely leading to new and additional demand compared to business as usual forecasts (see for food and land use issues: Obersteiner et al. 2016). The situation for energy fuels is less straightforward as climate policy will probably lead to restrictions for using fossil fuels, if political efforts succeed, although major suppliers may not join any future international agreement and have announced plans to expand production. If prices for fossil fuels stay low, efforts to curb demand will be difficult to achieve.

On the other hand, the SDGs also endorse the sustainable production and consumption agenda, and call for global increases in resource efficiency as well as for aims to achieve sustainable and resource-efficient infrastructures by 2030 (Goal 9) and sustainable management and efficient use of all resources by 2030 (Goal 12). Moreover, they aim to "improve progressively, through 2030, global resource efficiency in consumption and production and endeavour to decouple economic growth from environmental degradation (…)" (Goal 8).

The balance between such expected demand increases and other goals however is not entirely clear, in particular as key terms (such as sustainable management and efficient use of all resources) are insufficiently defined and will leave space for quite different implementation pathways.

Analysing and developing nexus innovations will thus have a key role to play in delivering the SDGs 2 (food), 6 (water), 7 (energy), 9 (infrastructure and industrialization), 12 (sustainable consumption and production) in a more integrated manner. Bringing nexus and eco-innovation closer together, it should clarify trade-offs and identify synergies. A joint approach of the nexus and eco-innovation will also be required to develop principles for a sustainable management of resources (SDG 12) and to understand future dynamics on resource markets and within societies.

12.3 What Can Policy Makers/Business/Other Stakeholders Do?

12.3.1 Improving Evidence Base and Policy Learning

The resource nexus requires a robust evidence based on interdisciplinary knowledge and diverse expertise actively brought in along the whole policy process. Policies need to encourage collaborative transdisciplinary research engaging various stakeholders in generating and validating context-specific evidence for policy intervention. The dynamic nature of the challenge requires that the evidence base needs to be continuously adapted based on an on-going policy learning process that interprets and prioritizes validated evidence in a transparent way. The complex and emerging nature of nexus requires that the evidence base provides data on critical resource interlinkages as well as other key data on footprints, and incorporates the long-term prospective view by engaging in inclusive foresight methods as well as establishing links with modelling. The establishment of an international open-access database would be a huge step forward. There is a need to integrate the notion of risk and uncertainty in such analysis, while integrating the pre-cautionary principle at the heart of the process.

12.3.2 Shared Understanding of the Nexus Challenges and Key Projects

Nexus platforms need to involve many stakeholders who bring with them diverse perceptions, understandings and interests that explain how they frame the problem. Various stakeholders will perceive the same problem through different lenses focusing on aspects of technology and infrastructure, environment, economic and business models, policy and regulations, as well as culture and values. All these perspectives are valuable for policy in the context of the nexus and decoupling. Policy processes should directly include nexus issues in a regional or national process to assess the resource base towards implementing SDGs and delivering green growth, a process in which these various frames are transparently presented, explained and supported with the use of evidence. It should help to make risk assessments and better planning, and facilitate key projects on such nexus innovations by comparing successful niches and scaling them up regionally and internationally. This process enriches the evidence base by bringing new stakeholders in, revealing the motivations and positions of stakeholders as well as prepares the foundations for the vision and strategy on how innovation alliances can tackle the nexus issues.

Such key projects may comprise:

- Transboundary river management with better planning for hydropower and co-benefits
- Development of urban green space and urban farming
- Development of business niches with local people at the Bottom of the Pyramid[3] towards eco-innovations with a potential to grow and become interconnected
- Enabling new alliances for collaborations with international companies seeking community involvement and eco-innovation across borders with local benefits
- Engaging with investors, large companies, and international organizations that are under pressure to serve long-term goals with more short-term returns.

12.3.3 Leadership, Participation and Shared Visions of the Future

The nexus challenges have the strength of being oriented towards a mid-term horizon of 10–30 years, which is in line with planning for water and energy infrastructures, land use, mining projects, and investments for producing capital goods. Key projects as outlined above are important for establishing ties between stakeholders, in particular when the institutions are weak (Acemoglu and Robinson 2012). Overall it requires a strong buy-in from both stakeholders affected by the challenges as well as those who may be instrumental in solving them. The latter may not be directly affected by the challenges, but they may consider their contribution beneficial, if evidence on short-term benefits and co-benefits can be established. Water management, food security, sustainable energy systems, transitions for resource-intensive industries, and sustainable urban development are key areas. Policy and think tanks need to establish and provide a platform for developing future scenarios of dealing with nexus and a shared vision of the future. Indeed, this should add a long-term view too, e.g., the year 2050. The main value added of the process is creating shared visions based on various perspectives, and combining potentially conflicting interests into a 'future project' by creating short- and long-term incentives for key actors.

12.3.4 Short - and Long-Term Scenarios and Transition Roadmaps

The overarching vision needs to be translated into more tangible strategies on how to both kick-start the process, identify potential asset losses and sunk investments, and follow up on it in the medium to long term. Planning will need to explicitly

[3] See e.g. http://www.bopglobalnetwork.org/about-us

recognize the interdependencies between uses of various resources and seek flexible solutions to overcome the current and avoid future lock-ins in the resource- and capital-intensive functional systems (e.g. energy and water supply) that contribute to unsustainable use patterns. Scenarios and roadmaps are examples of strategic tools well suited for dealing with complex challenges of winners and losers requiring both short- and long-term outlooks. They provide a practical framework for developing shared understanding of alternative innovation pathways (scenarios) and contribute a foundation for implementation of complex multi-actor innovation projects (roadmaps). It will be essential to bring new valuation perspectives in such scenarios and roadmaps, in order to identify risks and gains of eradicating poverty and enabling access to key resources for the world's poor. Research should support these processes via modelling efforts, potentially by soft linking bio-physical tools with macro-economic modelling and applying system dynamics as appropriate.

12.3.5 Systemic Policy for Nexus System Innovations

Policy makers are an important stakeholder in making system innovation possible; one may also consider an 'entrepreneurial state' (Mariana Mazzucato). Nexus challenges require an innovative, coordinated and coherent policy response based on a policy mix that both directly supports eco-innovation and also ensures that wider regulatory and policy frameworks favour the sustainability transition. The direct support can be delivered by dedicated market and financial instruments such as resource taxes or resource dividends (Thomas Pogge), allowing a degree of risk in the case of particularly promising investment, whereas the enabling environment can, step-by-step, develop a strong regulatory framework of 'inclusive institutions' (Acemoglu and Robinson 2012) as well as removing regulatory barriers, environmentally harmful subsidies and other forms of support that favours actors or technologies that contradict the direction agreed in the vision.

12.3.6 Governance for Nexus Innovations

The resource nexus requires revisiting existing governance structures and mechanisms. Conversely, it also should be applicable if the state is weak and institutions are weak too – as it is often the case in developing countries. Resource governance[4] thus is an attempt to provide lessons learned especially for resource-rich countries, with key areas such as transparency, accountability, diversification and revenue management. This angle is important and should be enriched with thinking about eco-innovations. Such governance will concern all dimensions of governance, including

[4] See the excellent work by the Resource Governance Institute at: http://www.resourcegovernance.org/, albeit it is yet weak at nexus eco-innovations.

leadership, strategic deliberation, participation, responsibility and accountability, implementation, and monitoring and evaluation. Resolving nexus challenges require envisaging alternative forms of governance that exist in parallel to, or even substitute, established institutions and organizations. On the one hand, the challenge is to design and implement viable governance approaches that make optimal use of existing capacities and power structures by stimulating collective action on the ground. On the other hand, the challenge may be to innovate and create new governance structures when existing settings do not suffice or are mobilized against the desired change. The latter suggests that design of policy and governance need to take into account the dimension of power and leadership as well as organizational capacity, technical competences and budgets. In the global perspective, the governance catered for the nexus challenges is likely to rely on the regionalized polycentric coordination of collective action towards a global coordination. This regionalized bottom-up perspective complements other planetary governance approaches, such as 'earth system governance' (Frank Biermann) that appear more top-down. Indeed global governance approaches need to combine both bottom-up and top-down.

12.4 Outlook

Amidst lively discussions about the nexus becoming a new buzzword, our contribution provides a nexus definition and suggests a number of core elements for nexus analysis, centred around critical interlinkages to integrate multiple sustainability goals like the SDGs. We suggest new conceptual underpinnings at the interface with human security and strategic choices. A broad scope, illustrated by our five node nexus of water, energy, food, land, materials, remains important for sustainability perspectives in general and for our notion of nexus innovation in particular. A nexus angle can help assess synergies and trade-offs of resource use across different spatial and temporal scales. It is potentially transformative of analytical and policy approaches to resource assessment and governance in the context of supply chain management and wider climate action, and also considering the fact of many countries being rich in some resources while depending critically on others. The strengths of the nexus approach may be seen in two regards: enabling a focus towards social issues, and helping to establish alliances beyond the usual environmental policy suspects. Clearly, for world-wide transformations both are indispensable assets.

Acknowledgements This contribution acknowledges funding from the EU project Inno4SD (www.inno4sd.net, grant agreement no 641974) and benefited from discussions with Will McDowall.

References

Acemoglu D, Robinson JA (2012) Why nations fail: the origins of power, prosperity, and poverty

Ackerman F, Fisher J (2013) Is there a water–energy nexus in electricity generation? long-term scenarios for the Western United States. Energy Policy 59:235–241

Andrews-Speed P, Bleischwitz R, Boersma T, Johnson C, Kemp G, VanDeveer SD (2012) The global resource nexus. The struggle for land, energy, food, and minerals. A Transatlantic Academy report, Washington, DC

Andrews-Speed P, Bleischwitz R, Boersma T, Johnson C, Kemp G, VanDeveer SD (2014) Want, waste or war? the global resource nexus and the struggle for land, energy, food, water and minerals. Routledge Publisher, Abingdon

Bazilian M, Rogner H, Howells M et al (2011) Considering the energy, water and food nexus: towards an integrated modeling approach. Energy Policy 39:7896–7906

Biggs EM et al (2015) Sustainable development and the water-energy-food nexus: a perspective on livelihoods. Environ Sci Policy 54:389–397

Bleischwitz R, van der Voet E, Spataru C, Hoff H, VanDeveer S (eds) (2017) Handbook of the resource nexus. Routledge/Earthscan (forthcoming)

Cairns R, Krzywoszynska A (2016) Anatomy of a buzzword: the emergence of 'the water-energy-food nexus' in UK natural resource debates. Environ Sci Pol. doi:10.1016/j.envsci.2016.07.007

Dobbs R, Oppenheim J, Thompson F, Brinkman M, Zornes M (2012) Resource revolution: meeting the world's energy, materials, food, and water needs. McKinsey Global Institute, New York

European Commission (2012) Confronting scarcity: managing water, energy and land for inclusive and sustainable growth. The 2011/2012 European Report on Development. Accessible at: http://ec.europa.eu/europeaid/index_en.htm

Giurco D et al (2014) Responsible mineral and energy futures: views at the nexus. J Clean Prod 84: 322–338. Available at: http://www.sciencedirect.com/science/article/pii/S0959652614006805

Glassman D, Wucker M, Isaacman T, Champilou C (2011) The water-energy nexus: adding water to the energy agenda. World Policy Institute, New York

Green J et al (2016) Research priorities for managing the impacts and dependencies of business upon food, energy, water and the environment. Sustain Sci. doi:10.1007/s11625-016-0402-4

Hertwich E et al (2010) Assessing the environmental impacts of consumption and production: priority products and materials. UNEP, Nairobi

Hoff H (2011) Understanding the Nexus. Background paper for the Bonn2011 Conference: the water energy and food security nexus. Stockholm Environment Institute, Stockholm

Howells M, Rogner H (2014) Water – energy nexus. Assessing integrated systems. Nat Clim Chang 4:246–247

Lawford R, Bogardi J, Marx S et al (2013) Basin perspectives on the water–energy–food security nexus. Curr Opin Environ Sustain 5(6):607–616

Lee B, Preston F, Kooroshy J, Bailey R, Lahn G (2012) Resources futures. Chatham House, London

Liu J et al (2015) Systems integration for global sustainability. Science 347(6225). doi:10.1126/science.1258832

Mo W, Zhang Q (2013) Energy–nutrients–water nexus: integrated resource recovery in municipal wastewater treatment plants. J Environ Manag 127:255–267

Mukherji A (2007) The energy-irrigation nexus and its impact on groundwater markets in eastern Indo-Gangetic basin: evidence from West Bengal, India. Energy Policy 35(12):6413–6430

Nature (2016) Editorial: scientific buzzwords obscure meaning: 'Nexus' is enjoying new-found popularity. But what does it actually mean? Nature 538(7624)

Obersteiner M et al (2016) Assessing the land resource–food price nexus of the Sustainable Development Goals. Sci Adv 2:e1501499

Rasul G (2014) Food, water, and energy security in South Asia: a nexus perspective from the Hindu Kush Himalayan region. Environ Sci Pol 39(0):35–48

Ringler C, Bhaduri A, Lawford R (2013) The nexus across water, energy, land and food (WELF): potential for improved resource use efficiency? Curr Opin Environ Sustain 5(6):617–624
Siddiqi A, Anadon LD (2011) The water–energy nexus in Middle East and North Africa. Energy Policy 39(8):4529–4540
Sharmina M et al (2016) A nexus perspective on competing land demands: Wider lessons from a UK policy case study. Environ Sci Pol 59:74–84
Wichelns D (2017) The water-energy-food nexus: is the increasing attention warranted, from either a research or policy perspective? Environ Sci Pol 69:113–123
Wilting H, Hanemaaijer A (2014) Share of raw material costs in total production costs. PBL Netherlands Environmental Assessment Agency, The Hague

Chapter 13
Germany's Resource Efficiency Agenda: Driving Momentum on the National Level and Beyond

Reinhard Kaiser

Abstract ProgRess—the first resource efficiency programme to be adopted by a European government—has been in place since 29 February 2012, with Germany embracing a trailblazing role once again. And that is a good thing, because only by being highly innovative both in terms of technology and society will we have a chance to compete globally and hold our ground as a prosperous country with a high-wage economy. We seek to use our innovative capacity very purposefully to enable as many people as possible throughout the world to permanently enjoy good material living conditions—without damaging nature or squandering our planet's irreplaceable resources. After all, this Earth is only on loan to us from our children.

The German government and Bundestag (lower house of parliament) passed resolutions requiring reporting on progress in resource efficiency and an update of the ProgRess programme every 4 years. We fulfilled this requirement on 2 March 2016 when the federal cabinet adopted ProgRess II, which is both an implementation report and an update of ProgRess. This article describes Germany's activities at the national level and Germany's involvement in European and international initiatives, and it outlines the broad range of work being done in Germany by the 16 Länder (states) and at local authority level. It concludes by looking ahead to the further discussion in Germany and developments towards ProgRess III, to be passed in 2020.

Keywords Resource Efficiency • National Resource Efficiency Programme • Circular Economy • Sustainability Strategy • ProgRess • NeRess • NaRess • BilRess • VDI-ZRE

R. Kaiser (✉)
Head of Division WR III: Resource Efficiency, Soil Conervation, Soil Conservation, Federal Ministry for the Environment, Nature Conservation, Building and Nuclear Safety (Germany), Stresemannstraße 128-130, 10117 Berlin, Germany
e-mail: reinhard.kaiser@bmub.bund.de

H. Lehmann (ed.), *Factor X*, Eco-Efficiency in Industry and Science 32,
https://doi.org/10.1007/978-3-319-50079-9_13

13.1 Introduction

In 2013, the editors of the current series of books—Michel Angrick, Andreas Burger and Harry Lehmann—pointed the way forward when they wrote in the "The Editor's View" in Volume 30: "Protecting natural resources means raising the issue of their efficient use and, far beyond that, the issue of our production and consumption patterns and therefore, of our lifestyles." (Angrick et al. 2013) Germany strives to meet this immense challenge with its resource policy and constant improvements to performance and processes. The German Resource Efficiency Programme ProgRess is a key factor in this.

Whatever action we, be it myself, you, the reader, or other stakeholders, take on resource efficiency will help to stabilize the global environment. But our actions will only make a quantitatively relevant, globally noticeable contribution if many actors take similar action in whatever way they can in their respective sphere of influence. It is vital that the actions we take complement and support each other. To this end, we must first know who is doing what. It is my hope that this book will go some way towards achieving that.

In this article, I would like to report on the activities in Germany, including on the work being done in the individual Länder and local authorities. However, first I would like to take a brief look at Germany's European and international initiatives. Maybe you have already come across one or two of them? And that is why I will start right there.

13.2 The G7 Is Backing Resource Efficiency

Germany held the presidency of the G7[1] in 2015, and with a great deal of effort and luck, and—to our great delight—with the help of the Federation of German Industries (Bundesverband der Deutschen Industrie, BDI), we managed to get the topic of resource efficiency included in the German proposal for the G7's agenda. This turned out to be quite momentous: at the summit of the G7 heads of state and government, held in Schloss Elmau on 7 and 8 June 2015, resource efficiency was declared to be one of the G7's ongoing tasks.

Specifically, the following decisions were adopted:

- To establish a **G7 Alliance on Resource Efficiency**, to be launched at an opening event in Berlin on 2 October 2015 (BMUB 2015a)
- To request that the UNEP International Resource Panel (IRP)[2] prepare a **synthesis report** on the most significant potential for resource efficiency at international level;
- To request that the OECD develop a **policy guidance paper**.

[1] G7=Group of 7: Germany, France, United Kingdom, Italy, USA, Canada, Japan.

[2] We continue to work in the Steering Committee there. http://www.unep.org/resourcepanel/

- And, as a general rule: **each G7 presidency** should in future organise **at least one event on resource efficiency** as part of the Alliance on Resource Efficiency (G7 Leaders Declaration 2015).

German business made a significant contribution to getting the issue off the ground, even in advance of the official meetings. Under the aegis of BDI, the **B7**—the business associations of the G7 countries—met on 29/30 May 2015. They adopted a comprehensive appeal for innovation with a separate chapter on promoting efficient use of resources and presented it to the summit (BDI 2015).

Work in the **G7 Alliance** got off to an encouraging start. It was prepared on 12 March 2015 (BMUB 2015b) and launched on 2 October 2015 (BMUB 2015d) at a major conference in Berlin. The **UK** carried out a Workshop on Industrial Symbiosis in Birmingham on 29–30 October 2015 (International Synergies Ltd. 2015), and the **USA** invited a conference for the automobile industry on 22/23 March 2016 (epa 2016). **Japan**, which holds the G7 presidency in 2016, continued the Alliance on an ambitious scale, starting by organising two meetings,[3] and then adding an Action Plan, which continues and expands the G8's 2008 Kobe 3R Action Plan.[4] And so, resource efficiency was once more a focus of discussion and the subject of resolutions in 2016 both at the G7 Environment Ministers' Meeting15–16 May (G7 Toyama Communiqué 2016)[5] and at the G7 summit, 26–27 May (G7 Leaders' Declaration 2016).[6] Japan will also host two more events to conclude 2016.[7]

We are already looking ahead to the time after 2016: in 2017, Germany will hold the **presidency of the G20** which, in addition to the G7 countries, also includes partners such as the EU, China, Russia, South Africa and India.[8]

Germany intends to take this opportunity to mainstream resource efficiency, embed it in a broader international context and set appropriate work processes in motion. The fact that the issue has gained a high level of support within the G7 is conducive to this, despite considerable resistance initially. Ultimately, it was a real learning process for us. The fact that **India** instigated an Indian Resource Panel not too long ago on 23 November 2015 is also conducive.[9] But the most encouraging

---

[3] 22./23.2.2016 in Pacifico/Yokohama, on international cooperation, especially with developing countries; and April 2016 on the IRP's synthesis report.

[4] This was highly influenced by the approach that originated in the waste management sector. Now the aim is to pursue the broader-based circular economy approach. https://www.env.go.jp/en/focus/attach/080610-a5.pdf

[5] Of particular interest to us in this Communiqué are pages 3-4 and 15 ff. on the Toyama Framework on Material Cycles.

[6] Of particular interest to us in this communiqué is p. 29.

[7] "Resource efficiency and Climate Change", Tokyo, 12–13 December 2016; "International Resource Recycling", Tokyo, 14th–15th December 2016.

[8] Germany's presidency began on 1.12.2016 and Argentina will take over on 1.12.2017.

[9] Incidentally, this was instigated by a GIZ project carried out under BMUB's International Climate Initiative, https://www.international-climate-initiative.com/en/news/article/new_indian_resource_panel_contributes_to_climate_protection/?iki_lang=en&cHash=ceb84b12ad053783c661c19e898469b6

thing is that resource efficiency is explicitly mentioned in several places in the **Sustainable Development Goals** (SDGs), which were adopted at the UN General Assembly on 25–27 September 2015, which took place in the form of a summit (UN 2015).[10] This gives us an excellent political framework and numerous starting points for practical initiatives.

Here again, we received a call to action from business: the B20—which comprises the major business federations from 15 of the G20 member states—drafted an appeal on 2 June 2016 in Paris, calling for the G20 to prioritise resource efficiency in its work (B20 2016). At the time of writing, we are, of course, still in the starting blocks with regard to the G20 initiative and currently in the process of devising a strategy and informally sounding out the possibilities.

13.3 Slow Progress in Europe—Impacts Germany Too

Unfortunately, developments at European level have not been as encouraging. Former EU Environment Commissioner Janez Potocnik engaged with the issue of resource efficiency very ambitiously, and it became one of the seven flagship initiatives under the Europe 2020 strategy[11] produced by the Barroso Commission (2009–2014). The Commission presented a roadmap for the strategy (EU-COM 2011)[12] and, as a result of the European Council of Environment Ministers conclusions in December 2011, which it felt were disappointing, set up a high-ranking stakeholder body—the European Resource Efficiency Platform (EREP) (EU-COM 2014a)—on 5 June 2012. Its members included Germany's environment ministers Altmeier and Hendricks. The EREP concluded its work with policy recommendations on 31 March 2014, (EU-COM 2014a)[13] which were passed on to the new commission that would be set up that autumn. At the same time, the Commission produced a Circular Economy Package, which contained communications on all areas of resource efficiency, along with a waste legislation package (EU-COM 2014b).

When a new Commission was created following the European elections of 25 May 2014, a far-reaching change of policy direction occurred. In the expectation that it would be widely applauded for "doing away with anti-business regulations that are unnecessary and infuriating" the new Juncker Commission withdrew the legislation part of the package in December 2014. However, this move was not applauded; on the contrary, it unleashed a storm of outrage among the Member States, interest groups and the European Parliament. The all-round policymakers at the head of the

[10] Here relevant primarily SDGs 8.4 and 12.2, and also 9.4, 11b.

[11] Links referring to this flagship initiative have apparently been removed from the Commission's website or are inactive (see, for example, the link in following footnote).

[12] The link "Go to the home page of the Resource Efficiency Flagship Initiative" given here was inactive on 12.11.2016.

[13] http://ec.europa.eu/environment/resource_efficiency/documents/erep_manifesto_and_policy_recommendations_31-03-2014.pdf p. 8 ff.

Commission had clearly misunderstood the language of the specialists in the field. Broad-based criticism of the Commission's proposal was indeed voiced in these circles—from Germany too—but this was seen as a completely normal part of the work process, and not as a neoliberal cry that "better regulation is no regulation."

The EU Commission felt obliged to fend off criticism with a voluntary commitment: within a year it would present an **"ambitious" new package on the circular economy**. In view of the massive amount of bureaucracy with which the new Commission hampered its own ability to initiate activities,[14] this seemed like an ambitious time frame. In fact, the new package was presented by the Commission as promised on 2 December 2015 (EU-COM 2015) Once again, its legislation package refers to waste management only and essentially revisits the former draft. A new component is an Action Plan in the area of product policy, the details of which will no doubt be made more specific in the course of further discussions. Many of the recommendations of the EREP were not taken up, or only partially, although they already had the character of a compromise—large companies and industry associations had participated in the EREP, and even Germany's contributions had been carefully agreed with the different ministries concerned and often incorporated feedback from the industry associations. It is particularly regrettable that the Commission made **no mention of a European target** for resource efficiency, as proposed by the EREP—something which Germany repeatedly advocated for.

It is therefore all the more encouraging that things have at least started to move in the area most closely connected to practice. Through the Association of German Engineers Centre for Resource Efficiency (VDI ZRE), we had instigated **collaboration across the resource efficiency agencies** in the Member States. From the very first meeting in Berlin on 18 October 2012, this proved to be an extremely interesting and fruitful approach; just how much specific experience was shared and the range of tools that people described to each other was astonishing.[15] From the very outset, our aim was to set the ball rolling and then pass it to the EU. And we succeeded: the Commission took the ball on—it organised a third meeting—dribbled it

---

[14] Proposals were only to be made when absolutely necessary and then only by several commissioners acting jointly. They would require the approval of the vice-president responsible before they could even be submitted to the Commission. The Secretariat-General would closely monitor the process and outcome at all times. The Secretariat-General alone was allocated some 200 additional posts to deal with its "additional" tasks. The Commission's activities were to concentrate rigorously on a growth agenda, the substance of which seems to consist in handing out funding. For more information on the ideology behind this policy, the Commission communication entitled "Better regulation for better results" is to be recommended and makes fascinating reading. https://ec.europa.eu/digital-agenda/sites/digital-agenda/files/dae-library/communication_br_package.pdf. I myself considered launching a soil conservation initiative to demonstrate "better regulation" in practice, but found it simply impossible to achieve anything following the ideas and guidelines of the cited communication.

[15] On 18.10.2012 10 ministries and agencies (M&As) from seven Member States participated in Berlin at the invitation of VDI ZRE. Other meetings took place on 17./18.02.2013 in Brussels at the invitation of WRAP UK with 20 M&As from 12 Member States; 15.01.2014 in Brussels with 20 M&As from 12 Member States at the invitation of the Commission; 10.11.2014 in Berlin (back to back with the European Resource Forum ERF), with 16 M&As from 10 Member States.

and then aimed for the goal. Initially it issued an invitation to tender for a smaller contract to develop an **Efficiency Tool for SMEs**,[16] then on 30 December 2015—after lengthy delays—it published an invitation to tender for a contract to organise a Europe-wide collaboration across different agencies within a **European Resource Efficiency Excellence Centre**. On 23 July 2016 the contract was awarded to a consortium,[17] of which the German VDI ZRE Centre for Resource Efficiency is a member. This should lead to important further improvements to the work of those agencies that are already working well, but more than that it could also be an incentive for those Member States in which little or nothing has happened to date with regard to resource efficiency. Our environmental concerns can only truly be addressed if they are not only pursued by a handful of pioneers but also enjoy support across the board.

13.4 Our Activities in Germany

We are proud to be one of the first countries in the world to have adopted a national resource efficiency programme, also known as ProgRess, on 29 February 2012.[18]

The aim of the German Resource Efficiency Programmes is to make the extraction and use of natural resources more sustainable and to minimise the environmental pollution associated with it to the furthest possible extent. In this way, we seek to create the basis needed to ensure a high quality of life on a permanent basis—also with a view to our responsibility for future generations.

The German government seeks to decouple economic growth from the use of resources as far as possible and reduce the associated environmental pollution, thereby making German industry more competitive and fit for the future, which in turn will promote stable employment and social cohesion. The resource efficiency policy is meant to help us shoulder our global responsibility for the environmental and social impacts of resource use. The aim must be to reduce the use of raw materials.

ProgRess deals with a selection of resources. ProgRess I focused on abiotic, non-energy resources, and also on biotic raw materials in material use. Energy resources were not included because their use was (and still is) the subject of comprehensive debate within the field of energy policy. The use of raw materials is connected with the use of other natural resources such as water, air, land and soil, biodiversity and ecosystems. However, since these resources are already covered by other programmes, processes or legislation, ProgRess does not address them in detail.

---

[16] I am happy to note that the contract was awarded to a consortium of which VDI ZRE was a member. Access to the tool at: https://ec.europa.eu/growth/tools-databases/resat

[17] The lead company in the consortium is the Brussels-based Technopolis Consulting Group Belgium.

[18] In fact, we assume that we were the very first country whose government adopted a programme of this kind. The author welcomes any information to the contrary. Austria's REAP, which is of great merit, was submitted a week before ProgRess, but it was a ministerial paper and was not presented to the Council of Ministers.

Fig. 13.1 ProgRess, 2012 (Source: © BMUB)

ProgRess I gives an overview of the wealth of activities that were already in place in 2012 and describes numerous approaches and measures for increasing resource efficiency (about 170 according to an internal list). It covers the entire value chain. It is about securing a sustainable raw material supply, increasing resource efficiency in production processes, making consumption more resource-efficient, expanding resource-efficient closed cycle management and using cross-sectoral instruments (Fig. 13.1).

Following the cabinet decision of 29 February 2012 on the first German Resource Efficiency Programme ProgRess (BMUB 2015e), a substantial decision on it by the Bundestag on 8 March 2012 (Bundestag 2012) and a decision of the State Secretaries' Committee for Sustainable Development of 8 October 2012 (BMUB 2012a, b), we have worked to **implement the programme** through a wealth of individual initiatives. There is not sufficient space to describe them in detail here, nor is it necessary because they can easily be found elsewhere: cabinet and Bundestag tasked us with reporting on progress in implementing the programme every 4 years and with updating it. The report and update document, **ProgRess II**, was adopted by the federal cabinet on 2 March 2016 (BMUB 2016a). And the Bundestag once again supported us in 2016, as it had in 2012, with a precise decision that injected many ideas and suggestions into our work (Bundestag 2016).

Structurally, our implementation work has worked out well. We have continually updated the activities of the **resource efficiency network,** which was established back in 2007 (NeRess 2016a). It now comprises 39 partner organisations (NeRess 2016b). The local resource efficiency events, which are organised by the network, are very important. They are primarily addressed to small and medium enterprises (SMEs) and aim to persuade them to engage in resource efficiency initiatives. In recent years, we have been able to hold events of this kind in almost all the German Länder, often in collaboration with the local chambers of industry and commerce. Moreover, biannual public network conferences, which take place in Berlin in early June and early December and focus on different priorities, create a regular rhythm for the exchange of ideas and experience on resources.

The subject of the 16th network conference on 7 December 2015, for example, was **resource efficiency's contribution to mitigating climate change**.[19] At the same time, the global climate conference was taking place in Paris.[20] The message of our conference was: unless we manage to significantly reduce our material usage and become significantly more efficient, it will be very difficult to mitigate climate change—because all material usage requires energy, and often the processes involved also release greenhouse gases. Raising levels of prosperity around the world—which is fully in line with the SDGs—while maintaining our current economic system will cause an explosion in demand for materials, which has severe implications for the environment. Tackling this issue is crucial in terms of climate policy and is also imperative for other environmental reasons—otherwise it is already possible to foresee that the planetary boundaries will be crossed (Wikipedia 2016).[21]

The **VDI Centre for Resource Efficiency (VDI ZRE)**[22] has also developed positively and with steadily increasing influence. It was established in 2009 with a joint initiative by what was then the Federal Environment Ministry (BMU) and the Association of German Engineers (VDI) and is financed under **BMUB's National Climate Initiative**.[23] VDI, which runs the centre,[24] is by far the largest association of engineers in Germany. It has approximately 130,000 members, many of whom are active in a voluntary capacity within VDI. This structure of voluntary cooperation represents a potential that is often useful for VDI ZRE. The VDI Centre for Resource Efficiency offers a broad range of recommendations, work resources, training activities and information about examples of good practice and sector-specific studies that are directed at **SMEs** in particular. They enjoy a high degree of take-up among SMEs and multipliers. Since 2012, the Centre for Resource Efficiency has also housed the resource efficiency network's offices.

We are delighted that the National Climate Initiative considers this to be a successful and important project and has therefore extended VDI ZRE's contract for the period from 1 June 2015 through 31 May 2019, following a Europe-wide tendering procedure.

---

[19] http://www.neress.de/termine/termin/1513-16-netzwerkkonferenz-in-berlin.html?cHash=79357ab25421cfa2fd1cf574f9fbf7b5&L=0 (in German)

[20] COP21, the 21st Session of the Conference of the Parties to the 1992 United Nations Framework Convention on Climate Change.

[21] This concept was developed by Johan Rockström et al. (Stockholm Resilience Centre) in 2009.

[22] http://www.resource-germany.com/

[23] http://www.bmub.bund.de/en/topics/climate-energy/climate-initiative/

[24] VDI ZRE is a limited company (Gesellschaft mit beschränkter Haftung), owned by the VDI.

Fig. 13.2 16th network conference, 7.12.2015, Neue Mälzerei Berlin; at the speaker's podium Dr. Martin Vogt, CEO of VDI ZRE (Source: © VDI ZRE GmbH/Leo Seidel)

And, on the subject of VDI: with its VDI Guidelines it also makes important contributions in the field of **standardisation. It has been working in the field of standardisation in general** since 1884 and has more recently made valuable contributions in the field of resource efficiency. It successfully concluded its work on VDI Guideline 4800 Blatt 1 "Resource efficiency—Methodical principles and strategies" in 2014 (VDI 2014), and its work on additional resource efficiency standards is making good progress (Fig. 13.2).

Germany's **Environment Agency (Umweltbundesamt, UBA)** started to work on resource efficiency long before the ministry.[25] It still plays an important role for us, putting forward many new ideas. All our research projects are supervised by UBA. UBA drafted ProgRess I and II (and will draft ProgRess III), and it has high public visibility. In 2013, it has set up a scientific Resources Commission, the **KRU**, at its headquarters (UBA 2016a). Among the UBA's activities that are very important

[25] To be precise: as early as 1999. As a result, UBA presented a substantial report entitled "Grundsätze und konkrete Schritte einer stoffmengenorientierten Umweltpolitik" (not published) to BMU in May 2001 (UBA (2001a)), and in the same year published the comprehensive study "Nachhaltige Entwicklung in Deutschland" with an extensive chapter (VII) on natural resources; this book is not freely available on the internet; short version: http://www.apug.de/archiv/pdf/uba_nachhaltige_entwicklung.pdf (in German), here p.18 ff UBA (2001b). And Harry Lehmann, co-editor of this book, was instrumental in getting UBA to organize its first "Natural Resources Day," held in Berlin on 16 September 2009. Its slogan was "Factor X: Beyond Climate Change." (UBA (2009)). https://sns.uba.de/chronik/de/concepts/t3e529bfc_12409c73a7a_-490b.html

for us are the large-scale events it stages regularly: along with many other diverse activities, UBA organised two "back to back" events in Berlin in 2012—the European Resources Forum (**ERF**) and the National Resources Forum (**NRF**),[26] repeating them on 10/11 November 2014 and 12 November 2014, again with very encouraging participant figures (ERF around 370, NRF 250) and above all with remarkable international interest: guests from over 40 countries. Here too the collaboration with VDI ZRE proved very valuable.

These large-scale events have become truly established with their fixed 2-year rhythm. In 2016 followed ERF 9–10 November[27] and NRF 11 November.[28] And they will be followed in 2018 by the 4th ERF on 27. -28.November (UBA 2016b) and the 4th NRF 29. November (UBA 2016c), both in Berlin. "Constant dripping wears away the stone," as the German saying goes.

We have also continued to hold regular **meetings with the Länder and their agencies**, the most recent, the 14th instalment, taking place on 14 June 2016. Here there are signs that the collaboration is intensifying. The Conference of Federal and Länder Environment Ministers (UMK) already sent out a strong signal to that effect on 13 November 2015 (UMK 2015), and on 17 June 2016 it passed a decision prompted by Baden- Württemberg to set up a permanent Länder Working Group on Resource Efficiency the LAGRE (UMK 2016).

All 16 Länder had already made their own contributions to ProgRess I, which were documented in the Annex to the programme.[29] Similarly, all the Länder described their work in ProgRess II. The activities in some Länder have experienced a huge upsurge in recent years. To mention just a few examples: the **Effizienz-Agentur NRW**[30] has been advising small and medium-sized businesses on resource efficiency for 17 years now; **EffNet Rheinland- Pfalz**, which pays special attention to synergies with energy efficiency, celebrated its 10th anniversary on 28 November 2015 and **Baden- Württemberg** staged its annual **conference on resource efficiency and circular economy** for the fourth time on 5/6 October 2016—always an inspiring event with contributions by the state premier and several ministers and with ever increasing numbers of participants (in 2016 well over 700). **Bavaria** opened a **resource efficiency centre (Ressourceneffizienz-Zentrum Bayern)** headquartered in Augsburg on 23 October 2016 (Bayern 2016), and from 31 May through 2 June 2016 it carried out its 1st European Resources Conference during IFAT, a leading trade fair for the waste management sector. **Hesse** has been working

[26] On 12.-13.11.2012 (ERF) and on 14.11.2012 (NRF).

[27] ERF: http://www.resourcesforum.eu/

[28] NRF: http://www.ressourcenforum.de/ (in German)

[29] When the cabinet decision was passed on 29.2.2012, no submissions had been received from Berlin and Brandenburg. However, this was quickly rectified as a result of parliamentary pressure from the Green party, part of the opposition in both Länder parliaments.

[30] On efa: http://www.ressourceneffizienz.de/ressourceneffizienz/startpage-en.html (efa 2016). For the 15th anniversary on 11.12.2013: http://www.ressourceneffizienz.de/aktuelles-termine/detailansicht-alle/news/detail/News/ressourceneffizienz-gestern-heute-und-morgen-1.html (in German) (efa 2013).

on its own resource conservation strategy since the end of 2014 (Hessen). A lot is happening at the state level in Germany, but we would still like to help bring momentum to those Länder that have been a little hesitant to date. Why should they not get bitten by the success bug too?

We have also developed lively collaboration with the **local authority associations** since 2015, which we intend to expand and develop to include further local initiatives. One of our principal ideas is to tap into local authorities' business development activities. We intend to explore and develop the options in a research project, scheduled to be up and running by the time this book is published.

The National Resource Efficiency Platform (NaRess) is a relatively new addition to our structures (BMUB 2016b). Established on 17 September 2013, it was designed initially to facilitate ongoing collaboration between the different federal government departments and the most important business associations. In fact, the idea for the platform originated at a discussion with some of these associations at the Federal Ministry for Economic Affairs and Energy (BMWi), and was promptly taken up by the environment ministry (BMU). It was agreed that biannual working meetings would be held, with the option of additional meetings where necessary on particular topics or with higher-ranking/political representatives. Former informal meetings at BMU and BMWi were included in NaRess, and we consult closely with BMWi about the management of the platform. The coalition agreement of 27 November/16 December 2013 (Coalition 2013) tasked us to expand NaRess to include trade unions, environmental associations, and the consumer association (Verbraucherzentrale). NaRess met for the first time with this new format on 17 March 2015. The local authority associations have also joined and, since the meeting on 27 September 2016, the Länder are also represented by two delegates from LAGRE.

Finally, the **Round Table on Resource Efficiency in the Construction Sector** should not go without mention. Launched on 3 April 2013 in what was then the Federal Ministry of Transport, Building and Urban Development (BMVBS), it is now continuing its work at BMUB. Associations and professionals from the building sector meet here twice a year to exchange ideas and experiences with actors from the ministries and government agencies that report to them. In the building sector, the Federal Institute for Research on Building, Urban Affairs and Spatial Development (BBSR) plays an important role. Questions concerning resource efficiency are combined with sustainable building strategies. The most fascinating question is how we can ensure that which is possible—technically and financially feasible—is actually put into practice. This is a persistent problem and it is not exclusive to the construction sector.

From the outset, we have seen education as playing an important role, because resource efficiency can only be achieved by changing people's mindset. One major research project known as BilRess for short (from the German for **Education for Resource Efficiency**) comprises a number of precise modules addressed to all levels of education from primary school through to adult further education. It culminated in the establishment of a **BilRess** network on 16 September 2015.[31] The third network conference was held in Berlin on 20 September 2016. We have also been able to house this network at VDI ZRE to ensure that it can continue to work in the long term after the original research project has been concluded.

After this—hopefully impressive yet admittedly incomplete—overview of our success in creating structures, the next question arises: where is progress being made in terms of content?

13.5 The Main Features of ProgRess II

The great milestone for us came on 2 March 2016 when the federal government passed the decision to continue and refine our Resource Efficiency Programme in the form of ProgRess II. This was preceded by a long phase of preparatory work, which was very important for the discussion on resource efficiency in Germany.

We received the first draft for ProgRess II from the Federal Environment Agency on 11 December 2014.[32] After what seemed like a never-ending round of consultation within the ministry, we were finally able to pass on a "BMUB draft" to the other ministries on 11 August 2015 and, thankfully with their agreement, we were able to circulate it for comment to the Länder on 14 August and on 17 August to the public, industry associations, institutes and any other interested parties (BMUB 2015c). We set the deadline for feedback at 15 September, which in view of the summer holidays was not very long.

We had always believed it was extremely important to involve the public, especially industry associations, trade unions and environmental organisations. A total of **23 associations** had already submitted **contributions to the Annex to ProgRess I**, in which they described their resource efficiency activities. The many years of preparatory work this represented were now paying off. The feedback we received surpassed our expectations: **66 comments** from industry associations and institutes, ranging from BDI, the German Trade Union Confederation (Deutscher Gewerkschaftsbund, DGB) the Association of German Cities (Deutscher Städtetag), the Länder and a number of well-known and less well-known individuals. Almost all of them were comprehensive and substantial, and gave us a great deal to read and process.[33] UBA also took its own position in the evolving discussion (UBA 2015).

In addition, back in the spring we had already launched something completely new (new for us, at least): an organised direct **public participation procedure**. We randomly selected about 50 people from the civil register at each of five locations throughout Germany and asked them to meet for a weekend and formulate their

[31] http://www.bilress.de/bilress-netzwerk.html (in German) (BilRess 2015)

[32] Not published.

[33] My thanks go to VDI ZRE, especially to Dr. Christof Oberender, for invaluable support here.

expectations of our Resource Efficiency Programme. In parallel to this, we organised an online dialogue. Delegates of these "5+1" groups summarised their thoughts in a **"citizens' input paper"**—which made for fascinating reading—and presented it to our environment minister on 6 November 2015 (BMUB 2015f). We incorporated the document in its unabridged form into the Annex to ProgRess II and have continued to keep it "on the table" as guidance for the debates that followed. We will once more review all the suggestions made in the "citizens' input paper" during our preparatory work for ProgRess III.

The basic structure of ProgRess II is similar to ProgRess I—why reinvent the wheel? I would just like to highlight four new aspects.

13.5.1 Addressing Energy Resources More Precisely

We had intended to extend the programme's scope to cover fossil energy resources—an obvious need, to say the least, since in Germany they account for the largest volumes of imported resources in the form of oil, coal and gas (Fig. 13.3).[34]

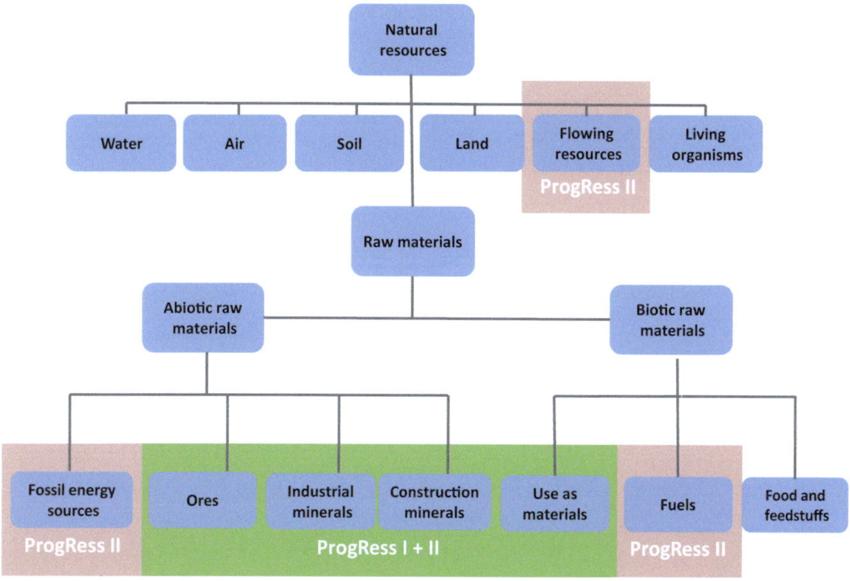

Fig. 13.3 Resource selection in ProgRess II in the BMUB draft version of ProgRess II 2015 (Source: © BMUB)

[34] Charts on this (and also on Sect. 13.5.2. below) can be found in my contribution in Thomé-Kozmiensky/Goldmann (eds.): Recycling und Rohstoffe, vol. 4. Neuruppin 2011, p. 26. The figures have changed very little since then.

Logically speaking, the biofuels have to be added to that and, of course, the "flowing" energy resources—water, wind and geothermal.

It must be said that our concern was in no way to redesign energy policy from the point of view of resource efficiency. We simply found it interesting to look at how Germany's energy transition impacts resource consumption. And additionally, energy resources are by no means used only for energy purposes; sometimes they are used for their material qualities. Our beloved plastic shopping bags, for example, are made from oil. Our primary concern is to identify the synergies between efforts to achieve energy efficiency and those to achieve resource efficiency and prevent clashes between them. It is not amusing when the German government sends an energy advisor to a company in the morning and a resource advisor in the afternoon, especially if the two know nothing about each other. Slight irritation on the part of actors about such already existing dual structures and others that are still being set up is becoming increasingly widespread and is completely understandable.

But first and foremost, a concern of the industry associations, especially the construction industry was that an "integrated" view should be taken: "We make thermal insulation, and in ProgRess you count the materials we use as a negative. But you dismiss the benefit—i.e., the energy resources saved as a result. That is simply not right!" And many other examples across all industries and their associations followed. We at BMUB believed this view to be both logical and expedient, which is why we integrated it into our thinking.

The outcome of the interdepartmental consultations was that we were able to word our objectives in a far clearer and unequivocal way. In fact, the term "energy resources" is misleading because, as I explained above, these resources are by no means used only to produce energy. Consequently, we now speak of "fossil resources" that can be used for their energy or material value. Their use in producing energy is the subject of energy policy, not ProgRess,[35] whereas their use as materials is part of our core concern. And the outcome looks like this (Fig. 13.4).

A positive spin-off from the intensive discussion on this—especially with the Federal Ministry for Economic Affairs and Energy, which has lead responsibility for energy policy in Germany—was the agreement to work together "to achieve an integrated view of energy and material efficiency." We included a new chapter (5.1.) on this in ProgRess II, and the German Energy Agency (dena) carried out a first workshop on the subject on behalf of BMWi on 27 June 2016.[36]

[35] To be precise: it is only marginally the concern of ProgRess, for example, regarding efficiency when designing or building energy installations. But in that case, the sort of questions we look at could be equally useful with regard to wind turbines or coal-fired power stations.

[36] http://www.dena.de/veranstaltungen/archiv/fachveranstaltung-rohstoff-und-energieeffizienz-synergien-und-zielkonflikte-im-kontext-der-plattform-energieeffizienz-pfee.html (in German).

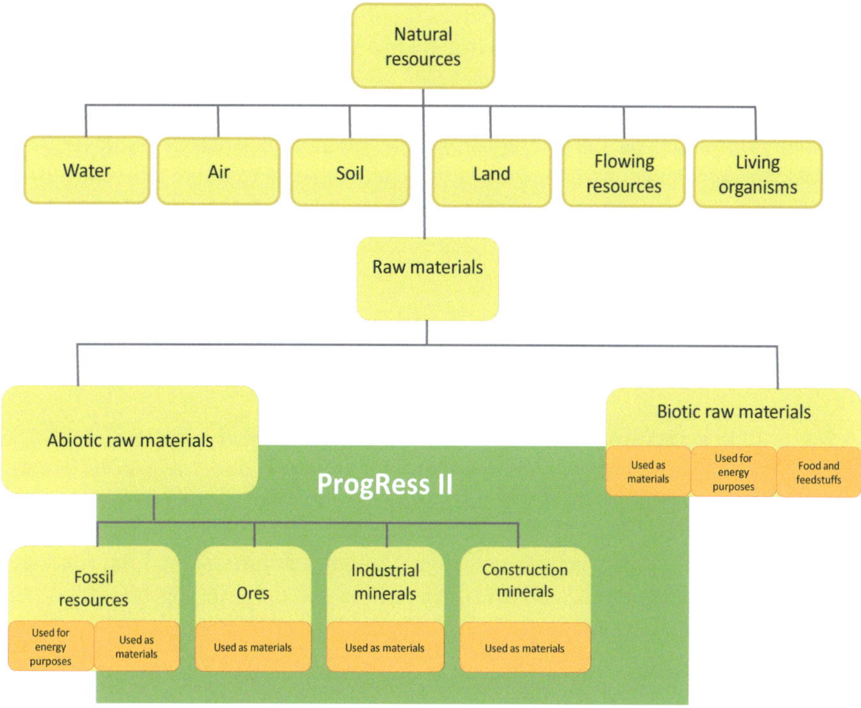

Fig. 13.4 Resource selection in ProgRess II, version adopted 2016 (Source: © BMUB)

13.5.2 The Building Industry Is One of the Top Priorities

In view of the fact that construction has been part of our ministry's portfolio since December 2013,[37] it is no surprise that all aspects of building have gained greater significance for our programme. It also makes complete objective sense because of all the resources produced domestically, building materials of all kinds absolutely dominate in terms of mass, which gives rise to serious problems. The focus on developing former East Germany following German reunification in 1990 meant that infrastructure in the West was to a certain extent neglected (although this had, in fact, begun earlier). The austerity policy of the last 15 years has caused further damage nationwide—a problem which now needs to be addressed. Since autumn 2015, we have also been facing huge challenges to build additional housing to cope with the influx of refugees from the Middle East and North Africa. This means that

[37] The ministry changed its name to the "Federal Ministry for the Environment, Nature Conservation, Building and Nuclear Safety". It acquired the building and urban development directorates-general from the former Federal Ministry of Transport, Building and Urban Development, and responsibility for renewable energy was transferred to the Federal Ministry for Economic Affairs and Energy (BMWi).

judicious construction that uses resources intelligently and sustainably is the challenge of the moment. Fortunately, Germany has a storehouse of ideas in the field of **sustainable building.** It is now crucial that we actually put into practice all those things that we know about in theory—and this is by no means a trivial task. We—especially our colleagues in the ministry's construction department—will continue to tackle this challenge, with the help of the round table on resource efficiency in the construction sector mentioned above.

13.5.3 Focus on Information and Communication Technologies

A new feature is that we have included a separate chapter on resource efficiency in and because of ICT. Here there are a number of major and minor questions that the ICT industry needs to address, ranging from the significant levels of energy consumed by computing centres to replaceable mobile phone batteries, but also the question of how ICT can improve efficiency in other industries, and the (likewise significant) problems connected with that. These are the gripping and important questions that affect our future and which BMUB, like BMU before it, has already worked on in a very focused way, including with the industry association BITCOM, for example,[38] and which we intend to devote a great deal of energy to in the years to come. This will be a priority topic at our network conference on 5 June 2017.[39]

13.5.4 New Indicators and Targets for 2030

Since adopting its national sustainable development strategy in 2002, Germany can boast to have three resource efficiency targets: increase energy efficiency, increase renewables' share in the energy mix (both applying to the energy sector), and finally "our" target: **double resource productivity** by 2020 compared with 1994. Resource productivity is the ratio of gross domestic product to (abiotic) raw material consumption (the unit of measurement is therefore €/t).

We feel it is very important to set quantitative targets because they necessitate annual explanatory reports, and also because issues that have quantitative targets attract far greater attention in the political public sphere. We are delighted and encouraged by the fact that our national sustainable development strategy and its

[38] This includes several annual conferences, which BMU carried out with BITCOM since 2007. They were organised by our product policy division (G I 4). http://www.bmub.bund.de/themen/wirtschaft-produkte-ressourcen/produkte-und-umwelt/produktbereiche/green-it/bmuubabitkom-jahrestag/ (in German).

[39] This was the planning status at the time of going to press. Updates (in German) are available at: www.neress.de.

targets have been included and explicitly reaffirmed in every coalition agreement since 2002, when the strategy was passed.

However, "our" resource productivity target and its indicator have continually been the subject of lively criticism since they were adopted in 2002. Two lines of criticism appear justified from an environmental point of view.

Environmental associations claim that the indicator permits an even greater increase in our **per-capita raw material consumption**, which by global comparison is already exorbitant, provided we further increase our economic output, measured in terms of gross domestic product, which itself is not without problems. This cannot be right; we in Germany should demonstrate the fact that a highly developed, innovation-oriented industrialised country is able to "achieve more with less" and should export our experience appropriately.

Industry associations claim that it is very easy to improve the indicator by **relocating our production abroad** and only importing finished products. This is because domestically we count all the raw materials used along the value chain whereas for imported products we count only the weight of the finished goods when they cross the border—and there is a world of difference between the two calculations. That cannot possibly be right. Actually, in our opinion, shifting production abroad does not improve the global environment.

In response to these criticisms, the indicators chapter in BMUB's draft of ProgRess II introduced **two new indicators**. To resource productivity we added **total resource productivity**,[40] which was designed to ensure that shifting production abroad no longer changes the calculations. This indicator also takes imports into consideration with the weight of their entire upstream chain, and it includes biotic raw materials. The second indicator is **per-capita raw material consumption (RMC)**, which reflects only domestic resource consumption caused by our lifestyle and does not worsen if we produce more for the world market.[41] So, as you can see, we cannot be accused of being anti-business; on the contrary, we believe that commitment to the environment is crucial to a forward looking economic policy.

Unfortunately, we were not able to get the per-capita RMC indicator through the interdepartmental consultation process; the opposition of large sectors of German industry to an "absolute target" was simply too great. Nevertheless, RMC data are reported in ProgRess II.[42] As **a target for 2030**—this is the time horizon for our ongoing work on updating the targets of the national sustainable development strategy—the BMUB draft for both indicators called for "**a continuation of the trends since 2000**." The federal cabinet in fact adopted this target for total resource productivity. In its resolution of 7 July 2016, the Bundestag translated this into a 30% increase from 2010 to 2030. According to new, revised figures from the federal

[40] Arithmetically it is (GDP + value of imports)/materials consumed, including upstream chains.

[41] BMWi introduced this indicator into the discussion in Brussels back in January 2014 in a background paper that had not been subject to interdepartmental consultation ("Resource efficiency targets and indicators," undated, not published). We remain grateful to them for this.

[42] In Sect. 3.3 of ProgRess II.

statistical office DeStatis, it even equates to a bit more than that.[43] And, in view of the numerous achievements with respect to "low hanging fruits" we have already chalked up and the new challenges in the construction sector outlined above, we believe that to be quite ambitious.

13.6 Future Prospects and Insights

Ultimately, it would be legitimate to ask: is that really all we have to offer **in terms of ideas**? More specifically:

What about the relationship of the Resource Efficiency Programme to the debates in society in recent years about growth, qualitative growth and **post growth economies**, about efficiency, effectiveness, **sufficiency,** material prosperity and **dematerialisation,** or about **rebound effects** and how to deal with them? What about the **term resources in the strictest sense and broader sense** and how different natural resources interact with one another—after all, the European Commission has always taken a broad, integrated view of natural resources? What has the German government learned from the tremendous study that the Bundestag's Enquête Commission on Growth, Prosperity and Quality of Life submitted at the end of the last legislative period (Bundestag 2013)?

It is my personal opinion that we in Germany had to start by doing a vast amount of grounding homework on resource efficiency. It was about putting the issue on a sound basis, mainstreaming it in a number of fields of practice and among a range of very different groups of stakeholders and acquiring useful experience. We did that with great success and joy, and ProgRess II is another milestone along this path.

It would seem that the time is now ripe to tackle the fundamental issues I have outlined. Thus far, we were able to simply make use of windows of opportunity wherever and whenever they appeared and our own resources permitted, jubilant at their great diversity and with a deliberate degree of randomness. In the past that was the right thing to do, but we cannot continue in this way. It was valuable in getting started, instigating action and gaining experience, but it goes without saying that it is not an effective approach. We need an **inner compass** to provide orientation for our further work and help us develop criteria for setting priorities and defining conflicts. A well-structured **fundamental debate** could help us to do this over the next few years. We are preparing that kind of debate and introducing it as I write. In 2017, we will systematically compile the experience gained by the different social stakeholders, try to identify obstacles, gaps and examples of success that we can build on, while at the same time refining our indicators and raft of instruments. We want the outcome to be reflected in **Progress III**, which the federal cabinet will adopt in February 2020. We are hoping to receive a wealth of ideas from external

[43] At the time of finishing this article, we are working on the basis of an annual augmentation of the indicator of 1.5658%, resulting in 36.4429% from 2010 to 2030. But these figures are still not stable; Destatis has already announced yet another revision of the data.

sources as we work on this concept—particularly from the international arena. Readers of this book are very welcome to take this as a personal invitation.

References

Angrick M, Burger A, Lehmann H (eds) (2013) Factor X: re-source—designing the recycling society. Springer, Dordrecht, pp 277–278

B20 (2016) http://www.b20coalition.org/coalition-activities/communiques/b20-coalition-calls-for-g20-to-address-resource-efficiency

Bayern (2016) https://www.lfu.bayern.de/abfall/ressourceneffizienz/index.htm (in German)

BDI (2015) http://bdi.eu/media/user_upload/Artikel_1_Broschuere_BDI_G7_Business_Summit_1_.pdf

BilRess (2015) http://www.bilress.de/bilress-netzwerk.html (in German)

BMUB (2012a) http://www.ressourcenpolitik.de/wp-content/uploads/2012/10/336-12-StA-NHK_PM.pdf (in German)

BMUB (2012b) http://www.bundesregierung.de/Content/DE/_Anlagen/Nachhaltigkeit-wiederhergestellt/2012-10-16-beschluss-sts-ausschuss-ressourcen.pdf?__blob=publicationFile&v=4 (in German)

BMUB (2015a) http://www.bmub.bund.de/en/service/events/details/event/opening-event-for-the-g7-alliance-on-resource-efficiency/

BMUB (2015b) http://www.bmub.bund.de/presse/pressemitteilungen/pm/artikel/deutschland-macht-ressourceneffizienz-zu-einem-schwerpunkt-seiner-g7-praesidentschaft/ (in German). Identical: www.bmub.bund.de/N51684

BMUB (2015c) http://www.neress.de/fileadmin/media/files/pdf/2015/ProgRessII_RA_pdf.pdf (in German)

BMUB (2015d) http://www.neress.de/news/news.html?tx_tknews_fe1%5Bpost%5D=1520&cHash=b00c698d3c2a86d16126c39d828b1920 (in German)

BMUB (2015e) http://www.bmub.bund.de/en/service/publications/downloads/details/artikel/german-resource-efficiency-programme-progress/

BMUB (2015f) http://www.bmub.bund.de/themen/wirtschaft-produkte-ressourcen/ressourceneffizienz/details-ressourceneffizenz/artikel/buergerratschlag-des-buergerdialogs-gespraechstoff/?tx_ttnews%5BbackPid%5D=289

BMUB (2016a) http://www.bmub.bund.de/en/service/publications/downloads/details/artikel/german-resource-efficiency-programme-ii/

BMUB (2016b) http://www.bmub.bund.de/themen/wirtschaft-produkte-ressourcen-tourismus/ressourceneffizienz/naress-nationale-plattform-ressourceneffizienz/ (in German)

Bundestag (2012) http://dipbt.bundestag.de/dip21/btd/17/085/1708575.pdf (in German)

Bundestag (2013) http://dip21.bundestag.de/dip21/btd/17/133/1713300.pdf (in German)

Bundestag (2016) http://www.neress.de/fileadmin/media/files/pdf/Politische_Dokumente/1937-16_EN_V03.pdf

Coalition (2013) http://www.kas.de/wf/doc/kas_36853-544-2-30.pdf?140820093605

efa (2013) http://www.ressourceneffizienz.de/aktuelles-termine/detailansicht-alle/news/detail/News/ressourceneffizienz-gestern-heute-und-morgen-1.html (in German)

efa (2016) http://www.ressourceneffizienz.de/ressourceneffizienz/startpage-en.html

epa (2016) https://www.epa.gov/sites/production/files/2016-09/documents/g7_us_workshop_summary_proceedings_final.pdf

EU-COM (2011) http://ec.europa.eu/environment/resource_efficiency/about/roadmap/index_en.htm

EU-COM (2014a) http://ec.europa.eu/environment/resource_efficiency/re_platform/index_en.htm

EU-COM (2014b) http://cor.europa.eu/en/activities/stakeholders/Documents/COM(2014)%20398%20final.pdf
EU-COM (2015) http://europa.eu/rapid/press-release_IP-15-6203_en.htm
G7 Leaders Declaration (2015) https://www.g7germany.de/Content/DE/_Anlagen/G7_G20/2015-06-08-g7-abschluss-eng.pdf?__blob=publicationFile&v=6
G7 Leaders' Declaration (2016) https://www.bundesregierung.de/Content/DE/_Anlagen/2016/05/2016-05-27-g7-eng.pdf?__blob=publicationFile&v=1
G7 Toyama Communiqué (2016) https://www.env.go.jp/earth/g7toyama_emm/english/pdf/160516_G7EMM%20Communique_FINAL.pdf
Hessen. https://umweltministerium.hessen.de/umwelt-natur/ressourcenschutzstrategie (in German)
International Synergies Ltd (2015) http://www.international-synergies.com/wp-content/uploads/2015/11/A4-portrait-G7-draft-programme-26-OCT-15-FINAL.pdf
NeRess (2016a) www.neress.de/ (in German)
NeRess (2016b) http://www.neress.de/netzwerk/partner.html (in German)
UBA (2001a) Grundsätze und konkrete Schritte einer stoffmengenorientierten Umweltpolitik. Report of the UBA to the BMU (not published), Dessau, May 2001
UBA (2001b) http://www.apug.de/archiv/pdf/uba_nachhaltige_entwicklung.pdf (in German)
UBA (2009) https://sns.uba.de/chronik/de/concepts/t3e529bfc_12409c73a7a_-490b.html (in German)
UBA (2015) http://www.umweltbundesamt.de/publikationen/elemente-einer-erfolgreichen (in German)
UBA (2016a) https://www.umweltbundesamt.de/en/topics/waste-resources/resources-commission-at-the-german-environment
UBA (2016b) http://www.resourcesforum.eu/
UBA (2016c) http://www.ressourcenforum.de/ (in German)
UMK (2015) https://www.umweltministerkonferenz.de/documents/endgueltiges_UMK-Protokoll_Augsburg_3.pdf (in German)
UMK (2016) https://www.umweltministerkonferenz.de/documents/UMK-Protokoll_Juni_2016.pdf (in German)
UN (2015) http://www.un.org/ga/search/view_doc.asp?symbol=A/RES/70/1&Lang=E. Thomé-Kozmiensky/Goldmann (Hg.): recycling und Rohstoffe, Band 4. Neuruppin (Germany) 2011 (in German)
VDI (2014) http://www.vdi.eu/nc/guidelines/vdi_4800_blatt_1-
Wikipedia (2016) https://de.wikipedia.org/wiki/Planetary_Boundaries

Chapter 14
Results of Three Cost-Effective, Innovative and Transferable Resource-Efficiency Instruments for Industries in the Basque Country

Ander Elgorriaga Kunze and Ignacio Quintana San Miguel

Abstract Driving resource efficiency in the industrial sector requires market and economic instruments that are flexible and easily accessible by companies. The Basque Country has used private-public partnership to develop three innovative instruments that generate an annual private turnover between €4 and €178 annually per public-sector euro spent, with a maximum dedication of 0.7 persons a year to manage each instrument. Driving product environmental innovation through the Basque Ecodesign Centre, reinforcing sustainable production with tax deduction for the Clean Technology List, and closing the loop by the Circular Economy Demonstration Projects Programme are critically analysed instruments to facilitate their total or partial replication in other countries or regions.

Keywords Circular economy • Ecodesign • Ihobe • Instrument • Cost effectiveness • Tax deduction • Clean technologies • Public private partnership • Basque country • Demostration • ISO 14006

14.1 Introduction

With a population of 2.17 million, the Basque Country is a region with a surface area of just 7234 km^2. However, its industrial sector makes an important contribution to the regional GDP of 23.5%, its productivity per worker is 30% higher than the European average, and the per capita income in the Basque Country is 30,459 €/year (Spri 2016). One aspect of the Basque Country that differentiates it from other regions is its power to collect the majority of taxes paid by citizens and to establish its own fiscal policy, a fact that bestows it with a high degree of self-government.

A. Elgorriaga Kunze (✉) • I. Quintana San Miguel
Ihobe, Basque Environmental Management Agency, Alda. Urquijo, 36, 48011 Bilbao, Spain
e-mail: ander.elgorriaga@ihobe.eus; ignacio.quintana@ihobe.eus

H. Lehmann (ed.), *Factor X*, Eco-Efficiency in Industry and Science 32,
https://doi.org/10.1007/978-3-319-50079-9_14

The environmental problems stemming from the Basque Country's highly industrialised and polluting past led to remedial environmental policies. These started to be seen as an opportunity from the turn of the new century onwards, with the Basque Environmental Strategy for Sustainable Development 2002–2020 (Basque Government 2014).

Ihobe, the environmental management agency of the Basque Government's Ministry of the Environment, Territorial Planning and Housing, was tasked with the design and deployment of instruments and innovative lines of action at operational level to convert sustainability into an opportunity, using a Life Cycle Thinking approach. Establishing such objectives required innovation regarding the "how" to adapt to the constant changes of the environment (legislation, corporate crises, etc.) and the internal changes (new guidelines, availability of human and economic resources). On the one hand, that means that the cost-effectiveness of the public action needs to be increased, which has involved innovating by means of a smooth development and deployment of instruments designed, in general, after a benchmarking process. On the other hand, excellent management is required in this publicly-owned company to achieve the objectives thanks to a high performance team.[1]

We have selected three public instruments that, using the life cycle approach, seek to accelerate environmental improvements in companies, specifically regarding the product design, production and end-of-life phases.

The three selected instruments have been operational for between 3 and 13 years, and have produced results and lessons learnt that increase the transfer potential to other interested European countries and regions.[2]

14.2 Life Cycle Thinking Implementation and the Basque Ecodesign Centre

14.2.1 Background

Drawing on Ihobe's systematic collaboration with the "European Roundtable on Sustainable Consumption and Production", an innovative ecodesign project emerged in 1998 in conjunction with Delft Technical University (The Netherlands). The aim was to ecodesign products in four Basque industrial companies, generate a method adapted for Basque SMEs and create an ecosystem where ecodesign could take shape.

[1] Ihobe obtained ISO 9001 certification 2000, followed by ISO 14001 in 2003, European Foundation Quality Model (EFQM) external assessment with over 400 points in 2005 and 2011, UNE 166.002 integral management of innovation certification and UNE 166006 Technology Watch certification in 2010 and the EMAS certificate in 2016.

[2] Taking into account the "CCC- Content and Critical Conditions" tool (Ecopol and Landes Energie Verein Steiermark 2014), developed as part of the European Pro-Inno Ecopol (Ecopol 2014) project in order to transfer successful ecoinnovation instruments from one European country or region to another. A simplified transferability analysis was performed for each of the experiences specified below (Table X.4).

The initial success of the release of an Ecodesign Guide (Ihobe 2000) with the four eco-designing companies was eclipsed by a lack of demand for ecodesign products on the market. This prompted the rethinking of public action (Jönbrink and Melin 2008), leading to two responses: creating a talent pool of eco-design engineers, and establishing a standard that rigorously recognises eco-designing companies on the market.

The first response was the creation of the Ecodesign Learning Centre in 2001 (Basque Country University et al. 2013), with the aim of providing constant mentoring and intensive training to engineers would then work on an ecodesign project in company as part of their thesis. Building on the accumulated experience from this initiative, the second response consisted of starting the development of a Spanish Ecodesign Management Standard by AENOR,[3] a process that ended with the approval of UNE 150.301 in 2003 (Aenor and Ihobe 2010).

The preliminary work in those two cases led to the launch of the "Ecodesign Promotion Programme" (Ihobe 2004) at the end of 2004, which established support services for Basque SMEs[4] in industrial sectors with the greatest potential for their application, such as electric-electronics, machinery, automotive, packaging, furniture, chemicals and construction materials. Those services were based on the AIDA+R strategic marketing scheme,[5] transferred to Ihobe by the Envirowise Programme (Defra 2011) and which subsequently became part of the British Wrap Programme, which allows the progress of the environmental commitment of companies to be managed and measured in time.

However, apart from internal benefits and material savings, support for companies offering ecodesign did not result in differentiation on the market leading to greater sales.

The implementation of the Programme coincided with the approval of the European Union's Ecodesign Directive (EuP, now ErP) in July 2005, which led to significant demand for information and support from the electric, household appliance and capital goods sectors of the Basque Country. In response to the request from companies, Ihobe started developing numerous methodological guides to facilitate the implementation of ecodesign.

The arrival of the economic crisis in 2010 shifted the corporate world's priority onto the survival of short-term businesses, which meant that industries focused nearly exclusively on meeting the short-term requirements of their customers.

[3] Spanish Normalization Association, integrated in European Normalization Committe (CEN).

[4] Small and medium enterprises.

[5] The AIDA+R scheme supports companies with standardised services aimed at creating Awareness (A), Interest (I), Demand (D) and Action (A), together with Recognition (R) of the work well done. The first Ecodesign Promotion Programme (Ihobe 2004) launched at the end of 2004 established a public catalogue of services to support ecodesign. General information through business channels for Awareness (A), an Ihobe-Line and different technical symposiums and conferences to respond to the Interest (I), detailed training and a in-factory technical visit to deploy the Demand (D) and both grants for ecodesign project and grants to implement the certifiable UNE 150.301 ecodesign management standard (now ISO 14.006), along with in factory projects by young engineers from the Ecodesign Learning Centre to drive Action (A). The cost of all these services, except for a small part of the action services, was assumed by the Basque Government through Ihobe.

In this context, where public coffers saw a drop in revenue, leading to a cut in the public services provided by Ihobe, the experiences existing in public-private partnerships as regards ecodesign worldwide were analysed in detail. This implied conducting in-depth benchmarking and a subsequent agreement with the Swedish Life Cycle Center (Palander and Wikström 2015). In November 2011, the Basque Ecodesign Center was launched. It is a public-private partnership involving Ihobe and multinational Basque companies and is thus a further commitment to Green Supply Chain Management from a product approach.

14.2.2 Description of the Instrument

The Basque Ecodesign Center is an organisation based in the Basque Country and structured pursuant to a partnership framework between firms in the private sector and the Basque Government. It aims to foster the design and execution of innovative ecodesign projects.

The Basque Ecodesign Center currently consists of Ihobe, Spri (the Business Development Agency of the Basque Country) and nine large companies: CIE Automotive, EDP Eroski, Euskaltel, Gamesa, Iberdrola, Ormazabal, Orona and Vicinay Sestao. With a total annual turnover of €36 billion, they employ 60,000 employees and have over 10,300 suppliers, 16% of whom are in the Basque Country. The clusters of the main value chains of the Basque Country (automotive, aeronautics, railway, maritime, machine tool, energy, environment, habitat and logistics) have recently joined this Center as associated partners to reinforce knowledge transfer to other SMEs.

The objectives of the Basque Ecodesign Center are to strengthen the competitiveness of the participating firms in the acquisition and application of cutting-edge ecodesign-related knowledge, to stimulate product eco-innovation through partnerships between companies, the Basque University and the top knowledge centres in the world and, most importantly, to ensure that the environmental factor is fully integrated into the supply chains of the participating firms.

The main activities of the centre (Fig. 14.1) are aimed at developing individual and collaborative technical ecodesign projects, supporting business competitiveness from the environment through new business models, training and building skills to improve technical expertise and supporting the SMEs of the Basque Country in ecodesign implementation.

14.2.3 Results

The commitment to ecodesign is long-term and the results analysis should therefore not only include the last 5 years of activity of the Basque Ecodesign Centre, with spending of €2.2 million, but also the related measures implemented since 2000,

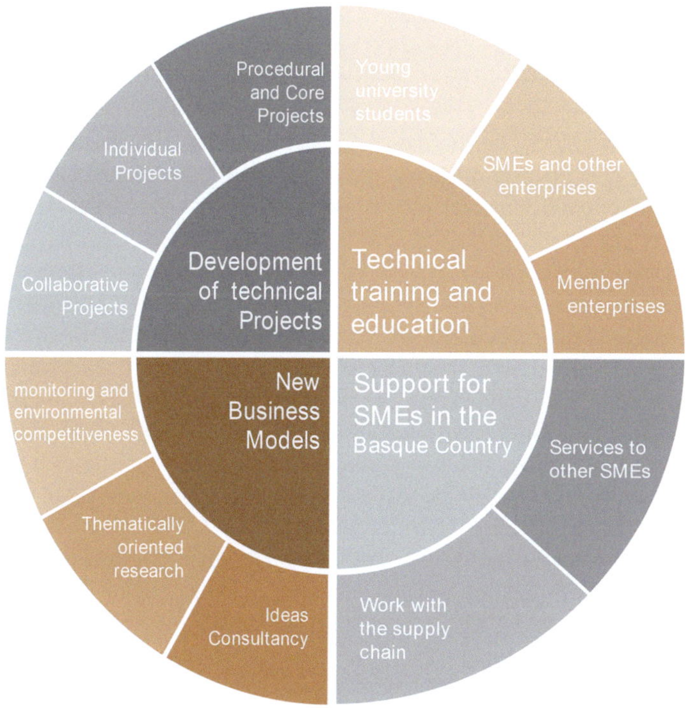

Fig. 14.1 Main lines of action of the Basque Ecodesign Center (Source: Basque Ecodesign Center 2016)

with spending by Ihobe totalling €5.5 million and average dedication of 0.7 people a year from Ihobe.

The overall results, presented in the report "15 Years of Product Environmental Innovation" (Ihobe 2014a), show that 156 Basque companies are eco-designing their products. A survey of these companies has shown that ecodesign products have generated €952 million in sales, and a detailed review of that survey is being carried out for early 2017. Forty-six per cent of these companies have found their ecodesign products to be profitable, having half of them a higher profit margin. The ecodesign of these energy-consuming companies have cut the generation of greenhouse gases by 11% in their life cycle (Pole Eco-Conception, Institut de Développement de Produits 2014).

Those results can be attributed to Ihobe's support of the creation of the UNE 150.301 ecodesign management standard which led to ISO 14.006 (Aenor and Ihobe 2010; Arana-Landina and Heras-Saizarbitoria 2011), thanks to the investment in skills-building through in-company projects for 100 undergraduates and graduates (Basque Country University et al. 2013) and to the development of 20 methodological guides to facilitate the deployment in the company of product environmental innovation.

The positioning of Basque companies on the global market in terms of ecodesign is advanced. Forty-seven per cent of ISO 14.006 certificates in Spain for ecodesign management systems are held by 100 Basque companies and 6% of all Environmental Product Declarations of the Global EDP System worldwide are from 22 Basque companies (Ihobe 2015). Furthermore, 40% of Basque industries declare that they have made product environmental improvements in the last 3 years and 50% forecast that product environmental innovation will be more important on the market (Ihobe 2011).

The initial results obtained in the last 5 years[6] can be summarised, on the one hand, as 31 ecodesign and/or life cycle assessment projects being conducted by the companies of the Basque Ecodesign Center, the majority of which involved companies of the value chain that make up the Life Cycle Thinking of their businesses.

On the other hand, there is the development of 21 innovative methodologies related to ecodesign, which have been piloted with the Center's companies and then disseminated to the value chains to lay the foundations for their launch in the near future. Special mention should be made of methodologies produced in the areas of Life Cycle Costing (Basque Ecodesign Center 2017b), Green Supply Chain Management (Basque Ecodesign Center 2014), Strategic Environmental Monitoring (Ihobe 2016c) or Servitization (Basque Ecodesign Center 2017a). The Competitive Environmental Monitoring reports prepared in different value chains, such as aeronautics (Hegan), automotive (Acicae), railways (Mafex), shipping (Maritime Forum), habitat (Habic) and machine tools (Afm), together with a Strategic Environmental Monitoring annual report for each of the companies of the Basque Ecodesign Center, create demand for new ecodesign projects and policies in the business sector.

The coordination of the ecodesign research and training activities through a new structure – the Basque Ecodesign Hub which involves the three Basque universities – is another of the results. Another important outcome was the participation of members of the Basque Ecodesign Center in multiple international working parties, such as the European Commission's Product Environmental Footprint pilot project, and the dissemination of Life Cycle Thinking at numerous conferences and symposiums held in the Basque Country, Spain, Europe and South America.

14.2.3.1 Lessons Learnt

Seventeen years of public work on ecodesign in the Basque Country has generated valuable lessons which can be summarised in five points.

First, opting for ecodesign is very cost-effective, with a return of close to €180 a year per euro of public expenditure in long-term private-public partnerships, according to currently available figures. The strategic nature of ecodesign requires consistency in time with clear messages and support tools, because it affects the core of business activity and due to the complexity of the knowledge involved.

[6] See Activity Report 2012–2015, Basque Ecodesign Center 2016.

Second, ecodesign has been focused in companies with greater potential for competitive and environmental improvement.

Third, the Basque Ecodesign Center has proven to be an effective private-public instrument of Green Supply Chain Management that uses its projects and the impetus of the Environmental Product Declarations in Europe to accelerate supplier drivers.

Fourth, the commitment to Strategic Environmental Monitoring is being consolidated. Assessing competitors, customers and future legislation allows companies to opt for ecodesign, plan ahead and be better positioned on the global market.

Fifth and finally, legislation and standardisation offer three driving forces with great potential that must be used intelligently. There is the European impetus of the Ecodesign Directive (ErP) reinforcing the durability aspects of the products by means of a prior standardisation process (Cen-Cenelec 2016) and reinforcing the inspection and control in Member States, while waiting for the allocation of sufficient resources to do so by the European Commission (Coolproducts 2013). Another driving force is the requirement of the new ISO 14001:2015 to integrate Life Cycle Thinking in the company. Finally, the development and piloting in Europe of the Product Environmental Footprint that, given the consensus with stakeholders, should foreseeably accelerate the presence of environmental products declarations rigorously comparable throughout Europe on the B2C and even B2B markets (European Commission 2016).

14.3 Best Available Techniques and Tax Deductions for the Clean Technology List

14.3.1 Background and Benchmarking

The Autonomous Community of the Basque Country and the Chartered Community of Navarra are the regions of Spain with the greatest tax powers, regulated by an agreement rooted in the Spanish Constitution and known as the Economic Agreement.

In the 1990s, the Basque Country established a system of corporation tax deductions of 15% for environmental investments by companies, in which case rigorous technical applications must be submitted. After an exchange with the Dutch Government, Ihobe decided in 2003 to embark on a project to adapt the MIA-VAMIL[7] system to the Basque Country. The simplified tax deduction system based on the Basque Clean Technology List (BCTL) with a deduction of 30% was introduced in 2004.

[7] MIA-VAMIL is the Dutch Government's fiscal instrument to foster material efficiency in companies using a standardized yearly updated innovative clean technologies list. MIA is the Enhanced Capital Allowance Scheme and VAMIL is the tax deductions instrument.

During a subsequent benchmarking process, conducted with the Government of Flanders, the latter's Ecologie Premium subsidy instrument was analysed. This instrument was likewise based on a better methodologically developed list of standardized best available technologies, which had been endorsed and approved by the European Commission, thus leading to an improvement of the Basque Clean Technology List.

14.3.2 Description of the Instrument

The Basque Clean Technology List (BCTL) consists of a list of industrial equipment developed and available on the market, with a low degree of implementation in the production sectors of the Basque Country and whose application is more efficient than the technologies conventionally used for that same purpose (Spri et al. 2016). It incorporates technologies that contribute to the mainly preventive objectives of the innovation, competitiveness, energy and environmental policy, with high application potential and direct impact on decreasing the consumption of resources and emissions. This instrument is managed by Ihobe, in coordination with the energy and competitiveness agencies (Spri and Eve) of the Ministry of Economic Development and Infrastructures, and with the Ministry of the Treasury and Economy of the Basque Government. The application of these technologies is aimed at enhancing the competitiveness of a large number of industrial SMEs that obtain tax deductions for investing in production equipment that is more efficient, less polluting and/or uses fewer energy and material resources.

In 2016, the BCTL includes 92 types of equipment. These include 32 renewable and energy efficient ones (e.g. a frequency inverter, a high performance heat pump and a geothermic heat pump), 26 aimed at preventing and reducing air emissions (e.g. a particle filter), 12 water-related technologies (e.g. the membrane bioreactor), 11 to minimise waste (e.g. grinding filter), and the 10 material efficiency ones, such as the minimal quantity lubrification (MQL) or the automatic mixing equipment.

The BCTL is reviewed every 3–4 years; new technologies are added, based on preliminary prospective studies, and others are removed if they fail to met the criteria. The selected equipment must be accessible, that is, available on the market, must be innovative or have a low degree of implementation on the market, and it must be environmentally better than the State of the Art and what is required by environmental legislation.

The tax incentives for the Basque Clean Technology List is in keeping with what is established in Commission Regulation (EU) number 1407/2013, on the Application of Articles 107 and 108 of the Treaty on the Functioning of the European Union to *de minimis* aid.[8]

[8]A programme run under European Unions "Minimis" scheme has a limit of €200,000 per company summing all different aids of this kind during 3 years time. This simplified scheme is used *de facto* to support SMEs.

Table 14.1 Comparative information based on 2013 data of each instrument

Country/region	Netherlands	United Kingdom	Flanders	Basque country
Instrument	MIA/Vamil	ETL energy tech list	Ecologie premium	LVTL clean tech list
Kind of instrument	Tax deduction/ECA	ECA	Subsidy	Tax deduction
Maximal yearly budget (MM €)	131	100	41	8
Direct revenue	1:12	1:14*	1:9*	1:4
Available since...	1995/2000	2001	2004	2004
Management costs (% budget)	1.3	1.2	1.3	0.3
Equipments and actives (n°)	304	55	171	92
Approved demands (n°/y)	10.000	8.000*	720	57

Source: Ecorys and Ihobe (2014)
Notice (*): estimative data

The main advantage of the Basque Clean Technology List is its simplicity and agility as this tax deduction is applied automatically in the annual corporation tax return, along with the extent of the tax deduction which is 30% of the corporation tax payable.

14.3.3 Results and Lesson Learnt

Based on currently available data, there is an estimated incentive effect of €4 from the private sector for each tax deduction euro. The costs of managing, controlling, updating and assessing the instrument stand at 0.3% of the total of tax deductions implemented. The applications approved annually stand at under 60 equipments.

The BCTL results gain more importance when compared with other similar systems in the European Union. After a detailed analysis of simplified tax instruments for companies based on standard equipment lists by Ihobe, benchmarking sessions were organised (Ecorys and Ihobe 2014) with representatives of the Dutch Government's MIA/VAMIL instrument (Mure 2012; Netherlands Enterprise Agency 2013), of the British Government's Energy Efficiency Technology List (Carbon Trust 2014; Department for Business, Energy and Industrial Strategy 2016) and with the managers of the Government of Flander's Ecologie Premium instrument (Flanders Entrepreneurship and Innovation 2016).

The main conclusions (see Table 14.1) were, first, that tax instruments are politically more stable, tenable and in keeping with the guidelines of the European Commission than direct aid or subsidies (Kosonen and Nicodème 2009). Second, it can be seen that the economic instruments based on standardised equipment and technology lists are adequate to drive tax deductions in a high number of SMEs due to their simplicity and agility. The aforementioned four instruments have jointly incentivised investment to a total of EUR 3,372 million a year for nearly 18,000 equipment units (Ecorys and Ihobe 2014).

Third, the incentive effect of this economic instrument stands, as the average result of the four instruments, at EUR 12.7 invested by the private sector for each tax deduction euro deducted. Four, it seems more effective to drive tax deductions for a smaller number of technologies. Thus, the British list with just 55 technologies achieves on average 145 deductions for technology per year, which is much higher than that of the Basque list. This is due to a great extent to a verification system that offers high legal certainty to companies and excellent promotion of the instrument.

The fifth conclusion was that success can be maximized by an ongoing collaboration of the developers and managers of the instrument with equipment manufacturers, distributors, potential users, business associations, financial entities and government departments.

The sixth and last lesson learnt was that those lists are key instruments to contribute to implement sectoral policies of the government. It is therefore essential to improve the alignment of objectives and measurement indicators. A clear example of this is the contribution of the list-based deductions to the British efficient lighting strategy (Lighting Industry Association 2014).

14.4 Closing Cycles by Circular Economy Demonstration Projects

14.4.1 Introduction

The Basque Country squanders 5.5 million tons of waste yearly, which means a loss of 8.6% of total material consumption. Non-hazardous waste, in general industrial, head this ranking with 58%, followed by municipal solid waste (20%), construction and demolition waste (16%) and hazardous waste (6%).

Ihobe conducted a study to establish the value of the materials contained in the waste sent to landfills (Ihobe 2016b). This was quantified at €44 million a year (Fig. 14.2), even though that does not presuppose a return under current conditions. Dumped secondary plastics with a potential value of €12 million per year head the list, followed by ferrous metals and sludge worth €7 million per year.

On the other hand, there is an indicative potential of economic savings in Europe of €633 billion in short life-cycle consumer products, such as packaged food, and a further €564 billion in medium-long life-cycle products, such as household appliances (Ellen MacArthur Foundation 2015). The European Commission estimates that product remanufacturing may generate between an additional €40 and €70 billion a year in turnover in Europe and create between 260,000 and 410,000 new jobs by 2030 (European Remanufacturing Network 2016). The Dutch Government calculated the additional business potential of the Circular Economy for products and technical materials containing metals to stand at €570 million a year and that it could generate up to 54,000 new jobs (Bastein et al. 2013).

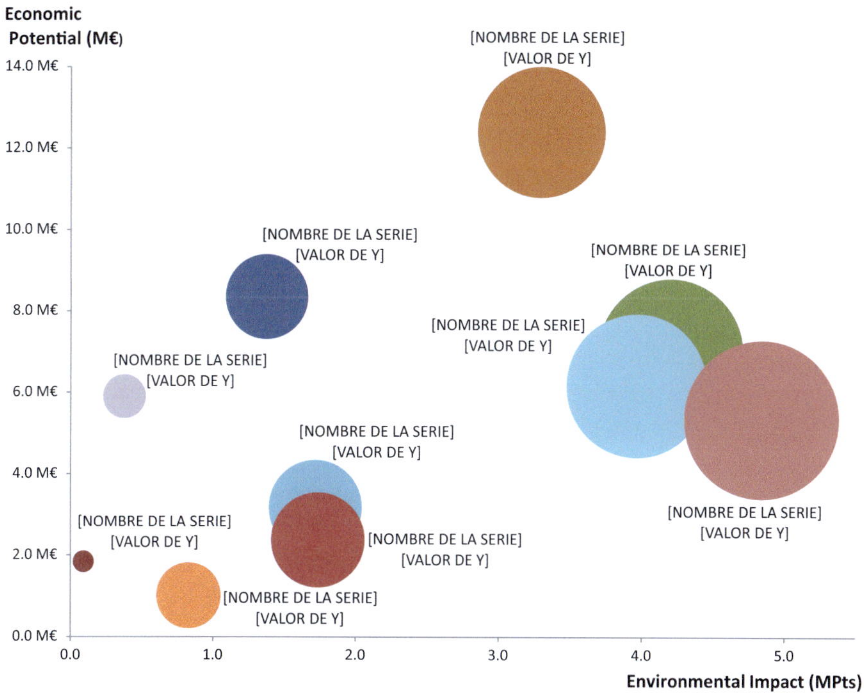

Fig. 14.2 Economic value (M€) and environmental impact (LCA approach: Mpoints according to Recipe/EcoInvent) of the materials contained in different waste streams to the landfill. The size of the balls reflects the environmental impact.(Source: Ihobe 2016b)

There are numerous circular economy technical alternatives developed in the European Union and in the Basque Country through R&D&I[9] projects. However, these solutions are not frequently available on the market due to multiple barriers that, in a simplified way, are known as the "valley of death".

14.4.2 Description of the Instrument

The fundamental objective of the "Circular Economy Demonstration Projects" call for grants is, similarly to the R+Impuls Programme (BMBF 2016), to facilitate new solutions on the market, which contribute to the circular economy goals of the Basque Country (Basque Government 2015), by conducting pre-industrial or industrial tests that confirm the environmental-economic-technical feasibility of innovative alternative to recover secondary materials, the manufacturing of products with high secondary material contents or to recover components and equipment.

[9] Research and development and innovation.

Table 14.2 Managing timeline and results of the call, participation in the annual calls and public funding for supported projects

Concept	2014	2015	2016	Total
Call full process period (days)	75	64	79	73
Call decision period (days)	20	26	27	24
Company overall satisfaction (0–10, max 10)	7.9	8.5	8.2	8.3
Preliminary ideas submitted (no.)	25	39	74	138
Projects submitted (no.)	11	19	39	69
Projects awarded and supported (no.)	5	12	19	36
Public grants (€)	100,000	188,000	429,750	717,500

Source: Ihobe (2016a)

The annual Demonstration Projects call for grants, piloted previously (Ihobe 2014b) and co-funded by the European Union through ERDF,[10] defines technical priorities according to environmental policies and the opportunities of new more circular markets, conducts projects on a short timeline (approx. 9–12 months) lead by a company that would exploit the new solution, finances projects with maximum grants of €25,000 per project, and provides technical mentoring by the Basque Government's Ministry of the Environment during the project.

Even though the project evaluation was performed in two stages, a major advantage of this call has been the short deadlines to approve submitted projects (Table 14.2).

In general, the selected projects (Fig. 14.3) tend to have a good degree of transferability to the rest of the industrial fabric (75%), intense inter-company cooperation, an important degree of proximity to the market with 86% of projects with a 6–8 baseline TRL[11] and public partnership needed, beyond funding, for the demonstrated solution to be established on the market.

Among the 36 projects implemented particularly in the case of those focused on the metal, construction, automotive, electric-electronic, plastic transformation, fashion & textile and paper sectors, special mention should be made of the following projects because of their high environmental performance and their relevant circular business opportunity: "Recovery of Zinc-Lead Oxides and Metallic Iron from Common Steel Mill Dust by means of a New Blast Furnace Concept", "Development of the Recycling and Recovery Model for Photovoltaic Solar Modules", "Reuse of Seats from End-of-Life Vehicles in the Furniture Sector", "Concrete Prefabs Produced Using Stainless or High Alloy Steel Aggregates", "Selective Production of Secondary Hydrate Aluminium Oxides to Develop New Functional Loads for Rubber Formulations" and "Recovery of the Sludge of the Paper Industry as Absorbent Material".

[10] European Regional Development Fund.

[11] Technology Readiness Level is a method used to estimate technology maturity. TRL 9 means high maturity, that is "actual system proven in operational environment" and TRL 1 low means "basic principles observed."

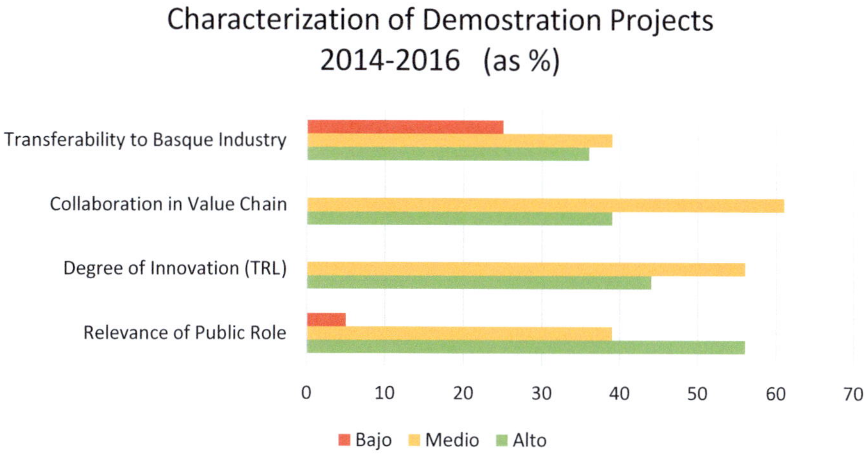

Fig. 14.3 Characterisation of the "36 Circular Economy Demonstration Projects" according to the degree of transferability to other Basque industries, to the collaboration existing in the value chain, to the degree of innovation and to the relevance of the public sector role to accelerate the availability of the solution on the market (Source: Ihobe 2016a)

14.4.3 Results

Ihobe has quantified the success of the concluded projects based on estimates of economic, social and environmental results carried out by the companies supported by the 'Circular Economy Demonstrative Projects' calls in the Basque Country between 2014 and 2016. The sample included up to 17 projects from a total of 36 that had finished by the end of 2016 (Ihobe 2016a). The preliminary results showed that the real market availability of the new circular business solutions would be 40% for the developed materials and/or products. Thus, if we extrapolate the former data to all the projects developed in the calls, we could estimate the potential results for a public budget that accounted for 0.7 million €. On average, each 100,000€ of public funding awarded resulted in 2 million € business turnover, 8.7 jobs and 15.3 ton/year of saved materials (Fig. 14.4). Hence, each public euro spent would lead to 21€/year of private turnover.

14.4.4 Lessons Learnt

Three years of managing this instrument leads to several conclusions, the first being that 56% of the demonstration projects require a high private-public partnership to achieve their leap to the market, with companies particularly appreciating the streamlining of environmental formalities.

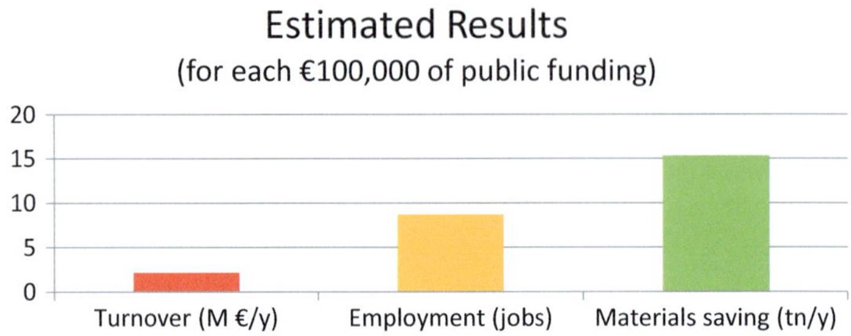

Fig. 14.4 Forecast results of turnover, jobs and materials savings for each €100,000 of public money spent for the 36 circular economy demonstration projects at 3 years after completion. The success factor of the projects has been established at 40% (Source: Ihobe 2016a)

Second, and given the limited timeline and budgets, 47% of projects need additional subsequent developments. Third, 61% of the projects interact with other value chains, which explains the difficultly to develop new solutions.

Fourth and finally, it was observed that recovering products and components generates greater value added than recovering of materials alone. The first demonstration projects underway to reuse and remanufacture components and products expect to charge between 1,500 and 10,000 €/ton and to create 30 jobs per thousand tons, while some key projects involving recycling materials aimed at the metal or construction value chain are under 10 €/ton and generate three hundredfold fewer jobs.

After 3 years of using this instrument, a new benchmarking process with similar European programmes (Ovam 2012; Hirschnitz-Garbers et al. 2015; BMBF 2016) is one of the actions indicated to detect new opportunities to increase the cost-effectiveness of the "Circular Economy Demonstration Project" funding of the Basque Country.

14.5 Final Conclusions

The analysis performed by the three eco-innovative instruments of the Basque Country allow us to reach the following conclusions:

- *The relevance of private-public partnerships and economic instruments as triggers of change.* The market instruments aimed at triggering demand, along with those aimed at supporting supply, are very flexible and offer a high return, even more so if they are always combined with legislative instruments (the so-called "policy mix") in the framework of a clearly results-aimed environmental strategy.
- *The cost-effectiveness of the three instruments considered is high.* Each public euro spent generates between €4 and €178 a year in private turnover or investment, depending on the instrument and with due caution relating to the quality of the data (Table 14.3). This figure is even more relevant if the few human resources

Table 14.3 Preliminary cost-effectiveness assessment of the three instruments considered

Instrument considered	**Ecodesign:** Basque Ecodesign Center	**Production:** Basque Clean Technology List (BCTL)	**End of Life:** Circular Economy Demonstration Projects
Public spending effect generated (€ per €/y)	178	4	21
Investment generated (M €/y)	952	10	17,4
Public spending (average M €/y)	0.4	2.5*	0.24
Minimum / maximum spending in x years (M €/y)	0.1-1.3 (x:17)	0.5-3 (x: 10)	0.1-0.4 (x:3)
People year	0.7	0,3	0.4

Source: Ihobe

Key: low data quality in red, average in orange and high in green

Table 14.4 Indicative assessment of the transferability of the instruments considered

Instrument Considered / Transfer Criteria	**Ecodesign:** Basque Ecodesign Center	**Production:** Basque Clean Technology List (BCTL)	**End of Life:** Circular Economy Demonstration Projects
Standardisation available			
Similar framework conditions			
Cost-Effectiveness			
Streamlined management			
Policy decision ease			

Source: Ihobe

Key: dark, high; light, low

available to design, drive and manage the instruments that on average vary between 0.3 and 0.7 persons a year are taken into account.

– *The transferability of those three instruments is average-high.* Even though all the instruments are well standardised in terms of their transference (Table 14.4), there are differences regarding their cost-effectiveness, in the simplicity of the operational management of the instrument by an operational unit, in the likelihood of similar framework conditions existing in other countries and regions, and in the simplicity of a favourable policy decision. The replication potential of the Grants for Circular Economy Demonstration Projects is high, but the impetus to ecodesign through a Green Supply Chain Management instrument such as the Basque Ecodesign Centre is average, particularly due to its complex operational management. The implementation of a simplified tax deduction system based on a Clean Technology List comes up against the limitation of tax powers in the majority of European regions and the ensuing complexity of policy decisions nationwide.

- *Collaborative, flexible and advanced public management enables a flexible design, effective implementation and constant improvement of the instruments.* The rapid adaptation and response capacity to changes to meet certain objectives with limited resources requires a good strategy, a focus on the needs of the companies, streamlined coordination between different government departments, a commitment to innovation as part of a technology watch and constant benchmarking, and appropriate management of people and equipment enable effective responses adapted to resources and environments that are constantly changing.
- *Life Cycle Thinking linked to the increase of business competitiveness must be present in the design, development and assessment of all market and economic instruments.* This facilitates the focus on more limited spheres of action and priority value chains, which are the ones that most contribute to business competitiveness and environmental results.

Lessons learned from 20 years of experience supporting material efficiency for businesses in the Basque Country at a regional level suggest that future work should focus on building stronger international collaboration and identifying circular economy opportunity niches. We strongly believe that once the strategy for a certain region is defined and the necessary tools are developed, a region should concentrate its efforts on increasing cost-effectiveness. It can be said that by standing on the shoulders of frontrunners, a region can advance very rapidly in its benchmark and concentrate its efforts in the collaborative development of the tools by building partnerships.

In the case of the Basque Country, after the definition of a framework strategy we have defined the following actions: (i) a prioritization of specific target topics at a regional level (e.g. eco-design of electrical equipment, near net shape technologies, process control system for material efficiency, remanufacturing or composites recycling); (ii) the development of public-private strategies for priority topics; (iii) a better contribution of demand driving instruments in specific policy mixes for each priority strategy and; (iv) a systematic measurement and better follow up of the cost-effectiveness of the applied instruments.

In this sense, due to the experiences and instruments developed in the Basque Country we believe that our region can become an ideal laboratory for other European initiatives to test innovative environmental instruments on a regional scale.

References

Aenor and Ihobe (2010) Guía para el desarrollo de la Norma de Ecodiseño UNE 150.301, base de ISO 14.006. Evaluación de Aspectos Ambientales de Producto, Bilbao

Arana-Landina G, Heras-Saizarbitoria I (2011) Paving the way for the ISO 14006 ecodesign standard: an exploratory study in Spanish companies. J Clean Prod 19:1007–1015

Basque Country University, Beaz and Ihobe (2013) 10 years of the Basque Ecodesign Classroom 2002–2012, Bilbao

Basque Ecodesign Center (2014) Comprometidos con la tracción ambiental en la cadena de suministro
Basque Ecodesign Center (2016) Activity report 2012–2015
Basque Ecodesign Center (2017a) Servitization and environmental opportunities for business
Basque Ecodesign Center (2017b) Guide for integrated LCA-LCC application
Basque Government (2014) Environmental Framework Programme of the Basque Country 2020, Vitoria-Gasteiz
Basque Government (2015) Plan for the Prevention and Management of Waste in the Basque Autonomous Community 2020. Executive Summary, Vitoria-Gasteiz
Bastein T, Roelofs E, Rietveld E, Hoogendoorn A (2013) Opportunities for a circular economy in the Netherlands. TNO for the Netherlands Ministry of Infrastructure and the Environment
Bmbf (2016) Vom Labor auf den Markt. R+Impuls setzt Impulse für industrielle Ressourceneffizienz, Bmbf, Bonn
Carbon Trust (2014) Energy-saving technologies. A guide to Equipment Eligible for Enhanced Capital Allowances, London
Cen-Cenelec (2016) Energy related products material efficiency aspects for ecodesign, JWG10. Brussels
Coolproducts (2013) International comparisons of product policy – final report. European Environmental Bureau, Brussels
Defra (2011) Business resource efficiency and waste programme disaggregated metrics results for 2007/08
Department for Business, Energy and Industrial Strategy (2016) Enhanced capital allowance scheme for energy efficient technologies. Energy Technology Criteria List, London
Ecopol (2014) Public innovation partnership for better policies and instruments in support of eco-innovation; Method for evaluating and benchmarking public policies
Ecopol-Landes Energie Verein Steiermark (2014) Content and critical conditions. A mindset for transferring EcoInnovation Policy Measures, Graz
Ecorys and Ihobe (2014) Cleaner technology lists for economic public instruments: four operating systems and common critical aspects to face the future – benchmarking meeting report, Brussels
Ellen MacArthur Foundation (2015) Delivering the circular economy: a toolkit for policymakers. Ellen MacArthur Foundation
European Commission (2016) Product environmental footprint pilot guidance for the implementation of the EU Product Environmental Footprint (PEF) during the Environmental Footprint pilot phase
European Remanufacturing Network (2016) Remanufacturing market study. ERN for the European Commission
Flanders Enterpreneurship and Innovation (2016) The ecology premium plus (EP-Plus). Flanders Government
Hirschnitz-Garbers M, Hinzmann M, Watkins E, Brink P, Milios L, Soleille S (2015) An analysis of support measures applied in the EU-28. Measure synthesis support for industrial symbiosis, Ecologic Institute, IEEP, Bio Intelligence Service and IVM for the European Commission
Ihobe (2000) Manual on ecodesign. 7 steps for implementation
Ihobe (2004) Programme for the promotion of the ecodesign in the Basque autonomous community 2004–2006
Ihobe (2011) Ecobarómetro industrial de la comunidad autónoma del País Vasco 2011
Ihobe (2014a) Ecodesign made in Euskadi. 15 years of product environmental innovation
Ihobe (2014b) Circular economy in Basque Country. Demonstration projects for reusing materials
Ihobe (2015) La declaración ambiental de producto. Un instrumento de información y comparación ambiental entre productos
Ihobe (2016a) Circular economy demonstration projects in the Basque Country. Results of 36 business initiatives

Ihobe (2016b) El valor de los materiales contenidos en los residuos. Oportunidades para una economía circular en el País Vasco, Fabricación Verde n° 4
Ihobe (2016c) Guía Práctica de cómo identificar oportunidades ambientales estratégicas en su empresa. Despliegue de la Vigilancia Ambiental Estratégica en los sistemas de gestión
Jönbrink A, Melin H (2008) How central authorities can support ecodesign: Company perspectives. Nordic Counil, TemaNord 2008:569, Copenhagen
Kosonen K, Nicodème G (2009) The role of fiscal instruments in environmental policy. Taxation Paper n° 19, European Commission
Lighting Industry Association (2014) UK lighting sector strategy. A development plan for a competitive and sustainable lighting industry
Mure (2012) Green investment and finance MIA, VAMIL, Ind-NL D03. Netherlands Enterprise Agency, Ministry of Economic Affairs
Netherlands Enterprise Agency (2013) Tax relief schemes for environmentally friendly investment. Vamil and MIA, Netherlands Enterprise Agency, Ministry of Economic Affairs
Ovam (2012) The flanders materials programme, Ovam
Palander S, Wikström A (2015) Swedish life cycle center summary report stage 7 October 2012 – December 2015. Chalmers University of Technology
Pole Eco-Conception, Institut de Développement de Produits (2014) Profitability of ecodesign: an economic analysis, Grenoble-Montreal
Spri (2016) BIGlittle. Invest in the Basque Country
Spri, Eve, Ihobe (2016) Basque List of Clean Technologies, Bilbao

Chapter 15
The Circular Economy Package of the European Union

Joachim Wuttke

Abstract The Circular Economy Package which the European Commission presented on 2 December 2015 aims to stimulate Europe's transition towards a circular economy, improve international competitiveness, foster sustainable economic growth and generate new jobs. The package fleshes out the EU Roadmap to a Resource-efficient Europe where it relates to the circular economy.

The package comprises legislative proposals on waste and an action plan (Communication from the Commission to the European Parliament, the Council, the European Economic and Social Committee and the Committee of the Regions, Closing the loop – An EU action plan for the Cirular Economy, COM(2015) 614 final). The legislative proposals on waste aim to expand recycling and reduce land filling, and the action plan for the circular economy contains proposals for "closing the loop" in the circular economy and including all phases of a product's lifecycle, from manufacture and use to waste management and the market for secondary raw materials.

The revised legislative proposal on waste, which encompasses amendments to four legal acts, sets waste reduction targets and is meant to create a long-term framework for waste management and recycling. To ensure effective implementation, the waste reduction targets in the new proposal are accompanied by various measures. However, the level of ambition of the measures included in the action plan offers countries with highly developed waste management systems such as Germany, the Netherlands, Austria or Denmark little new incentive, whereas they are ambitious for other Member States.

Keywords Circular economy • Circular Economy Act • German Resource Efficiency Programme • Waste prevention • Product design • Sustainable consumption • Obsolescence • Landfill ban

---

J. Wuttke (✉)
German Environment Agency (UBA), Wörlitzer Platz 1, DE-06844 Dessau-Roßlau, Germany
e-mail: joachim.wuttke@uba.de

H. Lehmann (ed.), *Factor X*, Eco-Efficiency in Industry and Science 32,
https://doi.org/10.1007/978-3-319-50079-9_15

15.1 Introduction

The growing level of global consumption requires us to rethink how we deal with natural resources. A circular economy – one which fully integrates all aspects ranging from product design, sustainable production methods and patterns of consumption to recycling – makes significant contributions to resource conservation.

Economic development in Germany has long been characterised by rising industrial production and an expanding service industry. Greater pressures on the environment have naturally accompanied this development. It has been possible to reduce or contain these environmental effects in many sectors, as environmental technology has become more advanced, and because organisations have contributed to protecting the environment within their companies and have become more efficient at dealing with energy and raw materials.

On a macroeconomic scale, many of the climate protection gains made in Germany have been reversed or even eclipsed because both the volume and variety of goods produced has risen simultaneously, and manufacturing has increasingly been transferred abroad. Electric devices are a good example – although each individual device has become significantly more efficient, many more devices are sold overall. This results in efficiency gains being neutralised or drowned out by the number of devices sold.

If we assume that more than 9 billion people will adopt the manufacturing and consumption patterns of the industrialised world in the future, the international consequences for nature and the environment would be catastrophic.[1] Increasing resource productivity and decreasing the burden on nature and the environment in relative terms per commodity is therefore insufficient – the absolute consumption of resources must diminish and thus be decoupled from economic growth.

15.2 Experiences in Germany

Elements of a Circular Economy have been discussed in Germany in the field of waste management for decades. Following the enactment of the first Waste Disposal Act[2] in 1972, the waste management system has been subject to consequent progressive development. The principles of modern waste management, which were developed in Germany as early as 1975 in the Federal Government's Waste Management Programme,[3] have influenced the legislation since that time. The latest phase of development in waste management was the enactment of the "Circular Economy Act" (CEA), which came into force at the beginning of 2012.

[1] vgl. Umweltbundesamt (Hrsg. 2012): Schwerpunkte 2012 – Jahrespublikation des Umweltbundesamtes. Ressourceneffizienz – Schlüsselkompetenz zukunftsfähiger Gesellschaften. Dessau-Roßlau.

[2] Waste Disposal Act of 7 June 1972, Federal Law Gazette, Part I, p. 873.

[3] Waste Management Programme of the Federal Government, Bonn 1975.

There is a range of environmental, economic and social arguments in favour of increasing the use of secondary raw materials obtained through recycling processes. Recycling rates in Germany for some materials are already very high. For example, since two decades, a greater portion of aluminium comes from secondary production than from primary production. This has led to a reduction of approximately 90% in the emissions relating to aluminium production.

Therefore the scope of discussion in the political, public and scientific spheres has widened. To this end, the German Resource Efficiency Programme (ProgRess) (n.d.) adopted by the German government has formulated a central goal: to reduce the consumption of resources and the environmental pressures associated with it in order to secure prosperity and opportunities for development.[4] The programme follows the model of an economy embedded in material cycles. Improved technical measures in all areas of industrial production and measures that affect finished products should allow us to have an economy that conserves natural resources and in the long-term stops the contamination of the environment with substances that are damaging to both our health and the environment itself, as far as possible.

The concept of the circular economy is of utmost significance in this regard. In a circular economy, materials are used more sparingly and efficiently during production, products are designed and used in an environmentally friendly way, and waste is preferably prevented or – at the very least – subjected to high quality recycling. This type of economy requires us to think in terms of what are known as material flows. Taking into account the entire lifecycle of raw materials and the products manufactured with them– starting with the mining of precious materials and rare earth elements for our mobile phones, for example, and going right through to the moment the products themselves are recycled.

15.3 The European Commission's Circular Economy Package

The action plan places a strong focus on recovery and recycling of end-of-life consumer waste and does not sufficiently address collection and management of production waste. Industrial symbiosis approaches are dealt with only briefly. It also fails to raise the problem of downcycling which is de facto practiced for virtually all material flows.

The package incorporates important aspects and topics and adequately identifies deficits, but the measures it proposes to overcome them are only – except for those for packaging – of an appellative, informational or descriptive nature. Proposals are often limited to vague statements of intent or very soft measures, like references to the work programme for the Eco-design Directive.

[4] Bundesministerium für Umwelt, Naturschutz und Reaktorsicherheit (Hrsg. 2011): Deutsches Ressourceneffizienzprogramm. Programm zur nachhaltigen Nutzung und zum Schutz der natürlichen Ressourcen. Berlin.

Overall, the proposals fail to convey a concept for how a circular economy can be fitted into a resource-efficient Europe. In that regard, the package could have built on the vision set out in the Roadmap to a Resource-efficient Europe.[5]

Concrete targets are set only in the field of waste, in the form of recycling quotas, while the key optimization parameters of a circular economy remain unclear. No reference is made to the resource use caused by an economy based on primary raw materials and the substitution effects from the use of secondary raw materials. The interdependencies between a circular economy and the energy sector are ignored. Overall, the Circular Economy Package is, from the German point of view, lacking in ambition and innovation.

In the following, some examples from the Package are discussed in more detail to consider how the European Commission's Circular Economy Package could provide impetus to the further development of the circular economy.

15.3.1 Resource Conservation

The action plan proposed by the Commission contains elements with a strong economic focus (boosting job creation, growth, competitiveness, security of raw material supply) while environmental aspects are, in many instances, mentioned only as a secondary argument. References to the EU 2020 Strategy, the flagship initiative "A Resource-efficient Europe" and the Roadmap to a Resource-efficient Europe are almost completely absent.

At the macro-economic level, the proposal does not include any targets, not even one for raw material efficiency is mentioned. Measures to reduce environmentally harmful and resource use-boosting subsidies are completely absent. The financial sector is not addressed. The financial sector should also do its part to draw more attention to raw material efficiency. Suitable indicators for the financial sector should be established as a first step.

Regarding production processes, the proposal is right in recognizing that mining will remain a major source of raw materials in the medium term even in a (growing) circular economy. It explicitly mentions the greater attention that must be given to the environmental and social impacts of the production of primary raw materials and rightly recognizes that even in a circular economy, mining will remain a major source of raw materials in the medium term. The statement that the Commission seeks to promote the sustainable sourcing of raw materials globally through international cooperation and its trade and development policy is very much to be welcomed. The view that responsibility should primarily be assigned to the industry, and not to consumers, is also in line with our assessment.

Even though it correctly identifies industry's role, the proposal implicitly excludes mandatory (legally prescribed) supply chain due diligence schemes by inviting the industry to make its own commitments to sustainable sourcing. The

[5] KOM (2011) 571 endg.

voluntary nature of this call is also evident from the fact that the list of measures annexed to the Communication and the legislative proposals do not include a specific measure on the topic of "environmental and social impacts of raw material sourcing".

Other processes, like those carried out by the European Innovation Partnership (EIP) on Raw Materials with its Strategic Implementation Plan (SIP) and other innovation partnerships (such as Directorate General Growth ((DG Growth), have in part already reached a higher degree of specificity. It would be consistent to acknowledge and connect with them. There should be procedural coherence with DG Growth. Much better linkage with the strategies of other Directorates-General would in any event be appropriate, since these strongly influence the framework conditions and potentials for development of a circular economy.

15.3.2 Waste Prevention

The Commission's proposal includes new measures to prevent food waste, incorporate reparability and durability into product design and strengthen the re-use of products.

These approaches to strengthening waste prevention as a strategy to be prioritised in the circular economy are seen as positive. The measures address relevant aspects along the product lifecycle which are assessed in the sections below. One important area is the prevention of food waste.

Reducing food waste all along the value chain (primary production, processing and manufacture, retailing, distribution and in food consumption at and outside the home) is one of the target areas of waste prevention measures.

The action plan contains a number of supportive elements to accompany the measures that need to be implemented at national level. They include the development of a common, EU-wide methodology and indicators to measure food waste, setting up a stakeholder platform on food waste at EU level, examining the use of date marking by actors in the food chain and supporting measures to enhance consumers' understanding of the "best before" label.

Overall, the measures proposed in the Package for reduction of food waste are seen as positive. A point worth noting is that the Communication on the action plan mentions the Sustainable Development Goals (SDG) target of halving per capita food waste by 2030. However, since it has not been incorporated in the legislative text, this target remains not legally binding. Developing common measurement methods and indicators is essential for determining the quantity of food wasted and the very prerequisite for measuring and assessing progress in preventing food waste. These measures are therefore very much welcomed.

A point that seems less convincing is that the Package still sees the consumer and the behaviour changes that can be induced on that end as the main focus for achieving relevant reductions and much less holds other actors responsible.

15.3.3 Product Design

For the important aspect of product design, the Commission's proposal includes the following elements in particular:

- Reparability, durability, upgradability and recyclability of products will be promoted under the Ecodesign Directive and material efficiency standards developed;
- Requirements concerning the availability of spare parts and repair information will be considered in future work on ecodesign;
- An independent testing programme will be prepared under Horizon 2020 to help identify issues related to possible planned obsolescence.

A set of measures of this kind for the action field of product design can make products more durable and easier to repair and upgrade. It can facilitate their disassembly and the identification of constituent materials for improved recycling. Therefore, the proposal envisages that reparability, upgradability, durability, recyclability and the identification of certain materials will be systematically examined under the Ecodesign Directive. As a first step, the Commission has included some of these aspects in the review of the regulation for televisions and the extension of the scope of this regulation to include all electronic displays.

In the preparatory studies conducted to prepare ecodesign regulations for specific product groups a material efficiency tool has been introduced. This tool has already created a fairly good basis for these measures. The request to develop standards on material efficiency under the Ecodesign Directive which was received and accepted by European standardization organisations will also contribute to the further development of relevant methods and standards.

It remains to be seen, however, whether product-specific implementing regulations will actually set requirements concerning the above aspects. Therefore, the implementation of the standardisation mandate referred to above and future preparatory studies should be monitored intensely in order to ensure that these aspects are adequately reflected in these implementing regulations.

A negative aspect is the fact that the Commission has not yet presented the Ecodesign working plan for 2015–2017, which it announced would be presented at the latest with submission of the Circular Economy Package. It remains to be seen, therefore, whether the Commission will sufficiently incorporate the aspects reparability, upgradability, durability, recyclability and identification of certain materials into the working plan.

Requirements for better recyclability and for identification of materials under the Ecodesign Directive are only enforceable and meaningful if they fit with primary-treatment and recycling practice. A close interaction exists here with the work to amend the German Electrical and Electronic Equipment Act, particularly with the planned treatment requirements.

15.3.4 Sustainable Consumption

Regarding the consumption phase, the proposal envisages improving consumer information. Among other measures, the Commission plans to consider the possibility of requirements on the provision of information on repair and spare parts (as part of the work on ecodesign).

In addition, the Commission plans to develop a European agenda for the collaborative economy. Its formulation under the Single Market Strategy is not sufficient to mobilise potential for reducing environmental pressures, new business models and social innovation towards sustainable consumption, since it only addresses aspects relating to new business models under competition and consumer legislation. It is not yet coupled with a European strategy for promoting the collaborative economy; instead, it is about creating a legal framework for further shaping the European single market. In view of the fact that the stance on commercial sharing platforms (such as ebay, Airbnb, etc.) has become increasingly critical of late, there is the danger that more non-commercial, civil-society-organised forms of the collaborative economy come under substantial pressure of being regulated and marketised. This would thwart current efforts to create legal structures enabling the promotion of social innovations for consumption as part of a green economy in Europe.

The proposal provides for consumer advice only via (improved) labelling, whilst actual advising of consumers as well as approaches for anchoring resource efficiency in education are left completely unaddressed.

The proposal encourages Member States to provide incentives (for consumers) and use economic instruments such as taxation to ensure that product prices better reflect environmental costs, but it does not mention more specific approaches. Economic instruments are, however, crucial for an effective policy, and possibilities for action in that regard also exist at EU level (e.g. VAT).

To raise public awareness of sustainable consumption and production patterns, it would be appropriate to provide for education and information measures and for the integration of resource efficiency and sustainable consumption into education curricula at all levels.[6]

15.3.5 Information Requirements on Life Span and Reparability

The fact that the Commission plans to specifically consider proportionate information requirements on durability under the Energy Consumption Labelling Directive (energy label) and on the availability of repair information and spare parts under the Eco-design Directive is a positive thing.

[6] cf. Environmental Research Plan project on education in resource conservation and resource efficiency (BilRess/FKZ 3712 93 103).

In order to choose products according to their respective needs, the product information that should be available to consumers before or while making their purchasing decisions should also include information on durability, reparability, the availability of spare parts and repair services. In regard to durability information, there are serious limitations to this, in that in the case of a number of products measuring durability is not possible and requirements can only be set at component level. On the other hand, measurements are very time-consuming and costly (e.g. lasting 9 months and more), depending on the product group, thus presenting a large hurdle for market surveillance authorities. There is a need for further research regarding this aspect.

Complementing the above, we see a need to improve the framework conditions for independent repair shops. Specifically, available spare parts, repair manuals and tools should always be available also to non-manufacturer-affiliated spare part distributors and repair shops. The arrangement that applies in the motor vehicles sector can serve as a model in this regard.[7] In that sector, diagnostic tools and detailed repair information are normally made accessible against payment of a nominal fee. We consider it necessary, in principle, that this arrangement be transferred to electrical and electronic equipment.

15.3.6 Obsolescence

We welcome the intended testing programme for detecting planned obsolescence. Results show that the first useful service life has decreased over the last years for most of the product groups investigated.[8] This is the result of material, functional, psychological and economic forms of obsolescence interacting to produce a highly complex pattern. Research into these issues should definitely be continued.

In addition to the "testing programme" on obsolescence, specific measures against obsolescence should be developed. Possible measures include e.g. requiring manufacturers to provide information on guaranteed product lifetimes and reparability to enable consumers to make a more informed choice.

15.3.7 Plastic Waste

The Commission proposes the preparation of a strategy on plastics, as a follow-up to the Green Paper on plastic waste.[9] We welcome its intention to improve the separate collection of plastics, increase the quantities recycled and strengthen recycling

---

[7] Regulation (EC) No 715/2007 as amended.

[8] Prakash et al.: "Einfluss der Nutzungsdauer von Produkten auf ihre Umweltwirkung – Schaffung einer Informationsgrundlage und Entwicklung von Strategien gegen Obsoleszenz"; im Auftrag des Umweltbundesamtes. Texte 11/2016. Dessau-Roßlau. Download at: http://www.umweltbundesamt.de/publikationen/einfluss-der-nutzungsdauer-von-produkten-auf-ihre-1

[9] GREEN PAPER On a European Strategy on Plastic Waste in the Environment, COM/2013/123 final, Brussels, 7.3.2013.

over landfill. Another positive aspect is its intention to launch work on EU-wide quality standards for secondary raw materials and to focus on product design as well as the waste phase. Thus, the Commission plans to examine aspects such as recyclability under the Ecodesign Directive and develop relevant requirements for products where appropriate. The package does not, however, include concrete targets (e.g. plastic-specific quotas for non-packaging uses) or measures to achieve them. It remains to be seen, therefore, whether such statements of intent will eventually be reflected in statutory requirements. Also missing are approaches for strengthening the demand for plastic recyclates. Likewise, statements on the issue of biodegradability of plastics remain vague, limited to pointing out that this aspect will be considered in the strategy on plastics that is due to be prepared.

15.3.8 Critical Raw Materials/Recovery of Precious and Specialty Metals

The Commission stated that it will foster high-quality recycling which through "voluntary certification of treatment facilities for certain types of waste (e.g. electronic waste, …)". This can have a positive effect on the efficiency of the recycling of precious and specialty metals along the entire waste management chain. However, this strongly depends on how "high-quality" is defined and how the certification is crafted.

Whether the stated further development of the "recently initiated Raw Materials Information System" as well as "EU-wide research on raw materials flows" will have a positive effect on the recovery of precious and specialty metals will depend on whether sound data on precious and specialty metal flows are (or can be) made available.

The description of the problems encountered in recycling of critical raw materials from waste electronic equipment is in line with available scientific findings:

- Poor recyclability of the products,
- No exchange of information between manufacturers and recyclers,
- Lack of data for all economic operators,
- Absence of recycling standards.

We see as positive the recommendation given to Member States in the Proposal to amend the Waste Framework Directive, which suggests that they should take measures to achieve the best possible management of waste containing precious and specialty metals, taking economic and technological feasibility and environmental benefits into account. The same is true for the recommendation to the Member States that they should promote the reuse of such products and include in their national waste management plans measures to improve the collection and recovery of relevant products.

15.3.9 Waste from Electrical and Electronic Equipment

In this area, the Commission proposes, in particular, requirements for facilitating the dismantling, reuse and recycling of electronic displays as well as the development of proposals regarding the reparability and upgradability of electrical and electronic products. Recovery of critical raw materials will also play an important role for reasons of security of supply.

Reparability and upgradability, on which the Package places a strong emphasis, should be supported, in addition to being promoted through repair information and availability of spare parts, through the development of repair concepts (including access to WEEE). Recovery of critical raw materials should be promoted not only for reasons of security of supply, but also for environmental reasons.

We welcome the development of previously absent standards for secondary raw materials and the promotion of non-toxic material cycles in order to increase the use of recycled materials. Where requirements are developed under the Ecodesign Directive, their impact on the treatment of waste products should be considered and the requirements adapted as appropriate.

Insufficient information exchange between manufacturers and recyclers, and a lack of data for economic operators on the potential for recycled critical raw materials are identified in the proposal as barriers to high-quality recycling. A possible approach that the Commission may wish to consider is the submission of waste management concepts for new products at the time they are launched on the market. Such concepts should also ensure suitable structures for collection and recovery.

15.3.10 Landfill Ban

The Commission's proposals include, in particular, a binding target to reduce landfilling to a maximum of 10% of municipal waste by 2030, a ban on the landfilling of separately collected waste, and the promotion of economic instruments to discourage landfilling.

The landfill ban is a central instrument for controlling waste management. The Proposal for amending the Landfill Directive assumes that separately collected fractions are subsequently sent to landfill and that this must be prohibited. This assumption is unrealistic and the provision is therefore unnecessary, because

- it cannot be expected that fractions are collected separately at great cost to be then sent to a landfill,
- this would run counter to the waste hierarchy prescribed by the Waste Framework Directive,
- separately collected fractions normally either have a positive market value (e.g. metals, paper) or at least can be consigned to further processing/recovery at less cost than to landfill disposal, and
- this provision would be hardly enforceable.

In addition, a practical problem lies in the fact that each separately collected fraction contains a relatively large proportion of extraneous material. The proposal says nothing about whether such material may also no longer go to landfill after having been sorted out, since it had previously been collected separately (unlike mixed residual waste).

It would be more effective to set requirements for the fraction of mixed residual waste for landfill in order to force treatment (sorting out of recyclable components, biodegradation, thermal conversion) up to defined target rates. This can stimulate more recycling/recovery and environment- and climate-friendly treatment.

By way of derogation from the currently applicable provision, the proposal does not call for a further reduction of the proportion of biodegradable fractions beyond the target for 2016 of 35% of the fraction produced in 1990. A possible target would be e.g. 25% of the biodegradable waste produced in 1990 by the year 2025, coupled with requirements for the residual organic component (determined as e.g. TOC or ignition loss) and/or biodegradability activity parameters (determined e.g. as breathing activity or gas formation rate).

The binding target of 10% in the Proposal to amend the Landfill Directive is far more definite than what was proposed in the package tabled and withdrawn in 2014, and is therefore welcomed. It would be helpful, however, if there was a clear and final definition of the term "municipal waste", based on waste code numbers (e.g. group 1501 and Chap. 20 of the EU list of waste[10]) instead of verbal descriptions.

This 10% target is a very ambitious requirement. Without the waste pretreatment measures taken (mechanical and biological treatment, incineration), even in Germany some 35% of municipal waste would still have to be landfilled today. It is completely incomprehensible, therefore, why there are warnings of future overcapacities even though many Member States have the needed plants either not at all or not at sufficient capacity. In many Member States, over 75% or even 90% of municipal waste is still sent to landfills without prior treatment.

15.4 Conclusion

Based on the waste hierarchy circular economy instruments have been derived. Whereas some like recycling have been used successfully for decades others like waste prevention and reuse were revitalized and complemented, e.g. by sustainable product design and sustainable consumption concepts, in a new context and in the light of new findings. They help us complete the paradigm shift away from remedial waste management to a preventive circular economy.

In order to come closer to achieving the goal of closed material cycles in the long term, new concepts and instruments as described above must be established. To reach this goal, a resource-efficient material flow economy will also take those impacts of resource use into account that occur abroad, e.g. in the extraction of raw materials.

[10] Commission Decision 2014/944/EU.

We must advance our technologies and logistics if we want to better harness our wastes' potential as a source of raw materials. In the future, whether the achieved benefits justify the cost should be judged less by current market prices and much more by the resource use and environmental impacts incurred for producing comparable materials from primary raw materials. Much of the burden that we place on the environment to produce primary raw materials has merely constituted an external cost to date, which is not reflected in the raw materials' market prices. Materials whose provision is associated with substantial, and in many instances growing, environmental impacts and incorporated into products using high-tech production methods should also be recovered by the use of high-tech processes at the end of the chain of utilization in order to prevent new environmental impacts and return the materials to the cycle.

Disclaimer This paper does not necessarily reflect the opinion or the policies of the German Federal Environment Agency.

References

German Resource Efficiency Programme (ProgRess). Decision of the Federal Cabinet of 29 February 2012, published by Federal Ministry for the Environment, Nature Conservation, Building and Nuclear Safety (BMUB), Download at: http://www.bmub.bund.de/en/service/publications/downloads/details/artikel/german-resource-efficiency-programme-progress/

Schwerpunkte (2015) Jahrespublikation des Umweltbundesamtes, Dezember 2015

UBA – Umweltbundesamt (Hrsg.) (2002.) Nachhaltige Konsummuster – Ein neues umweltpolitisches Handlungsfeld als Herausforderung für die Umweltkommunikation. Mit einer Zielgruppenanalyse des Frankfurter Instituts für sozial-ökologische Forschung. Erich Schmidt Verlag GmbH

UBA – Umweltbundesamt (Hrsg.) (2014). Soziale Innovationen im Aufwind – Ein Leitfaden zur Förderung sozialer Innovationen für nachhaltigen Konsum. Download at: https://www.umweltbundesamt.de/publikationen/soziale-innovationen-im-aufwind

Chapter 16
Saving Natural Resources Through Conversion and Constructional Densification in Urban Areas: Ecological Potentials and Limits

Daniel Reißmann and Matthias Buchert

Abstract Considering the increasing urbanization in already dense agglomerated regions, constructional conversion and densification could be promising options to achieve exceptional savings on natural resources – more particularly of construction materials and land. However, challenges that relate to compact city structures have to be considered in an integrated view to avoid conflicts between the saving of scarce resources and further targets like enabling a high quality of urban life. This article will focus on resource saving potentials through the measures of constructional densification and conversion in urban areas in Germany under consideration of challenges and conflicting goals to other relevant issues of urban development. The main finding show that considerable savings of construction materials, energy or land could be reached through dense building structures and the conversion of building stocks into alternative uses.

Keywords Natural Resources • Constructional Densification • Conversion Activities • Urban Development • Ecological Potentials and Limits

16.1 Introduction and Background

More than 75% of the German population lives in cities and agglomerated areas today. Globally, the share is about 50% with an increasing trend (BMUB 2016a). This concentration of people induces high demands on natural resources like water, raw materials and land (UBA 2012), and thus cities and metropolitan areas are

D. Reißmann (✉)
German Federal Environment Agency (UBA), Sustainable Spatial Development, Environmental Assessments, Wörlitzer Platz 1, 06844, Dessau-Roßlau, Germany
e-mail: daniel.reissmann@uba.de

M. Buchert
Oeko-Institut, Resources & Transport, Rheinstr. 95, 64295, Darmstadt, Germany
e-mail: m.buchert@oeko.de

H. Lehmann (ed.), *Factor X*, Eco-Efficiency in Industry and Science 32,
https://doi.org/10.1007/978-3-319-50079-9_16

major hot spots of the global consumption of natural resources. It is even expected that about 80% of the global resources consumption will take place in urban areas by 2030 (McKinsey Global Institute 2016). In addition, the growing metropolitan areas of Germany in the South (Stuttgart, Munich, Frankfurt etc.), in the West (Cologne etc.), in the East (Berlin, Leipzig, Dresden) and in the North (Hamburg) will be confronted with further population growth in future which will induce high demand for additional living space and infrastructure and the corresponding building materials (Geppert and Gornig 2010).

With these conditions of urbanization, constructional conversion and densification in urban areas could be promising options to achieve exceptional savings on natural resources – more particularly of construction materials and land. This is why these concepts have gained rising attention in the last few years. Also, current strategic urban policy frameworks on the international and national level highlight the promotion of these concepts, which shows their relevance for urban development policy (e.g. UN HABITAT (2016), BMUB (2016a), BMUB (2015a) and EC (2007)).

A recent study which analyzed the potential for the creation of affordable housing space through the addition of floors on building stocks (vertical densification) computed an ascertainable potential of 1.1 million additional and cost-efficient apartments in Germany on already sealed urban land (Tichelmann et al. 2016). This is only one of several measures of constructional densification and conversion, so a far higher potential is conceivable through a portfolio of various constructional actions.

Besides this study, other research projects have also already investigated possibilities and threats of constructional densification and conversion in urban areas. For example, a study of the German Environment Agency (UBA[1]) from 2010 highlighted the urban land saving potential of constructional densification and conversion as one solution for the environmental challenge of land consumption (UBA 2010). Another UBA study from 2013 focused on possible solutions for the dilemma of constructional densification in urban districts under assurance of sufficient urban green and open spaces as basis for a high urban quality of life (UBA 2013). Solving this dilemma is the main focus of the concept of dual development of inner city planning ("doppelte Innenentwicklung" in German). It aims to utilize urban land reserves in a constructional useful way under consideration of conservation, development, linkage and qualitative improvement of urban open spaces (Böhm et al. 2015). For this, green areas, roof gardens, house gardens or lagoons are integrated into the constructed environment to balance the presence between buildings and urban green and blue. Additionally, the aforementioned UBA study from 2013 also highlighted the importance of the revitalization of living space (e.g. through conversion) and a functional mix in urban districts.

Elements of constructional densification (e.g. vertical densification or brownfield revitalization) and conversion are often seen in the context of the so-called compact city. This concept aims to revitalize inner city districts, enable short distances and counteract urban sprawl (Dieleman and Wegener 2004). This should reduce

[1] Umweltbundesamt.

(and ultimately stop) additional land take, induce a more efficient use of existing inner city infrastructures and improve inner city districts through a mix of urban functions with short distances and corresponding environmentally friendly mobility. However, more compact city structures are often criticized, especially in regard to the vision of a healthy, climate friendly and livable city. For example, air circulation could be significant reduced through a dense city structure, which could compromise the urban ecology and micro climate (e.g. urban heat islands) (Stone et al. 2010).

For this reason, a broader vision of the compact and mixed use city is promoted today. The typical elements of this concept are short distances, a dense and land-saving building development as well as a structural balanced mix of compatible urban uses and green spaces which deliver ecologic services, recreational functions and promote meeting and cohabitation within the urban districts (BMUB 2015a).

Even today, constructional densification and conversion is applied in cities and urban regions with a high population growth and the associated high demand on housing and infrastructures. For example, in 2009 the city of Munich agreed to initiate a project which should promote constructional densification of settlements from the 1950s and 1980s and the conversion of former only commercially used areas into areas that are mixed used for commercial and housing purposes. According to its own statement, the project named "Langfristige Siedlungsentwicklung (LaSie)" (long-term residential development) has motivated several urban stakeholders to deal with tolerable constructional density considering a lack of sufficient urban land and the high pressure on the housing market (Stadt München 2015).

Beside the already mentioned challenges and potentials of constructional densification and conversion, considerable potentials for the saving of natural resources are connected with these concepts. In addition to the saving of scarce land, densification and conversion could also help save raw materials like minerals (e.g. stones) and metals (e.g. ores) as well as energy (e.g. coal).

This article will point out what are the driving forces, objectives and expected benefits of constructional densification and conversion in urban areas. It will also be shown which resource saving potential for land, minerals and metals is considerable through these concepts. Challenges and limits of constructional densification and conversion with regard to urban micro climate functions, urban quality of life and urban green and open spaces will also be addressed. As a summary, the authors suggest some recommendations for a resource saving urban development under consideration of aspects like urban climate, health and quality of life.

16.2 Why Urban Land Conversion and Constructional Densification? Driving Forces, Objectives and Expected Benefits

The population in most large cities in Germany is increasing since 2010. Due to the lifestyles in urban environments the growth rate of private households in cities is even higher than the growth rate of the urban population (BBSR 2016). Currently Frankfurt am Main is growing by 15,000 inhabitants per year. Local authorities expect 800,000 residents in Frankfurt in 2027 and even 830,000 in 2040, which means a 17% increase compared to the current population (Frankfurter Allgemeine Zeitung (FAZ) 2015). The city expects a demand for 90,000 additional living units. Due to limited areas available for new residential buildings in this city, re-densification is regarded as an important part of the solution (FAZ 2016).

Berlin's population is also growing. Germany's capital registered more than 50,000 additional residents in 2015, but just 10,877 new living units were finalized in the same year. And this trend seems to continue in the upcoming years. Hence, the tensions in the housing market are increasing in Berlin (Tagesspiegel 2016).

The third example is Munich. The city's authorities expect an increase of the population by 230,350 residents for the period 2013 to 2030, which means a total growth of 15.4% (München 2015). The Bavarian state capital wants to address the growing population with an active housing policy which resulted in 6595 new living units in the city for the year 2015 (München 2016). Nevertheless, the increasing number of residents in Munich will require an on-going effort for satisfying the enhancing need for residential buildings and necessary urban infrastructure.

The main reasons for the large scale trend of growing urban population in Germany are the attractiveness of cities for younger people due to universities and job opportunities, and often better options for culture and leisure which attract different groups of people.

The clear trend of an increasing number of residents in cities and agglomerated regions calls for resolute measures to mitigate potential negative effects for the urban environments. The strategy of conversion and constructional densification in urban areas is a promising way to achieve this goal. Nevertheless, the limits of this concept must be explored and discussed with the responsible actors and the citizens in the growing German cities.

16.3 Densification and Conversion Instead of New Construction: Potential for the Protection of Natural Resources and the Environment

Just in Germany, 147,304 new buildings were constructed in 2015 according to the national statistics for building permits in construction (residential and non-residential buildings). In total, 222,280 construction activities were accounted with new buildings having a share of about 66%. A very similar relation was also observed for the previous years as Fig. 16.1 shows.

Especially the construction of new buildings generates a high demand on materials like ferroconcrete, brick, sand-lime brick, cellular concrete, lightweight concrete and wood (Destatis 2016c). Because these building materials consist of raw materials like limestone, clay, sand, gravel, grit and iron, the natural resource basis is considerable claimed (Destatis 2014). A study from 2009 that analyzed the consumption of construction materials of non-residential buildings in Germany, shows

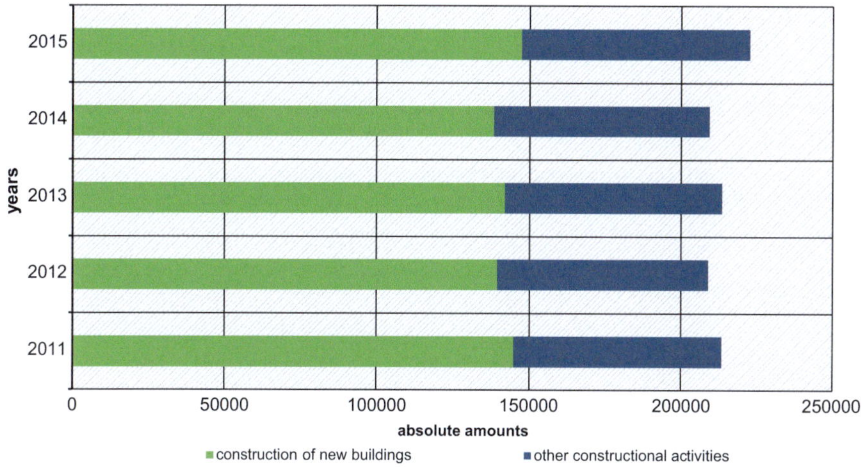

Fig. 16.1 Construction activities in Germany from 2011 to 2015 (building permits) (Source: Adapted from Destatis 2016b)

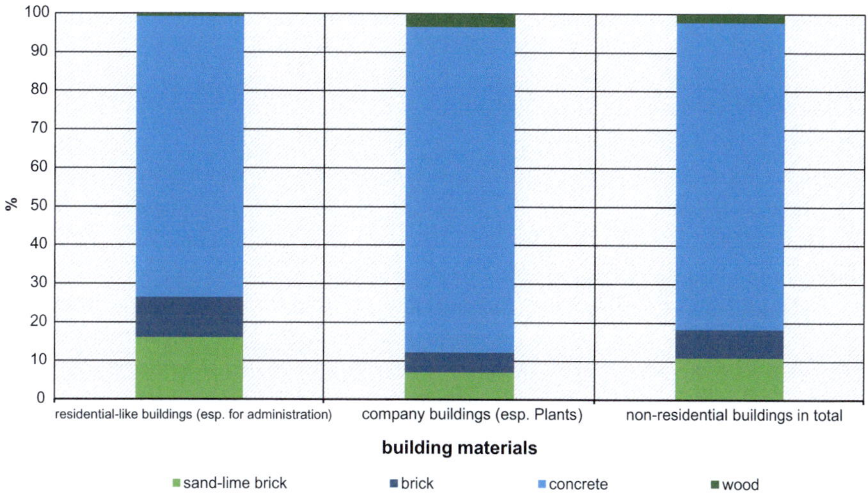

Fig. 16.2 Consumption of major material groups for non-residential building activities in Germany (survey year: 2009) (Data Source: BDZ 2009)

that cement – a major element of concrete – is a main part of most non-residential buildings. Figure 16.2 shows the shares of different materials. Obviously, especially concrete and thus its main raw materials limestone, clay, sand and iron ore were consumed through construction activities (BDZ 2009).

In regard to environmental protection, the production of building materials like cement or steel requires high amounts of energy and water and produces significant greenhouse gas emissions (during the exploration of raw materials, manufacturing

Table 16.1 Example of environmental effects caused by the production of 1 ton of cement (IBU 2012)

Parameter	Unit	Caused amount
Global Warming Potential (GWP)	(kg CO_2-eq.)	691.7
Potential for the abiotic degradation of non-fossil resources	(kg Sb-eq.) (Antimony-equivalent)	0.0013
Potential for the abiotic degradation of fossil fuels	(MJ)	1901.4
Potential for eutrophication	(kg PO_{43-}-eq.) (Phosphate-equivalent)	0.12
Potential for acidification	(kg SO_2-eq.) (Sulfur dioxide-equivalent)	0.83

of the product and through transport and distribution) even before the construction and use of the building itself (Schirmer 2015). To illustrate this, Table 16.1 shows the exemplary environmental effects from the production of just 1 ton cement in Germany.

To put this in context, in 2015 the cement consumption of Germany was about 26,638,000 tons of which 30.7% were used for residential construction, 34.1% for non-residential construction and 35.0% for civil engineering (VDZ 2016). That means that Germany's construction sector induced greenhouse-gas emissions of about 18,425,000 tons CO_2-equivalent for the production of cement of one exemplary year. The total greenhouse-gas emissions for Germany in 2015 were 908 million tons CO_2-equivalent (UBA 2016) which shows that just the usage of cement for construction induced 2% of the total amount. These numbers clearly show which negative environmental consequences construction activities can cause, although this example was limited to cement.

The high amounts of building materials that are used for constructional activities also induce considerable waste streams that occur during the construction process itself and through demolition activities. According to the European Environment Agency (EEA), the construction sector generates annually about 33% of Europe's waste (EEA 2010). In Germany, the share is even higher and was about 53% in 2014. In total, 197,735,000 tons of construction and demolition waste was generated in Germany in 2014 (Destatis 2016a). Currently, approximately 70% of construction waste is recovered. However, most of these materials were recovered for a low quality usage in civil engineering and landscaping which is a down-cycling. Only a minor share is recycled for the more valuable usage as secondary material for building construction, especially because most construction materials are not recyclable for these purposes and buildings are mostly not planned with regard to sustainable deconstruction and material recycling. However, demolished concrete as substitute for gravel or gypsum from used plasterboards has promising potentials and is already recycled in low amounts (Schindhelm et al. 2012).

Based on all this information it is clear that measures to minimize the resource consumption through construction activities are urgently needed to save natural resources, decrease energy consumption, counteract land consumption and minimize other negative environmental impacts like greenhouse gas emissions or eutrophication. Two examples of measures of densification and conversion will be illustrated to show how they can contribute to save constructional materials, land and energy as well as to minimize negative ecological effects in comparison to new construction.

Example 1: Vertical Densification Through the Addition of Floors on Building Stocks

There are considerable differences between urban forms of residential construction in regard to land take. For instance, dense forms like block or line structures induce a significantly lower land take than detached housing development. According to Tichelmann et al. (2016) the detached housing development stock claims 40.7 ha for 1000 housing units. In contrast to this, the line structure inventory needs only 9.1 ha and block structure inventory even just 3.8 ha for 1000 housing units. Tichelmann et al. (2016) computed that by adding floors on these building stocks, the additional land take occupied for buildings and associated open spaces can be reduced by 3 to 10 ha per day for Germany and 1.1 million additional and cost-efficient apartments could be realized assuming a 10 years period to skim off the overall potential. These computations focused on regions where a compact urban structure is useful because of scarce living space. Regions that have enough vacancies which could be reactivated were not considered as suitable for vertical densification, and were therefore excluded from the calculation.

Obviously, vertical densification could significantly contribute to saving land. Next to this, Tichelmann et al. (2016) have also shown that the addition of floors on inventory buildings could reduce the energy demand of apartments that are downstairs by up to 50%. Vertical densification could thus be an energy-saving alternative to conventional measures like roof renovation for buildings in need of restoration.

Example 2: Conversion and Renovation of Building Stocks Instead of New Construction

Germany has a large inventory of relatively old buildings. Most German cities consist of building stocks from the mid-nineteenth century and early twentieth century as a result of the population growth during the industrialization. However, as many buildings were destroyed during the world wars, buildings stocks from the 1950s up to the 1970s are very present in most German cities (Grassnick and Hofrichter 2013). In rural areas, relatively old agricultural buildings like houses with stalls, barns and farmhouses are very present (LfL 2008). Most of these building stocks are possibly usable for other purposes why they can be converted or reused. A study from 2008 which was prepared for the Agricultural Agency of Saxony (Sächsische Landesanstalt für Landwirtschaft (LfL)) computed that through the conversion of an exemplary German farmhouse up to 36% of the specific need of constructional materials of an average single-family house could be saved considering the legal minimum standards for energetic renovation (LfL 2008). Based on calculations of Markova et al. (2010) for the average material effort of a massive constructed

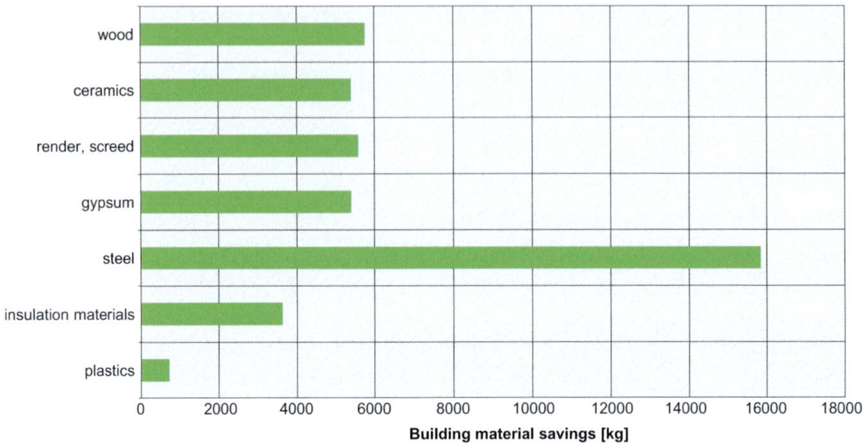

Fig. 16.3 Exemplary projected savings of selected building materials for the conversion of a German farmhouse in comparison to a new massive constructed single-family house (Data Source: adapted from Markova et al. 2010)

single-family house, this conversion measure could possibly lead to the exemplary savings of selected building materials that are illustrated in Fig. 16.3.

Only considering the production of galvanized construction steel – including credits through the recovery of scrap metal for primary production – approximately 12,600 kilogram CO_2-equivalent could be saved through this conversion measure, just for one building (IBU 2013).

Both examples show that considerable potentials for the saving of resources like materials, land or energy are conceivable through densification and conversion measures. Through vertical densification, high potentials for the saving of land and energy could be reached. In addition, existing infrastructures are used for new living spaces which also minimizes the need for resources (e.g. materials, energy, water) to develop new infrastructure as it is often the case for new constructed buildings. In addition, the conversion of building stocks into new uses could also reduce the resource consumption – especially regarding construction materials – significant. All this helps to decrease negative environmental impacts like greenhouse gas emissions.

Given that the overall potential for inner urban development for Germany is estimated at 165,000 hectares (Schiller et al. 2013) – which is four times the area of Cologne – an additional significant land saving potential by reactivating urban brownfields and empty sites[2] is possible.

[2] Brown fields and empty sites could be reactivated for constructional purposes (e.g. as building land) or other land usages (e.g. green areas) if these fields are not hazardous contaminated through e.g. high concentrations of heavy metals in soil.

16.4 Ensuring Quality of Life: Challenges and Limits of Compact Urban Structures

Since compact urban structures became part of urban planning concepts, they were seen as solutions for a portfolio of problems that occur as results of decentralization, rapidly growing individual mobility and population growth in already dense urban areas. The compact city concept which counteracts urban sprawl and additional urban land take was seen as a sustainable way for urban development (de Roo 2000).

Nevertheless, this enthusiastic view on compact city structures ignores the environmental dilemmas that could occur with a highly dense settlement structure. For example, dense urban structures have a higher negative impact on the regional climate in terms of so named heat island effects which are caused by absorption and retention of solar energy and exhaust heat from urban energy infrastructures and usages. Simultaneously, denser structures have less impact on global climate change due to shorter distances and less motorized traffic. Such environmental conflicts can also be observed for other aspects like the lower ecological footprint per capita in dense urban areas versus usually higher water and air pollution levels per acre in such regions in contrast to lower dense areas (Kelbaugh 2012).

When dealing with potentials for the saving of natural resources through urban densification, such environmental dilemmas associated with compact city structures must be considered. Taking into account that cities should serve a high quality of life for their inhabitants, cultural, social and economic aspects also have to be considered.

A major problem of high density cities is that the inner city infrastructures can be overloaded. Although one aim of compact city concepts is to enable short distances and hence promote short distance travel like walking or biking, medium distance travel must also be guaranteed to have a high urban quality of life. Public transport is the most dominant form of environmentally friendly medium distance travel in urban areas, with increasing tendency in most growing cities (BMVBS 2010). However, in many rapidly growing urban areas the rising passenger numbers exceed the capacity of public transport infrastructures, especially in major traffic hot spots and during rush hours. Figure 16.4 shows the rising number of public transport passengers for three German cities that are already highly populated and confronted with further population growth.

A potential solution for this increasing pressure would be to develop the public transport infrastructure as a whole. But the current situation shows that most cities are not able to develop their public infrastructure as fast as passenger numbers are increasing, which leads to much more pressure. Figure 16.5 illustrates this for Berlin. Obviously, the growth rate for passenger numbers is considerable higher than the growth rate of public transport routes.

Considering this, it is necessary to create urban mobility concepts that avoid such public transport overloads. One major advantage of compact city structures is that they already avoid long distance travel and reduce the need for individual motorized mobility. Although mixed functions – which means the mixture of activities like

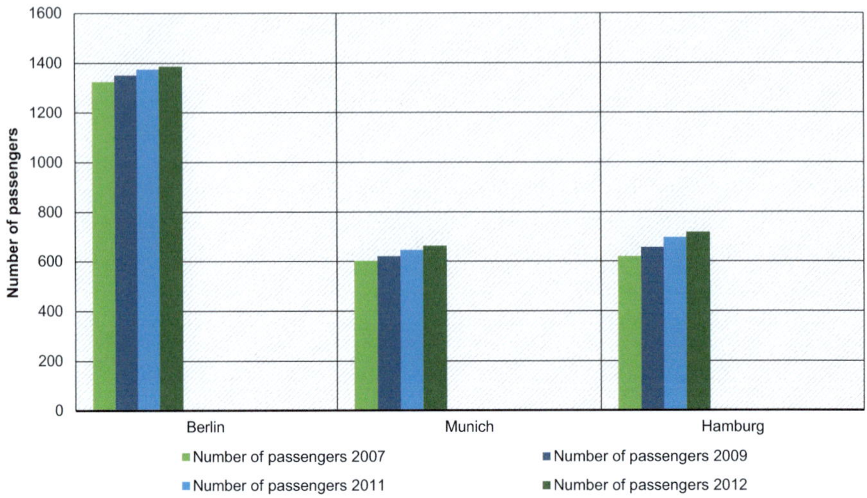

Fig. 16.4 Number of passengers of public transport in Berlin, Munich and Hamburg from 2007 to 2012 (Data Source: Senatsverwaltung für Stadtentwicklung und Umwelt Berlin 2013; HVV 2012; MVV 2015)

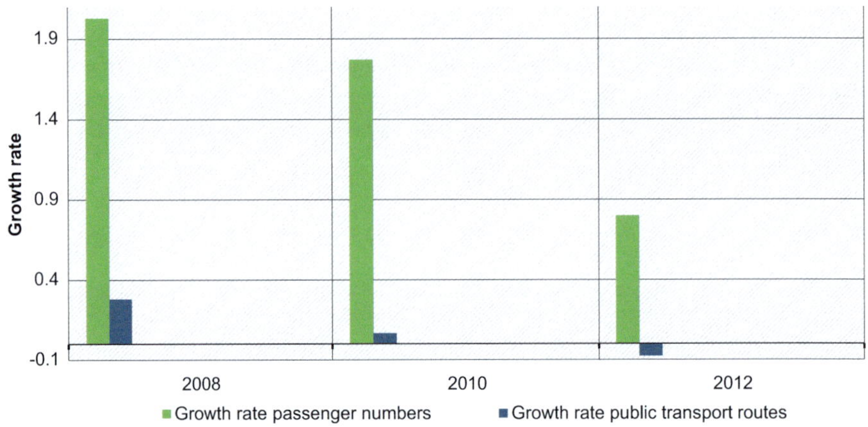

Fig. 16.5 Comparison of annual growth rates of passenger numbers and routes of public transport in Berlin for 2008, 2010 and 2012 (Data Source: Senatsverwaltung für Stadtentwicklung und Umwelt Berlin 2013)

living, working and recovery within one urban district – are one central aim of compact structures, there is no guarantee that all inhabitants live and work within short distances that can be reached by foot or bike. Hence, medium distance travel within the city is needed. Thus, a mix of urban medium distance travel is seen as promising approach to avoid possible mobility and infrastructure problems. For example, among other national and international activities the Directorate-General for

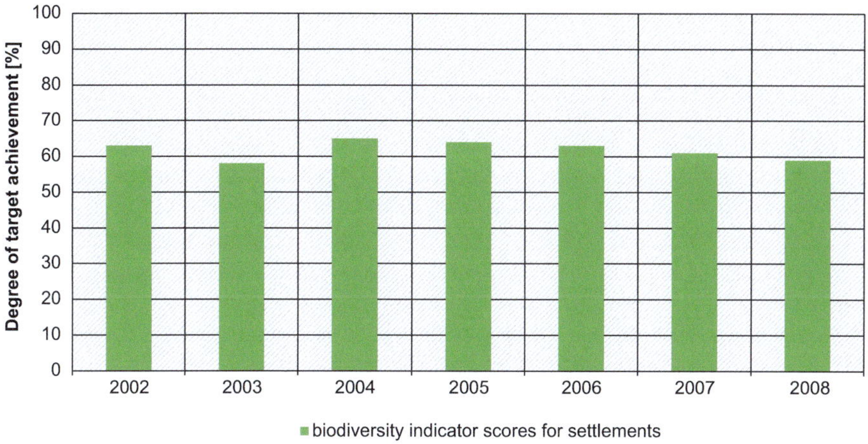

Fig. 16.6 Development of the indicator for biodiversity in settlement areas for the years 2002–2008. (Data Source: BFN 2014)

Mobility and Transport of the European Union has launched an urban mobility program that promotes a mixture of different medium distance mobility concepts like intermodal transfers, park and ride systems, demand responsive transport, car sharing concepts, electric bicycles and others (EU COM 2012).

In addition to infrastructure overloads, compact urban structures could also lead to high rates of land sealing which is in conflict with the preservation and development of urban green spaces which are major habitats for several species and thus contribute to save the natural resource biodiversity (Böhm et al. 2015). Considering that a score of 100% is aimed, the trend of the indicator scores for biodiversity in settlement areas – illustrated in Fig. 16.6 – shows the need for more urban green to preserve the biodiversity in settlement areas.

Besides biodiversity conservation, urban green spaces have various other functions like the absorption of air pollutants, CO_2 sequestration, evaporative cooling and shadowing. All these functions promote the quality of urban habitats, the health of the urban population, a comfortable urban microclimate and urban biodiversity. Additionally, urban green and open spaces are places that promote neighborly cohabitation, nature experience and environmental education (BMUB 2015b).

The guiding principle of dual inner urban development deals with these issues. It promotes an integrated approach of reactivating urban brownfields to save land and a simultaneous development, linkage and qualitative improvement of urban green and open spaces (Böhm et al. 2015).

Noise is also often mentioned as a problem of dense urban structures (City of Amsterdam 2008). Nevertheless, this problem is already partly counteracted through short distances which should contribute to reduce noise exposure from individual motorized mobility. However, the functional mix in compact cities leads to the problem that settlements and commercial activities move spatially closer. Some commercial activities (e.g. carpenters) generate noise exposure which could lead to problems regarding the fulfillment of noise thresholds for settlement areas. High noise pollution reduces the quality of life why these thresholds are normally very strict for settlement areas. To solve this conflict between mixed functions in compact structures that enable short distances and higher noise exposure through commercial activities near to settlement areas, a new type of building area is discussed in Germany. The so-called "urban area" could enable higher noise thresholds for compact urban structures (BMUB 2016b). These thresholds are close to thresholds for commercial areas, which makes this legal initiative highly controversial (Wuerfel et al. 2016) and thus further constructional solutions to enable a quiet living space in compact city areas are in development.

16.5 Conclusion and Recommendations

In the previous chapters it was shown that the strategy of conversion and constructional densification in urban areas is very promising to save natural resources. Addressing the significant population pressure in many cities and dense regions in Germany this strategy will become a key for satisfying the growing demand for living space and urban infrastructure as well as for preventing negative environmental impacts. In some cities like Frankfurt am Main the population pressure seems to be as dramatically that efforts for conversion and constructional densification have to be intensified.

But this also means that quality of life and health have to be ensured for the growing urban population. Therefore urban green spaces have to be preserved for the people in urban surroundings and also within the inner city. Vertical densification through the addition of floors on building stocks is a promising way to address both objectives following a recent study which assessed the potential of this option for Germany. Additional living units can also be generated in cities through the conversion of unused non-residential buildings to residential buildings, the revitalization of brownfields and empty sites as well as the reallocation of parking areas to construction land for buildings. The last option requires a noticeable shift of the modal split in the urban areas for transportation. Therefore a substantial extension of the public transport in the urban areas is an important pre-condition to reduce the number of vehicles in cities and to gain space for additional residential buildings.

Disclaimer This paper does not necessarily reflect the opinion or the policies of the German Federal Environment Agency.

References

BBSR (2016) Körner-Blätgen N, Sturm G: Wandel demografischer Strukturen in deutschen Großstädten, Bundesinstitut für Bau-, Stadt- und Raumforschung im Bundesamt für Bauwesen und Raumordnung, BBSR-Analysen KOMPAKT 04/2016, Bonn

Böhm J et al (2015) Urbanes Grün in der doppelten Innenentwicklung. Bonn

Bundesamt für Naturschutz (BFN) (2014) Artenvielfalt und Landschaftsqualität: Teilindikator "Siedlungen" in: http://biologischevielfalt.bfn.de/fileadmin/NBS/indikatoren/diagramme2014/Teilindikator_Siedlungen_barrierefrei.pdf, 26.10.2016

Bundesministerium für Umwelt, Naturschutz, Bau und Reaktorsicherheit (BMUB) (2015a) Grün in der Stadt – Für eine lebenswerte Zukunft. Grünbuch Stadtgrün, Berlin

Bundesministerium für Umwelt, Naturschutz, Bau und Reaktorsicherheit (BMUB) (2015b) Neues Zusammenleben in der Stadt. Bonn

Bundesministerium für Umwelt, Naturschutz, Bau und Reaktorsicherheit (BMUB) (2016a) Den ökologischen Wandel gestalten. Integriertes Umweltprogramm 2030. Berlin

Bundesministerium für Umwelt, Naturschutz, Bau und Reaktorsicherheit (BMUB) (2016b) Entwurf eines Gesetzes zur Umsetzung der Richtlinie 2014/52/EU im Städtebaurecht und zur Stärkung des neuen Zusammenlebens in der Stadt. Berlin/Bonn

Bundesministerium für Verkehr, Bau und Stadtentwicklung (BMVBS) (2010) Mobilität in Deutschland. Ergebnisbericht: Struktur – Aufkommen – Emissionen – Trends. Bonn/Berlin

Bundesverband der Deutschen Zementindustrie e.V. (BDZ) (2009) Baustoffverbrauch in Nicht-Wohnbauten 2009: Ergebnisse einer Erhebung zum Baustoffverbrauch nach Bauteilen. Berlin

City of Amsterdam (2008) Sustainable living in a compact city. City of Amsterdam Sustainability Report 2008–2009. Amsterdam

De Roo G (2000) Environmental donflicts in compact cities: complexity, decisionmaking, and policy approaches. Environ Plann B Plann Des 27(1):151–162

Dieleman F, Wegener M (2004) Compact city and urban sprawl in built environment 30(4):308–323

European Commission (EC) (2007) Leipzig Charter on Sustainable European Cities. Leipzig

European Commission (EU COM) Directorate-General for Mobility and Transport (DG MOVE) (2012) Action Plan on Urban Mobility – State of Play. Brussels

European Environment Agency (EEA) (2010) The European Environment State and Outlook 2010. Material, resources and waste. Kopenhagen

Frankfurter Allgemeine Zeitung (FAZ) (2015) Wir werden enger zusammenrücken müssen. 5th August 2015, http://www.faz.net/aktuell/rhein-main/bevoelkerungswachstum-frankfurt-2025-wird-es-enger-13731614.html. Accessed 14 Nov 2016

Frankfurter Allgemeine Zeitung (FAZ) (2016) Schwierige Suche nach Wohngebieten in Frankfurt. 25th October 2016, http://www.faz.net/aktuell/rhein-main/frankfurt-stellt-frankfurt-2030-fuer-stadtentwicklung-vor-14496555.html. Accessed 14 Nov 2016

Geppert K, Gornig M (2010) Mehr Jobs, mehr Menschen: Die Anziehungskraft der großen Städte wächst in Wochenbericht Nr. 19/2010 vom 12.Mai 2010, Deutsches Institut für Wirtschaftsforschung (DIW). Berlin

Grassnick M, Hofrichter H (2013) Stadtbaugeschichte von der Antike bis zur Neuzeit. Springer-Verlag, Heidelberg/Berlin

Hamburger Verkehrsverbund (HVV) (2012) Bericht 2012. Hamburg

Institut Bauen und Umwelt (IBU) (2012) Umwelt-Produktdeklaration nach ISO 14025: Zement. Königswinter

Institut Bauen und Umwelt (IBU) (2013) Umwelt-Produktdeklaration nach ISO 14025 und EN 15804: Feuerverzinkte Baustähle: Offene Walzprofile und Grobbleche. Berlin

Kelbaugh D (2012) The environmental paradox of cities: Gridded in Manhattan vs. Gridless in Dubai. J Sustain Dev 9(1):84–96

Markova S et al (2010) Voraussetzungen zur Prognose der materiellen Zusammensetzung zukünftiger Baurestmassen als Grundlage zur langfristigen, zielorientierten Bewirtschaftung von Baurestmassen. In Konzept zur nachhaltigen Nutzung von Baurestmassen basierend auf der thematischen Strategie für Abfallvermeidung und Abfallrecycling der EU. Wien: TU Wien

McKinsey Global Institute (2016) Urban world: the global consumers to watch. Brussels
München (2015) Demografiebericht München – Teil 1, Analyse und Bevölkerungsprognose 2013 bis 2030, Landeshauptstadt München, Referat für Stadtplanung und Bauordnung, Mai 2015
München (2016) Wohnen in München V – Bericht zur Wohnungssituation 2015, https://www.muenchen.de/rathaus/Stadtverwaltung/Referat-fuer-Stadtplanung-und-Bauordnung/Stadtentwicklung/Grundlagen/Wohnungspolitik.html, accessed at 14th November 2016
Münchner Verkehrs- und Tarifverbund (MVV) (2015) Verbundbericht 2015. München
Sächsische Landesanstalt für Landwirtschaft (LfL) (2008) Umnutzung versus Neubau im Dorf. In Schriftenreihe der Sächsischen Landesanstalt für Landwirtschaft, Heft 13/2008
Schiller G et al (2013) Innenentwicklungspotenziale in Deutschland – Ergebnisse einer bundesweiten Befragung. In Meinel et al. (ed.). Flächennutzungsmonitoring V: Methodik – Analyseergebnisse – Flächenmanagement. IÖR Schriften 61:51–59
Schindhelm S et al (2012) Recycling of demolishing wall material consisting of fine grained mineral by agglomeration. Chemie Ingenieur Technik 84(10):1798–1805
Schirmer S (2015) Der ökologische Plan B. Nachhaltig leben und bauen im Schatten des Klimawandels. o.O
Senatsverwaltung für Stadtentwicklung und Umwelt Berlin (2013) Mobilität in der Stadt: Berliner Verkehr in Zahlen 2013, Berlin
Stadt München, Referat für Stadtplanung und Bauordnung (2015) Perspektive München. Langfristige Siedlungsentwicklung: Statusbericht 2015. München
Statistisches Bundesamt (Destatis) (2014) Umweltnutzung und Wirtschaft. Bericht zu den umweltökonomischen Gesamtrechnungen, Wiesbaden
Statistisches Bundesamt (Destatis) (2016a) Abfallbilanz (Abfallaufkommen/–verbleib, Abfallkennzahlen, Abfallaufkommen nach Wirtschaftszweig).. Wiesbaden
Statistisches Bundesamt (Destatis) (2016b) Bautätigkeit und Wohnungen: Bautätigkeit. Fachserie 5 Reihe 1. Wiesbaden
Statistisches Bundesamt (Destatis) (2016c) Bauen und Wohnen: Baugenehmigungen von Wohn- und Nichtwohngebäuden nach überwiegend verwendeten Baustoff. Lange Reihe ab 1980. Wiesbaden
Stone B et al (2010) Urban Form and Extreme Heat Events: Are Sprawling Cities More Vulnerable to Climate Change than Compact Cities? Environ Health Perspect 118(10):1425–1428
Tagesspiegel (2016) Berlin hat so viele Bewohner wie seit Jahrzehnten nicht mehr, Tagesspiegel, 20th July 2016, http://www.tagesspiegel.de/berlin/bevoelkerung-und-wohnungsnot-berlin-hat-so-viele-bewohner-wie-seit-jahrzehnten-nicht-mehr/13903866.html, accessed at 14th November 2016
Tichelmann KU et al (2016) Deutschland Studie 2015. Wohnraumpotentiale durch Aufstockungen. Darmstadt
Umweltbundesamt (UBA) (2010) Nachhaltiges Bauen und Wohnen. Ein Bedürfnisfeld für die Zukunft gestalten, Dessau-Roßlau
Umweltbundesamt (UBA) (2012) Glossar zum Ressourcenschutz.. Dessau-Roßlau
Umweltbundesamt (UBA) (2013) Szenarien für eine integrierte Nachhaltigkeitspolitik – am Beispiel: Die nachhaltige Stadt 2030. Band 4: Der Szenario-Prozess – Dokumentation der Prozessergebnisse.. Dessau-Roßlau
Umweltbundesamt (UBA) (2016) Nationale Treibhausgas-Inventare 1990–2014 (Stand 01/2016) und Zeitnahprognose 03/2016 in: https://www.umweltbundesamt.de/daten/klimawandel/treibhausgas-emissionen-in-deutschland, 17.11.2016
United Nations Human Settlements Programme (UN HABITAT) (2016) HABITAT III. New Urban Agenda. Draft Outcome Document for Adaption in Quito, October 2016. New York
Verein Deutscher Zementwerke (VDZ) (2016). Zahlen und Daten: Zementindustrie in Deutschland. Düsseldorf
Wuerfel W et al (2016) GSK-Update: Das Urbane Gebiet – Ein neuer Baugebietstyp erleichtert dichtes Bauen mit hohem Wohnanteil. München

Chapter 17
The Path to Degrowth for a Sustainable Society

Serge Latouche

Abstract The word degrowth has been recently used in the ecological, economic and social debate; it started being used in 2002 as a provocative slogan to denounce the mystification of the ideology of sustainable development. It now designates a complex alternative project, with undeniable analytical and political significance. It is about first escaping from consumption society and then building a durable society of prosperity without growth or frugal abundance. Thus degrowth is not the alternative, but rather a matrix of alternatives that opens the human adventure to a plurality of destinies and space for creativity, while removing the lead blanket of economic totalitarism. Degrowth therefore offers a general framework that provides meaning to many sectorial initiatives or local resistances favoring strategic compromises and tactical alliances. This project may seem like a pipe dream today but it is extremely realistic if we want to avoid the collapse of human society in the future.

Keywords Bioeconomy • Degrowth • Development • Decolonization (of the imaginary) • Sustainable development

17.1 Why This Neologism?

Portraying degrowth as a version of sustainable development, as has been done by some authors (Zaccai 2007), makes little sense historically, theoretically and politically in regard to the meaning and significance of the initiative. The motto of degrowth was almost accidentally launched by a pressing need to break with the wooden language of sustainable development that was felt by whole streams of political ecology and development critics (Revue Silence 2002). Thus the phrase is not originally a concept (at least not symmetrically to growth) but rather a defiant political slogan with the main objective of reminding us of the meaning of limits; more specifically, degrowth is neither recession nor negative growth.

S. Latouche (✉)
Université de Paris-Sud, Orsay, France
e-mail: serge.latouche@free.fr

H. Lehmann (ed.), *Factor X*, Eco-Efficiency in Industry and Science 32,
https://doi.org/10.1007/978-3-319-50079-9_17

The word should not be interpreted literally: degrowing to degrow would be as absurd as growing to grow. Of course, *degrowthers* want the quality of life, air and water to grow, along with several other things that growth for the sake of growth destroyed. To speak more rigorously, we should probably employ the word a-grow, with the Greek privative prefix "a", as we say a-theism. And it is, as a matter of fact, the abandon of faith and religion: those of progress and development. We should become atheists of growth and the economy. The rupture of degrowth is therefore related to both words and things, it implies the *decolonization* of the imaginary and the implementation of another possible world.

17.2 History, Causes and Meaning of Degrowth

While chance did play a role in the sequence of events, it was a need that can be rightfully called historic that led to the apparition of a radical movement offering a real alternative to the consumption society and the dogma of growth. Facing the triumph of ultra-liberalism and the arrogant proclamation of Margaret Thatcher's famous TINA (there is no alternative), small anti-development freemason groups could no longer get by with a quasi confidential, theoretical criticism for *Third Worldists*. However, the other face of the triumph of single-thought ideology was no other than the consensual "sustainable development" slogan, a nice oxymoron that was launched by UNEP (United Nations Environment Programme) to try to save the religion of growth in the ecological crisis and which seemed to be well accepted by the anti-globalization movement. It became urgent to oppose the capitalism of a globalized market with another civilization project or, more specifically, to give visibility to a plan that had been in formation for a long time, but progressed underground. The rupture with *developmentism*, a form of productivism for the use of so-called developing countries, was thus the foundation of this alternative project.

Are we therefore dealing with another economic paradigm contesting neoclassical orthodoxy, similar to what Keynesianism did back in the day? That is how some try to label it following Nicholas Georgescu-Roegen's bioeconomy project. Clearly, other political economies are possible in a growth society. The so-called "trentes glorieuses"[1] period (1945–1975) that saw the triumph of Keyneso-Fordist regulation is proof. Nonetheless, in a growth society without growth – which is the current situation of industrialized countries – alternatives to neo-liberal inspired policies seem impossible without calling into question the economic system and/or exacerbating the ecological crisis.

In fact, it is a defiant slogan to break the soft consensus of submission to the dominant productivist order. The eponymous movement was born during the colloquium "Défaire le développement, refaire le monde[2]" (Undoing development, redoing the world) which was held in March 2002 at UNESCO, an intellectual

[1] English: The Glorious Thirty.

[2] Proceedings published as *Défaire le développement, refaire le monde*, Parangon, Lyon, 2002.

adventure that was reinforced a few months later by the birth of the newspaper that was its echo chamber. Degrowth quickly became the rallying cry of all those aspiring to the construction of a true alternative to an environmentally and socially unsustainable consumption society, and from then on it was a *performative fiction* representing the need for a rupture with growth society and the birth of a frugally abundant one.

The degrowth project does not aim for *another* growth, nor for another kind of development (sustainable, social, fair, etc.), but rather for the construction of *another* society, a society of *frugal abundance*, a *post-growth* society (Paech 2012), or one of *prosperity without growth* (Jackson 2009). In other words, it is not an economic project from the outset, not even of another economy, but a *societal* project that implies an exit of economy as a reality and as imperialist discourse. The word degrowth from now on designates a complex alternative project with an undeniable analytical and political reach.

17.3 The Imposture of Growth as an Organic Metaphor

Growth is a natural phenomenon and, as such, is undisputed. The biological cycle of birth, development, maturation, decline and death of life and its reproduction are also a precondition for the survival of the human species, which needs to metabolize with its vegetal and animal environment. Mankind first celebrated the cosmic forces that ensured their well-being in the symbolic form of gratitude for this interdependence and their debt towards nature in this regard. The problem begins when the line between symbols and reality disappears.

While all human societies rightfully worshipped growth, only modern western civilization made it its religion. The product of capital, resulting from a trick or market deception and most often from exploitation of workers and predation of nature, is assimilated to the recovery of plants. The economic organism, that is the organization of the survival of society, is no longer in symbiosis with nature but rather exploits it mercilessly and must indefinitely grow, just like its fetish, capital. The reproduction of capital/economy combines fertility and regeneration, the interest rate and the growth rate. This climax of economy/capital leads to consumer society's fantasy of immortality. This is how *growth societies* are lived.

Growth society can be defined as a society dominated by a growth economy, and that tends to be absorbed by it. Hence growth for the sake of growth becomes the main or even the only objective of the economy and life. It is not about growing to satisfy recognized needs, which would be a good thing, but just growing to grow. Consumer society is the normal outcome of a growth society. It rests on triple unlimitation: unlimited production and thus of the harvesting of renewable and non-renewable sources, unlimited production of needs – and thus superfluous production – and unlimited waste generation which leads to the emission of waste and pollution (of air, land and water).

In order to be sustainable and lasting, every society needs to set limits. But our society takes pride in being free of all constraints and chose excess. Admittedly, something in human nature motivates individuals to surpass themselves. This is both a strength and a threat. Thus all societies except for ours tried to channel this aspiration and to harness it for the common good. In fact, it is not detrimental when it is invested in things such as non-commodified sports. However, it becomes destructive when the impulse of greed (the constant quest for more) is left unchecked, as well as accumulation of merchandize and money. It is therefore necessary to find a sense of limits again, in order to ensure the survival of humanity and the planet. The point of degrowth is to escape from a society that was absorbed by the fetishism of growth. The *decolonization of the imaginary* is primordial to do so.

17.4 The Decolonization of the Imaginary

The idea and project of the decolonization of the imaginary have two main sources: on one hand, Cornélius Castoriadis' philosophy, and on the other hand, the anthropologic critic of imperialism. These two sources are naturally also at the origins of degrowth, just like ecological criticism. Castoriadis understandably places the emphasis on the imaginary, whereas the anthropologists of imperialism focus on decolonization. Coming back to these two sources allows us to coin the exact meaning of the expression "decolonizing the imaginary".

The use of this performative expression goes without saying in Castoriadis' approach even though he never directly used it, to my knowledge. For the author of *The imaginary institution of society* (Castoriadis 1975), social reality is the implementation of *imaginary meanings*, in other words representations with mobilize affects. However, if growth and development are beliefs, and hence *imaginary social meanings*, just as progress and all the foundation categories of the economy, we need to change the imaginary in order to escape them, abolish them and move past them (the famous Hegelian *Aufhebung*).

The achievement of a degrowth society requires the decolonization of our imaginary to really change the world, before changes in the world condemn us to it painfully. This is the strict application of Castoriadis' teaching. "What is required," he notes, "is a new imaginary creation of previously unseen importance, a creation that would place at the center of human life meanings different from the expansion of production and consumption, that would set new life objectives that can be perceived as worthwhile by human beings […] This is the enormous difficulty we need to face. We should want a society where economic values are no longer central (or unique), where economy is relegated to its place as a simple means to human life rather than an ultimate end, where we refuse the race to constantly increased consumption. This is not only necessary to avoid the final destruction of the terrestrial environment, but also and most importantly to escape the psychic and moral misery of contemporary humans." (Castoriadis 1996, p. 96). In other words, this much needed exit from *surmodern* consumer and show society is also extremely desirable.

"But in order to have such a revolution," adds Castoriadis, "deep changes need to occur in the psychosocial organization of the western man, in his attitude towards life, in short in his imaginary. The idea that the only goal in life is to produce and consume more – an idea both absurd and degrading – must be abandoned; the capitalist imaginary of a pseudo-rational pseudo-mastery, of unlimited expansion, must be abandoned. This can only be done by men and women. A single individual or an organization can only, at best, prepare, criticize and sketch possible directions" (Castoriadis 2005, p. 244).

Before attempting to plan an exit from the dominating imaginary, we must first return to the way we entered it, that is to the process of the *economization* of spirits simultaneous to the *commodification* of the world; in other words, an analysis of the way the economy was instituted in the modern western imaginary.[3]

The uprooting of a belief can readily be formulated through a *decolonization* metaphor in the analysis of North/South relations. The colonization term, commonly used by anti-imperialist anthropology about mentalities, can be found in the titles of several works. Octave Manonni probably led the way with his psychology of the colonized, but Gérard Althabe, a follower of Georges Balandier, more explicitly names his studies on Madagascar *Oppression and liberation in the imaginary* in 1969. Most importantly, in 1988 Serge Gruzinski published *The colonization of the imaginary*, with a subtitle referring to the same westernizing process (Gruzinski 1988). Nonetheless, when Gruzinski mentions the colonization of the imaginary, he is still referring to the colonization process in the strict sense of the term, in this case to the conversion of indigenous people by missionaries. The change of religion is both a deculturation of spirits and an acculturation to christianism and western civilization within the frame of the imperialist project. This refers to true oppression in the imaginary, implemented not only with symbolic means if we remember the stakes of inquisition that were often used by Spanish conquerors in the new world.

With growth and development, we are facing a process of conversion of mentalities, thus of an ideological and quasi-religious nature, aiming to institute the imaginary of progress and the economy, whereas the *rape of the imaginary*, to use Aminata Traoré's beautiful expression, stays symbolic (Traoré 2002). With the colonization of the imaginary in the West, we are dealing with a mental invasion of which we are both the victims and the agents. It is mainly an auto-colonization, partly voluntary slavery.

The expression "decolonization of the imaginary" creates a semantic slip. Its originality lies in the accent placed on the particular shape taken by the inverse process from the one analyzed by anthropologists. It consists of a *software* change or paradigm change or even a true revolution of the imaginary, as claimed by West Indian writer Edouard Glissant. It is first and foremost a cultural revolution, but also goes beyond that. It is about escaping the economy, changing values and thus *decolonizing* ourselves. This is exactly the program that degrowth "supporters" are developing for the post-development project.

[3] We tried to develop this in the book *L'invention de l'économie* (Latouche 2005).

Getting out of the dominating or *colonial* imaginary is a central question both for Castoriadis and for anti-imperialist anthropologists, but it is a very difficult one because one cannot decide to change their imaginary and even less the imaginary of other people, especially if they are "addicted" to the drug of growth. The first though goes to education, *paideïa*, which plays a crucial role according to Castoriadis: "What is the meaning of, for example, liberty or the possibility for citizens participation, he asks, if the society we are discussing is lacking something – which is disappearing from contemporary discussions [...] – which is *paideia*, the education of the citizen? It is not about teaching him arithmetic, it is about teaching him how to be a citizen. Nobody is born a citizen. And how do we become it? By learning to be it. We learn, first, by looking at the city we are in. And certainly not the television we watch today" (Castoriadis 2010, p. 96).

However, the withdrawal cure is only possible if degrowth society has already been achieved. We should first have exited consumer society and its "civic moronization" regime, which traps us in a cycle we must break. Denouncing aggressive advertising, the vehicle of today's ideology, is certainly the starting point to escape what Castoriadis calls "televisual and consumerist onanism" (Castoriadis 2010, p. 194). The fact that the newspaper "La décroissance"[4] came from the organization "Ad breakers" is not really due to chance, since advertisement is the main spring of growth society, and the growth objection movement is greatly and naturally linked to the resistance to aggressive advertising.

17.5 The Way Forward

Finally, degrowth is not *the alternative*, but rather a matrix of alternatives that reopens the human adventure to a plurality of destinies and the space of creativity, by lifting the lead blanket of economic totalitarianism. This is bout exiting the paradigm of *homo œconomicus* or Marcuse's one-dimensional man, the main source of planetary homogenization and the suicide of cultures. Consequently, the a-growth society will not be established the same way in Europe, in sub-saharan Africa or in Latin America, in Texas and in Chiapas, in Senegal and in Portugal. It is crucial to favor or rediscover diversity and pluralism. Therefore we cannot suggest a turnkey model of a degrowth society, but only an outline of the essentials for any non-productivist sustainable society, in addition to concrete examples of transition programs.

Nonetheless, transition programs will necessarily be reformist. Thus many "alternative" propositions that do not explicitly claim to support degrowth can find their place. In this way, degrowth is a general framework that gives meaning to many sectorial or local struggles and favors strategic compromises and tactical alliances. Leaving the economic imaginary with nevertheless require concrete ruptures. It will be necessary to set rules that frame and limit the lack of control and

[4] Degrowth in French.

greed of agents (chasing profit, always more): ecological and social protectionism, labor legislation, limitation of company sizes, etc. And first and foremost, the "decommodification" of the three *fictitious* merchandises, which are work, the earth, and currency, according to Karl Polanyi. Their removal from the globalized market would mark the starting point of a reincorporation/reembedding of economic matters in social ones, while simultaneously fighting against the *spirit* of capitalism.

The redefinition of happiness as frugal abundance in a fair society that corresponds to the rupture created by the degrowth project requires exiting the vicious circle of unlimited creation of needs and products and the increasing frustration it creates. Self-limitation is the condition to achieve prosperity without growth and thus avoid the collapse of human civilization.

The recent debates on the relevance of wealth indicators served as a reminder of the inconsistency of GDP (gross domestic product) as an indicator to measure well-being, while it is the untouchable functional fetish of growth society. We did not sufficiently notice in this occasion that it is the ontological inconsistency of the economy itself that is thus underlined. When we criticize GDP, we shake the very foundations of the belief in the economy or the economy as a belief. Economy as a discourse presupposes its object, economic life, which only exists as such because of it. In fact, regardless of the definition we give political economy, be it a classical one (production, distribution, consumption) or a neo-classical one (optimal allocation of rare resources for alternative use), the economy online exists when it presupposes itself. The specific field of practice and targeted theory can only be limited if wealth and the allocation of resources only concern the economy. Garry Becker is more coherent by stating that everything that is the object of a human desire is rightfully part of the economy, but if everything is economical, then nothing is anymore. In this case the omni-quantification of social matters and the obsession with calculation it illustrates are only the result of a power grab, the one of the institution of capitalism as *omnicommodification* of the world.

The degrowth movement intends to oppose this transformation of the world into a commodity which was largely caused by globalization.

> *"But in order to have such a revolution," adds Castoriadis, "deep changes need to occur in the psychosocial organization of the western man, in his attitude towards life, in short in his imaginary. The idea that the only goal in life is to produce and consume more – an idea both absurd and degrading – must be abandoned; the capitalist imaginary of a pseudo-rational pseudo-mastery, of unlimited expansion, must be abandoned. This can only be done by men and women. A single individual or an organization can only, at best, prepare, criticize and sketch possible directions."* (Castoriadis 2005, p. 244)

> *"What is the meaning of, for example, liberty or the possibility for citizens participation, he asks, if the society we are discussing is lacking something – which is disappearing from contemporary discussions […] – which is paideia, the education of the citizen? It is not about teaching him arithmetic, it is about teaching him how to be a citizen. Nobody is born a citizen. And how to we become it? By learning to be it. We learn, first, by looking at the city we are in. And certainly not the television we watch today."* (Castoriadis 2010, p. 96)

References

Castoriadis C (1975) L'institution imaginaire de la société. Seuil, Paris
Castoriadis C (1996) La Montée de l'insignifiance. Les carrefours du labyrinthe IV. Points, Paris
Castoriadis C (2005) Une société à la dérive. Seuil, Paris
Castoriadis C (2010) Démocratie et relativisme : Débats avec le MAUSS. Mille et une Nuits, Paris
Gruzinski S (1988) La Colonisation de l'imaginaire: Sociétés indigènes et occidentalisation dans le Mexique espagnol. Gallimard, Paris
Jackson T (2009) Prosperity Without Growth? The Transition to a Sustainable Economy. Sustainable Development Commission, London
Latouche S (2005) L'invention de l'économie. Albin Michel, Paris
Paech N (2012) Liberation from excess: The road to a post-growth economy. oekom verlag, Paris
Revue Silence (2002) La peur de la décroissance. Lyon
Traoré A (2002) Le Viol de l'imaginaire. Fayard, Paris
Zaccai E (2007) Sustainable Consumption, Ecology and Fair Trade. Routledge, New York

Part III
Examples of Good Practice

Chapter 18
Social Innovation Repair – The R.U.S.Z Case: A Systemic Approach Contributing to the Unplanned Obsolescence of Capitalism

Sepp Eisenriegler and Greta Sparer

Abstract Repair and Service Centre R.U.S.Z (Reparatur- und Service-Zentrum R.U.S.Z) is a social business and a service for consumer protection and sustainability. In 1998 it started as a work integration social enterprise for long-term unemployed persons creating and using the business model of a sustainable repair shop. R.U.S.Z founded the success story Repair Network Vienna with some 80 SME members. It was among the initiators of the Austrian umbrella organization RepaNet and its EU equivalent RREUSE, which gather social enterprises with activities in re-use, repair and recycling. These networks help making advocacy work successful. Today, R.U.S.Z is Austria's biggest independent repair centre for electrical and electronic appliances for all kinds and brands and a centre of excellence for the fight against planned obsolescence, for consumer protection and social businesses. Moreover, R.U.S.Z is working for the transformation of the current linear economy into a circular economy.

R.U.S.Z' primary objectives are resource efficiency and social inclusion. R.U.S.Z provides repair services for household appliances, consumer electronics and IT. It sells certified, high-quality used equipment as well as new washing machines that were diagnosed in the in-house R&D department as particularly durable and easily repairable. From 1998 to 2007, R.U.S.Z was commissioned by the Public Employment Service Austria (AMS). R.U.S.Z was successfully transformed into a not-for-profit private enterprise in 2008. Today, it operates on a cost-recovery basis and employs more than 20 former long-term unemployed people. R.U.S.Z has been leading many initiatives to replicate its model and also lifts the barriers it is facing, both in Austria and Europe by initiating changes in national and EU policies.

Keywords Growth driven economy • Planned obsolescence • Resource depletion • Circular economy • Repair services • Product service systems

S. Eisenriegler (✉) • G. Sparer
Reparatur- und Service-Zentrum R.U.S.Z, Lützowgasse 12-14, 1140 Wien, Austria
e-mail: sepp.eisenriegler@rusz.at; greta.sparer@rusz.at

H. Lehmann (ed.), *Factor X*, Eco-Efficiency in Industry and Science 32,
https://doi.org/10.1007/978-3-319-50079-9_18

18.1 The Beginnings

R.U.S.Z (Reparatur- und Service-Zentrum) was founded in 1998 in Vienna as a social economy enterprise[1] for reintegrating long-term unemployed persons and other people at risk into the labour market. Initially, the project was commissioned by the Austrian Public Employment Service (AMS) and supported by the Vienna Adult Education (VHS), the Environmental Counselling Vienna (Umweltberatung Wien), the Viennese Municipality (MA 48) and the European Union. R.U.S.Z started its activities with 12 people at risk who got working contracts limited up to 12 months (temporary workers), two administrative employees and one qualified social education worker. The temporary workers included long-term unemployed persons, persons with disabilities and former offenders. Most of them were men over 45 years of age.

All repair activities were carried out by the temporary workers. This was only possible because some of them were already trained technicians who could train the others, who in turn would train newcomers. Therefore, it was one of the important gains of R.U.S.Z when the first technicians got permanent contracts in 1999. For every temporary worker who became a permanent staff member, four new temporary workers had to be employed, following the guidelines of AMS.

Regarding operational activities, the original intention was to offer repaired second-life washing machines and dishwashers to socially deprived households and to service and repair these devices exclusively. But the initiative was so well received by the citizens of Vienna, that they soon started lining up in front of the repair shop with their vacuum cleaners, TVs, hairdryers and other appliances that they had stored in their garages and attics. They all appreciated the unique offer of having electronic and electrical appliances of all kinds repaired at reasonable prices and the R.U.S.Z team was more than happy to meet this demand.

Coming from a public environmental counselling agency (Umweltberatung Wien) which he had initiated in 1988, R.U.S.Z founder and CEO Sepp Eisenriegler had put the focus on prolonging the lives of electrical and electronic appliances and thereby conserving resources and reducing the e-waste of private households. The support of the previously mentioned institutions allowed R.U.S.Z to operate at prices below the market price. Therefore, R.U.S.Z not only created job opportunities for the long-term unemployed, but also offered repair services for those who would not have been able to afford it otherwise. The latter would have been forced to throw away possibly repairable devices, and in exchange they would have probably chosen a new and affordable, but low-quality device of short durability, hereby creating more waste of resources and e-waste in the near future.

[1] In Austria, a social economy enterprise is a project commissioned by the Public Employment Service Austria (AMS) that serves the purpose of taking over outsourced AMS duties and responsibilities by offering trainings to people at risk at the labour market aiming at reintegrating them into regular jobs. Social economy enterprises are an important instrument of active labour market policy in Austria.

After 1 year of activity, the business was flourishing, but R.U.S.Z was also overwhelmed by the volume and variety of repair-requests; for example, the increasingly famous repair shop received a request to repair an auxiliary motor of a sailing yacht. As a solution, R.U.S.Z founded the Vienna Repair Network (ReparaturNetzWerk Wien) together with 12 smaller repair businesses in the area in 1999. Today, the network contains more than 80 small and medium enterprises (SMEs) which repair all kinds of electrical and electronic devices and other household goods.

R.U.S.Z was among the initiators of the European umbrella organization RREUSE (1999) and its Austrian equivalent RepaNet (2004) two lobby organisations for durability, reparability and re-use of products like electrical and electronic equipment, textiles, furniture and building materials.

In its early years R.U.S.Z already became a respected institution for repair services and repair standards. In 2005, R.U.S.Z created the "Ö3 Wundertüte", the most successful mobile phone collection system in the world, in cooperation with Ö3 (the most widely heard radio station in Austria) and Caritas Austria. From 2006 to 2007, experts from R.U.S.Z developed the Austrian standard ONR 192102 "Sustainability mark for electric and electronic appliances designed for easy repair (white and brown goods)" (Austrian Standards 2006) in cooperation with the Ministry for Agriculture, Forestry, Environment and Water Management, the Austrian Standards Institute and other partners.

18.2 Transformation into a Private Limited Enterprise and Years of Struggle

By the end of 2007, R.U.S.Z had become a well-known repair shop and a retailer of second-life electrical and electronic devices. By then, R.U.S.Z had grown to a remarkable size employing 124 temporary workers and permanent staff (88 temporary and 36 permanent). Over the years it had employed around 450 temporary workers. More than 300 of them were placed into regular jobs.

The year 2008 brought some big changes to R.U.S.Z. The long-term partnership with the Austrian Public Employment Service (AMS) ended, which was not only a major change in the organisation's setup but also heavily influenced R.U.S.Z in its today's shape. Ten one-year contracts with the AMS had been signed and fulfilled. The main reason for the end of the cooperation was a crucial change in strategy by the AMS. At this point it was communicated that cutting the costs of the temporary work places would be prioritized over the employment of people at risk. As a consequence, R.U.S.Z would have had to reduce the overall costs and additionally, the maximum training period was cut down from 12 to 6 months. The short amount of time that a temporary worker would now have spent in R.U.S.Z would not have sufficed to train a person effectively and making him or her employable.

Instead of giving up its activities, as many other social economy enterprises did at the time, R.U.S.Z was transformed from a VHS project into an association and its activities were carried out on a not-for-profit basis. Since then, R.U.S.Z has taken part in numerous public and private social initiatives and projects. R.U.S.Z received several awards, including the Energy Globe Austria Award 2008, the Austrian Climate Protection Award 2009 (Klimaschutzpreis), the Innovation Price "Ideas Against Poverty" 2009 ("Ideen gegen Armut"), the Viennese Award for Environmental Protection 2012 (Wiener Umweltschutzpreis) and the famous Austrian CSR award TRIGOS 2017 in the category Social Entrepreneurship. In 2014, the CEO was decorated with the Golden Merit-Medal of the Federal State of Vienna.

In 2011 R.U.S.Z Ltd., held by CEO Sepp Eisenriegler (52%) and three employees (48%), was founded as a not-for-profit enterprise with the core business of repairing electrical and electronic devices and the sale of second life products. Since then the association R.U.S.Z – Association for the Promotion of Social Economy (R.U.S.Z – Verein zur Förderung der Sozialwirtschaft) is in charge of research and development, product testing and the development and acquisition of new projects.

The first years after the privatization of R.U.S.Z were also the hardest in the business's history. After having to give up the public funding of the early years as a work integration social economy enterprise, a severe increase in the prices for repair services and second hand products was unavoidable in order to cover costs. Since then, R.U.S.Z' CEO has intensified his media campaigns and is, as an expert for reparability of electronic and electrical devices and planned obsolescence, a frequent guest and interview partner for various Austrian TV and radio channels, newspapers and magazines. As a consequence, R.U.S.Z is now recognized by many citizens as a centre of excellence not only for repair, but also for political activism against planned obsolescence and for resource efficiency. The number of sympathisers and the customer base have grown. In 2014, the seventh year after privatization, the company performed cost-effectively for the first time.

Even in the years of struggle, R.U.S.Z stayed faithful to its initial mission and exclusively employed formerly unemployed persons, and it still does today with few exceptions. Long-term unemployed can join R.U.S.Z as trainees for 3 months. If possible and feasible, they are then employed. The second part of R.U.S.Z' social mission is to find the economically best solution for every customer while saving resources. R.U.S.Z employees always advise customers to repair a device rather than throw it away and buy a new one, which is usually more expensive in the long run. One of the broader goals has always been to raise awareness amongst customers and the general public regarding the so-called planned obsolescence of electronic and electrical devices. Today, R.U.S.Z is the biggest independent and most well-known repair service in Vienna, and its awareness raising media campaigns reach out to all of Austria, and even enters other European member states and Switzerland at times. CEO Sepp Eisenriegler is member of various EU committees and is frequently invited by consumer initiatives as an expert for reparability and longevity of electronic and electrical devices and planned obsolescence.

In 2014, when the first surplus became available, a position for product testing was created with the objective of testing the reparability of various electrical device categories. The same year, the above mentioned unique Austrian standard on durable, easily repairable electrical and electronic equipment was updated and redefined (ONR 192102:2014 Label of Excellence for Durable, Repair-friendly Designed Electrical and Electronic Appliances) (Austrian Standards 2014). This is an internationally unique standard for household appliances and consumer electronics. It still is the first benchmark for durability and also includes guidelines for the construction of appliances that allow repair services at reasonable prices. It consists of 93 mandatory and voluntary criteria. For example, an appliance must be easy to open and disassemble. Instead of gluing and welding joints together, screws and snap connections must be used in order to fulfil the criteria. The components must be standardised and spare parts must be available for an additional 10 years. The access to blueprints and other appliance service documents must be easily available to independent repair shops. Otherwise the search for information will be too time-consuming and can make the repair much more expensive or only possible through a small number of manufacturer approved service partners. The standard was used to test 28 new washing machines in 2015 and 40 vacuum cleaners in 2016. The results of these tests show that newer products tend to be less repairable than older generation products of the same brands. The trend of the last 10–15 years of new electronic and electrical appliances being less durable than their older counterparts becomes obvious in R.U.S.Z' everyday activities.

In the first years of the initiative, social inclusion was the main focus of R.U.S.Z and the long-term environmental approach was secondary to this. Today the social side of the business focuses on training people (mostly former long-term unemployed) and employing them, if possible, for an unlimited period of time, while at the same time advocating for resource efficiency and the circular economy on the national and the EU levels. In this way, R.U.S.Z has turned into a centre of excellence for the economic interests of repair shops, e.g. shifting taxes from labour-intense work to critical resources (environmental tax reform), fighting planned obsolescence, lobbying for reparability of new electronic and electrical devices and the disclosure of diagnosis software codes which are necessary for certain repairs, but also for resource efficiency and the circular economy.

R.U.S.Z also advocates for a mandatory warranty that reflects the promise made by the producers regarding the longevity of devices. Producers of high end appliances would gain market shares if they communicated, for example, a five-to-ten-year full guarantee to consumers. As a consequence, producers of low-quality products would have an incentive to improve the durability and reparability of their products (Alt et al. 2015). Together with an industrial supply of product service systems this range of requirements would lead to a sustainable, resource efficient economy.

18.3 R.U.S.Z as a Social Innovation and Its Socially Innovative Projects

R.U.S.Z is regarded as a social innovation (CASI 2016; Brunauer and Schartinger 2016). Several aspects of the business contributed to this perception.

18.3.1 Services and Processes

R.U.S.Z aims to connect social and ecological requirements by bringing back to life technical goods and re-integrating people at risk into the labour market. It creates jobs for unemployed people and qualifies them as professional technicians. Moreover, R.U.S.Z contributes to resource conservation and prevention of (hazardous) waste (CASI 2016).

18.3.2 Degree of Innovation

Customer services of manufacturers today are hardly considered anything more than an external marketing unit. Various TV contributions with hidden cameras[2] have shown that very often, a service technician will come into a customer's home, take a look at the product to be repaired, and state that it would cost way too much to repair it and that, if they buy a new product instead, they can even save the costs of this diagnosis. In this market situation a serious, forceful repair service is a quite radical change and even a threat to the status quo.

R.U.S.Z has had a high success rate for the reintegration of long-term unemployed people back into society from 1998 until 2007. These individuals were trained to be service technicians and received assistance for financial and social issues, e.g. declaring private bankruptcy and finding new housing. Most of the people trained at R.U.S.Z were men over the age of 45, and in some cases immigrants and ex-offenders were trained while finding themselves in very challenging life situations. These people got the chance to start over again, which in the experience of the R.U.S.Z team is very empowering: 71 percent of R.U.S.Z temporary workers have been re-integrated into the labour market and thus into society. Nowadays R.U.S.Z does that at a smaller scale: the business provides practical training for 16 long-term unemployed people and apprentices per year, free of charge (CASI 2016).

---

[2] Reparaturdienste im Test:

PULS 4: SUPERNOWAK! Waschmaschinentest 10.5.2016: (https://www.youtube.com/watch?v=2bWDUkocLnM)

PULS 4: SUPERNOWAK! Geschirrspülertest 05.11.2015 (https://www.youtube.com/watch?v=gtDijkwwnSg)

ATV: DER GROSSE ÖSTERREICH TEST: Waschmaschinentest 02.03.2015 (https://www.youtube.com/watch?v=U9Py3pKlrXo)

ATV: DER GROSSE ÖSTERREICH TEST – Reparaturtest Flat-TV 25.11.2013 (https://www.youtube.com/watch?v=F2VZqHeD06Y).

18.3.3 Recent Social Innovations

Since 2010, R.U.S.Z makes use of the know-how it gained from the mobile-phone-collection project "Ö3-Wundertüte" and collects household appliances donated by private households. The resource efficiency project Donate Your Old Washing Machine – The Environmentally Sound and Social Reallocation of Household Appliances[3] enables the in-house accredited Re-Use Center to refurbish and sell some 1.000 high end, second-life products per year. The products are good for another 10 years, and sold at prices comparable to new low-quality devices which in most cases are dead after 3–5 years. In addition to contributing to resource efficiency and climate protection (53 percent of environmental impact is due to production and distribution), the donation program enables R.U.S.Z to offer additional jobs for former long-term unemployed.

In 2010 and 2011, the in-house R&D unit developed the so-called "Tuning of Washing Machines", a technical innovation to increase the energy efficiency of old washing machines from category C to A. This helps save energy and resources and creates new jobs in the EU, instead of enhancing the shift of production to the Far East, where much of the fully automated production takes place. Moreover, it shows that resource efficiency and energy efficiency do not contradict each other. The disputable alternative is to buy a new supposedly energy efficient device and to dispose of the old one.

In 2012 R.U.S.Z started a media campaign against planned obsolescence. In the following years, some 500 awareness-raising newspaper articles and TV contributions were launched.

The Austrian Public Employment Service (AMS) commissioned R.U.S.Z in 2013 to perform "StarGate", a vocational training program at R.U.S.Z for mechatronic engineers who have quit their primary education before graduation. This innovative pilot project proposed by AMS ended after 1 year, as AMS could not provide enough trainees. Nevertheless 55 percent of the trainees succeeded in finalizing their education within 1 year.

Since January 2014, the in-house repair-café "schraube14" takes place on a weekly basis and empowers consumers to repair their electrical and electronic appliances. On average, 25 customers are helped free of charge to generate DIY-repairs with a success rate of 65 percent. The customers get professional advice, tools and coffee for free.

The product-service system "Clean Laundry" is offered since February 2016. Thirty-five washing machines and dryers are currently rented on an unlimited basis to private households and a refugee camp. A product-service system is a competitive system of products, services, supporting networks and infrastructure. The system includes product maintenance, parts recycling and product replacement, which sat-

[3] "Spenden Sie Ihre alte Waschmaschine – Die ökosoziale Umverteilung von Haushaltsgeräten."

isfy customer needs competitively and with lower environmental impact over the life cycle. If producers of household appliances made their profits by renting their products, consumers could rely on the intrinsic incentive of manufacturers to design durable, repairable products.

18.4 Conclusion and Outlook

Since 1998, R.U.S.Z has been able to build its reputation as a reliable partner in repairs of electronic and electrical appliances, re-use and social entrepreneurship both in the eyes of customers as well as in those of other organisations and the general public. For many years, R.U.S.Z has now been a pro-active actor in the field of resource efficiency. Its corporate responsibility is therefore not only directed towards its close sphere of activity, but to a growing sphere of influence e.g. in national and EU politics.

In the years to come, R.U.S.Z will be part of the shift in paradigm of the economic system of the EU towards a circular economy. Currently, the R.U.S.Z CEO is leader of the Austrian delegation to the CEN-CENELEC[4] Joint Technical Committee 10 "Energy related products – Material efficiency aspects for eco-design". The technical committee is commissioned by the European Commission until 2019 to develop standards for durable, repairable and re-usable products which will be crucial to put the Circular Economy Concept in place. These standards will find their way into European legislation and improve the Eco-Design Directive which currently is a mere energy efficiency directive.

Social franchising of the general business model of R.U.S.Z or parts of it are going to be offered soon. At the moment R.U.S.Z is working on the reality check of its social franchising handbook by developing a new branch in the city of Graz in southern Austria.

References

Alt M, Bretschneider W, Gawel E, Schlacke S, Tonner, K (2015) Stärkung eines nachhaltigen Konsums im Bereich Produktnutzung durch Anpassungen im Zivil- und öffentlichen Recht. Reihe UBA-Texte 72/2015 [Online]. Available from: http://www.umweltbundesamt.de/publikationen/staerkung-eines-nachhaltigen-konsums-im-bereich. Accessed 15 Nov 2016

Austrian Standards (2006) ONR 192102. Nachhaltigkeitssiegel für reparaturfreundlich konstruierte Elektro- und Elektronik-Geräte (Weiß- und Braunware). Vienna, Austrian Standards

Austrian Standards (2014) ONR 192102:2014. Gütezeichen für langlebige, reparaturfreundlich konstruierte elektrische und elektronische Geräte. Vienna, Austrian Standards

[4]CEN and CENELEC are two recognized European Standardization Organizations. CEN stands for the European Committee for Standardization, CENELEC stands for the European Committee for Electrotechnical Standardization.

Brunauer A, Schartinger D (2016) Report: SI drive – WP environment and climate change. R.U.S.Z. Repair and service center. A case study of a social innovation. Austrian Institute of Technology, Vienna

CASI (2016) Reparatur- und Servicezentrum R.U.S.Z. [Online]. Available from: http://www.casi2020.eu/casipedia/cases/751. Accessed 15 Nov 2016

Eisenriegler S (2016) Konsumtrottel. Wie uns die Elektro-Multis abzocken und wie wir uns wehren, Vienna. edition a

Chapter 19
Resource Efficiency in the Building Sector

Klaus Dosch

Abstract Endeavours towards resource-efficient and climate friendly construction seem to be reaching their limits. The current approaches, which focus on continual reductions in heating energy consumption, are leading to increasingly elaborate insulation measures and complex air-tight buildings. Further improvements can only be realized in the scope of the building's lifecycle. Factor X offers a pragmatic approach by reducing the consumption of non-renewable raw materials, non-renewable energy and the emission of greenhouse gases throughout the building's entire life cycle, resulting in significant improvements in comparison with other strategies employed to date. This concept has already been put into practice in two small residential areas in Inden and Eschweiler (North Rhine-Westphalia, Germany), and the first positive experiences with building contractors and related companies have been gathered.

Keywords Building sector • Residential areas • Energy Saving Ordinance • Construction costs • Marginal costs • Life cycle • Marginal energy savings • Ecological rucksacks • Factor 4 house • Resource-efficient construction • Klimaexpo NRW • Faktor X Agentur

19.1 Background

Building and habitation result in the highest rate of material exchange between nature and the anthroposphere worldwide, cf. UNEP (2011), Herczeg et al. (2014). Raw materials are extracted, transported and processed to make construction materials. During the extraction process the natural environment is destroyed, energy is consumed and waste products are expelled to the environment. Green spaces are converted into residential areas and energy is consumed when the buildings are inhabited.

K. Dosch (✉)
Faktor X Agentur der Entwicklungsgesellschaft indeland GmbH, Düren, Germany

Aachener Stiftung Kathy Beys, Aachen, Germany
e-mail: dosch@faktor-x.info

H. Lehmann (ed.), *Factor X*, Eco-Efficiency in Industry and Science 32,
https://doi.org/10.1007/978-3-319-50079-9_19

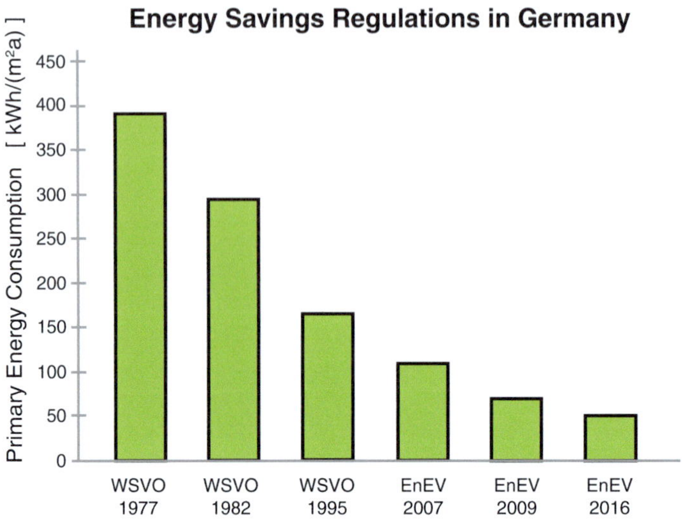

Fig. 19.1 Energy conservation regulations in Germany since 1977 (Source: Author)

Reducing building energy consumption in the building sector has been a goal of in German politics for some time.

The first thermal insulation regulations in Germany (Wärmeschutzverordnung, WSVO) came into force as early as 1977. This saw the first capping of energy consumption specifically for the building's use phase. Since then the permissible upper consumption limits have been continually adjusted. The latest Energy Saving Ordinance (*Energieeinsparverordnung, EnEV*) from 2016 reduces the required energy consumption during the building's use phase by almost 87% of the theoretical maximum reduction (cf. Fig. 19.1). The European Unions (EU) directives on Energy Efficiency and Energy Performance of Buildings Directive have a similar goal, cf. EU (2010), EU (2012).

However, in view of the material expenditure involved in sufficiently insulating the outer shell of a building in order to reach this goal, the continual reduction of energy consumption limits proves to be inefficient. At approximately half of the theoretically possible energy savings, the insulation effort required increases disproportionally to the benefits. The marginal energy savings from each centimetre of additional insulation begin to decrease drastically above an insulation thickness of 5 cm (cf. Fig. 19.2).

Furthermore, the construction costs increase dramatically for reaching this degree of insulation in the building. The necessary thermal insulation can only be reached by installing mechanical ventilation systems and only if the building is planned and built to be completely air tight. The actual behaviour of the user also plays a crucial role as windows must not be opened during the heating phase.

In this context a further tightening of building energy consumption requirements should be viewed extremely critically. In other words: if buildings are to reach

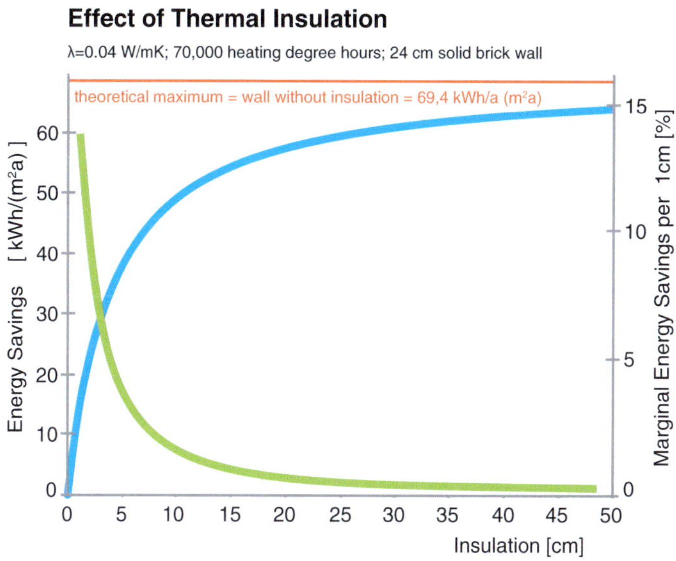

Fig. 19.2 Theoretical energy savings of thermal insulation (Source: Author)

energy savings beyond those currently attainable in new buildings, other potential sources of energy savings must be unlocked.

In addition to reductions in energy consumption, the production of renewable energy through photovoltaic or solar-thermal collectors can also be pursued. The current energy saving regulations only consider these energy sources in terms of the energy balance. The energy produced is deducted from the building's annual energy consumption. However, this does not reflect the actual energy flow. During the sunnier periods, from April to September, the building's energy requirement can be sustained by the energy yielded from photovoltaic or solar-thermal collectors. Surplus energy is produced but this cannot yet be stored to any significant extent for the low-radiation times of the year. External energy sources are thus required for heating and warm water in the low radiation months from October to March. It is therefore possible for an energy neutral or even a plus energy house on the balance sheet to still require an additional energy supply in low-radiation months. This fact is largely ignored in current legislation (Fig. 19.3).

Measures to improve energy efficiency currently only consider the building's use phase. But before the building is put into use, building materials need to be produced which requires the extraction, transport and manufacturing of raw materials into construction materials, which subsequently need to be transported to the construction site. In modern buildings this phase of a building's life cycle uses 30–75% of the total energy consumed during the building's entire life span. If the building needs to be renovated due to a change in its intended use or a defect technology needs to be replaced, the energy expenditure is even higher. At the end of the life cycle the building may be demolished, which also requires energy.

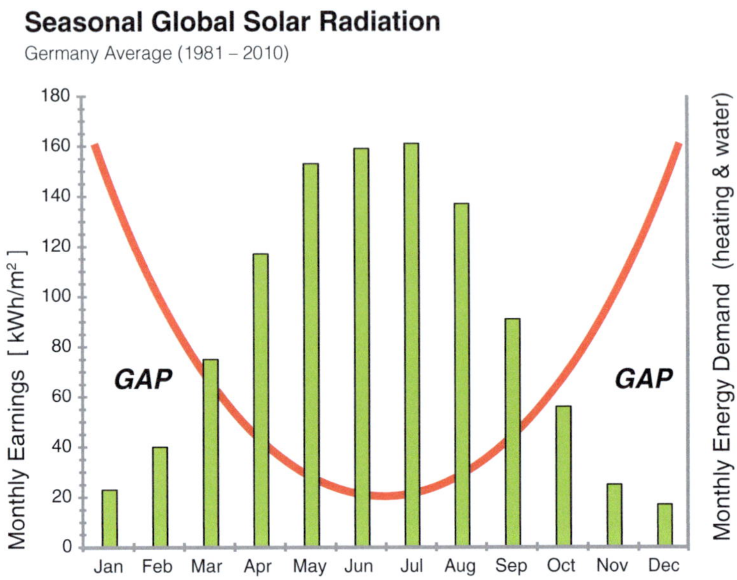

Fig. 19.3 Seasonal global solar radiation in Germany (Source: Modified after DWD (2016))

19.2 The Factor X Approach

The decreasing marginal energy savings associated with additional insulation measures and the lack of options for storing surplus energy in high-radiation periods indicate that it is necessary to develop and utilize other potential methods of climate protection. If construction work and the renovation of existing buildings are to contribute to improved climate protection, then the consumption of non-renewable energy, the emission of greenhouse gases and the consumption of non-renewable raw material must be reduced over the complete building life cycle. As only low marginal energy savings result from improvements during the building's use phase, the boundaries of the system must be are widened considerably (cf. Fig. 19.4).

The Energy Saving Ordinance considers only the use phase of the building. Factor X extends the system boundaries to include energy production, the production of construction materials, the construction phase, maintenance, repairs and renovation as well as building demolition and all related transportation.

Within the extended system boundaries, non-renewable primary energy, greenhouse gas emissions and the consumption of non-renewable raw materials as well the related ecological rucksacks are considered.

In order this extension of the system boundaries to be widely applied, the system needs to be easily accessible for users from the construction industry. It is crucial to consider reductions in resource consumption early on in the planning process. Due to the high amount of effort involved in assessing a building's environmental

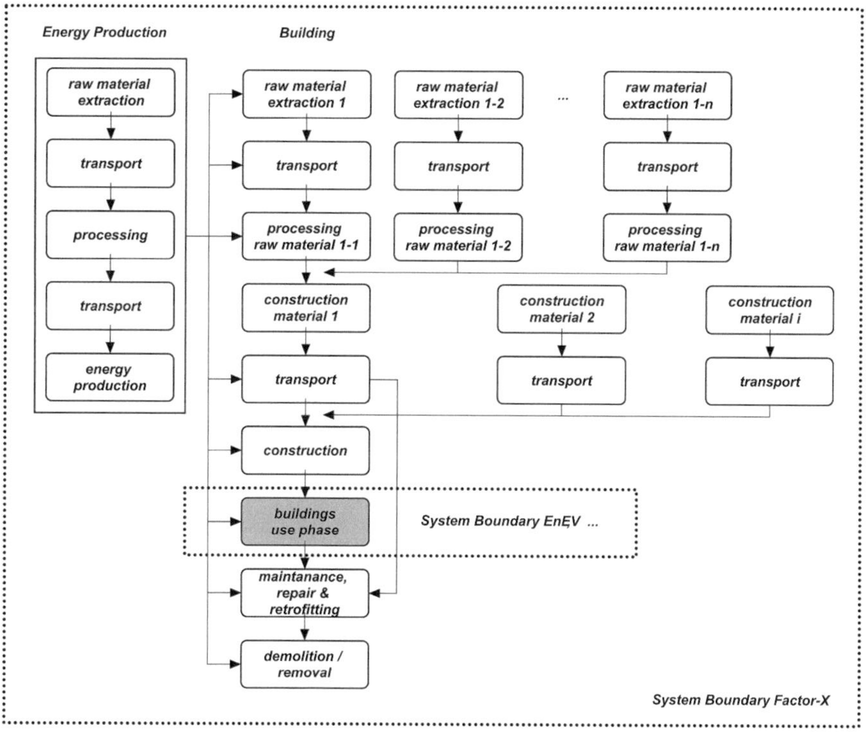

Fig. 19.4 System boundary of factor-x (Source: Author)

balance—as required by the sustainable building process of the Deutschen Gesellschaft Nachhaltiges Bauen (DGNB) (German Society of Sustainable Construction) or the German Federal Institute for Research on Building, Urban Affairs and Spatial Development—an assessment is generally not drawn up until finalising the planning details or after the building work has been completed for the purpose of applying for a sustainability certificate.

Since the summer of 2016, the Factor X approach has been tested in two new build residential areas between Aachen and Düren. In Inden-Seeviertel and Eschweiler-Dürwiß two building sites each with approximately 40 structures are currently under construction. The only criteria (apart from the 2016 legally applicable Energy Saving Ordinance) is the application of a Factor 2. Over a defined time period of 50 years, the buildings may only consume 50% of the non-renewable energy and non-renewable resources and may only emit 50% of the greenhouse gases compared to conventional construction. The system boundaries were extended as shown in Fig. 19.4.

In order to make this method accessible to a wide user group, ranging from architects to interested building contractors, an Excel tool was developed to help estimate resource consumption involved in the main features of the house at a very early stage in the planning.

Fig. 19.5 Factor 4 House in Inden, Architecture by Prof. Jörg Wollenweber, Düsseldorf (Source: Jörg Wollenweber, unpublished illustration)

A Factor 4 house was also constructed. In the three resource categories non-renewable embodied energy, non-renewable raw material and greenhouse gases, this optimised building achieves a maximum consumption of 25% compared to a similar conventional building. The fundamental principles for resource-efficient construction, such as lightweight construction, abstaining from concrete materials, recycling-oriented building, the use of renewable construction materials and allowances for easy changes in usage through a modular and marketable design, were all implemented (Fig. 19.5).

All three projects have been or are being considered for an award from the Klimaexpo NRW (www.klimaexpo.nrw.de) as pioneering climate protection projects.

From January 2017 the Faktor X Agentur located at the indeland Entwicklungsgesellschaft in Düren will focus on promoting this holistic approach to climate and resource conservation on a regional and national scale.

References

DWD (2016) https://www.dwd.de/DE/leistungen/solarenergie/download/aktueller_jahresgang_einstrahlung.pdf?view=nasPublication&nn=16102. Accessed 29 Oct 2016

EU (2010) Directive 2010/31/EU of the European Parliament and of the Council of 19 May 2010 on the energy performance of buildings

EU (2012) Directive 2012/27/EU of the European Parliament and of the council of 25 October 2012 on energy efficiency

Herczeg M, McKinnon D, Milios L, Bakas I, Klaassens E, Svatikova K, Widerberg O (2014) Resource efficiency in the building sector, Final report to DG Environment. Rotterdam

UNEP (2011) Decoupling natural resource use and environmental impacts from economic growth, A Report of the Working Group on Decoupling to the International Resource Panel. Fischer-Kowalski M, Swilling M, von Weizsäcker EU, Ren Y, Moriguchi Y, Crane W, Krausmann F, Eisenmenger N, Giljum S, Hennicke P, Romero Lankao P, Siriban Manalang A, Sewerin S

Chapter 20
Eco Efficiency and Circular Production: Cases from the Netherlands' Eastern Region

Frank A.G. den Butter and Harry A.A.M. Webers

Abstract This chapter discusses three case studies of circular production in the eastern region of the Netherlands, where the Dutch Hanse tradition of craftsmanship and family business links up with modernity and care for the environment. The main goal of the firms in our case studies is to prevent waste and to use recycled and non-fossil resources in their own production processes. Moreover, attention is paid to the circularity of the whole supply chain by acting as orchestrator of that chain. Government should support these firms by facilitating knowledge spill-overs, providing proper education and internalising the positive externalities that the strategies of these firms bring about in terms of the circular economy and sustainability.

Keywords Hanse tradition • Circular economy • Sustainability • Eco efficiency • Waste reduction • Supply chain orchestration • Recycling and reuse

20.1 Introduction

The report 'Limits to Growth', published in 1972 by the Club of Rome, created awareness that further exponential economic and population growth with finite resources would eventually be harmful to welfare. The vision of the Brundtland Commission in 1987 shifted the focus to the environment and emphasised the trade-off between current and future economic growth, or alternatively the trade-off between people, planet and profit. The argument is that environmental degradation should be stopped in order to make (economic) development sustainable. The

F.A.G. den Butter (✉)
Department of Economics, VU University Amsterdam, RITM and Tinbergen Institute, De Boelelaan 1105, 1081 HV, Amsterdam, The Netherlands
e-mail: f.a.g.den.butter@vu.nl

H.A.A.M. Webers
Witteveen-Bos Consulting Engineers, The New Hansa, Onder de Linden 3, Deventer 7411 SK, The Netherlands
e-mail: haam.webers@gmail.com

H. Lehmann (ed.), *Factor X*, Eco-Efficiency in Industry and Science 32,
https://doi.org/10.1007/978-3-319-50079-9_20

Brundtland Commission (1987, Ch 2) defined sustainable development as '*development that meets the needs of the present without compromising the ability of future generations to meet their own needs*'. This definition introduced intergenerational equity as a point of concern. Later the concept of sustainability was operationalised in various policy proposals and treaties. For instance, on 25 September 2015, the United Nations launched a new Sustainable Development Agenda with 17 Sustainable Development Goals to end poverty, protect the planet and ensure peace and prosperity for all. To reach these ambitious goals all groups in society need to take responsibility: government, the private sector, industry and civil society.

A practical approach to bring about sustainability is to aim for circularity in the supply chain. Here circularity means that all resources used in the production process are recycled and used again, so that no waste is left to pollute and degrade the environment. At the macro level such circularity would lead to a 'circular economy'.

In September 2016, the Dutch Government published the report 'Rijksbrede Programma Circulaire Economie', which sets forth the aim that by the year 2050 the Dutch economy should be completely circular (Dutch Ministry of Infrastructure and Environment 2016). It is an ambitious policy perspective for a future-proof and sustainable economy and a liveable planet. The ambition of the Dutch Government is to reduce the use of natural resources (minerals, fossils and metals) by 50% in the year 2030. The circular economy is regarded to provide a solid basis for resource efficiency, increasing the lifetime of products and materials and fostering ecological values. The Dutch government is not unique in its ambition for a circular economy. The European Commission (2015), the United Nations (2015), Scotland and Finland have made similar policy plans for circularity in their economies (Scottish Government 2016; SITRA Studies 100 2015).

This chapter discusses how circular production is implemented in the eastern region of the Netherlands. The remainder of the chapter is organised as follows. The next section provides an overview of economic activity in this part of the Netherlands, with special emphasis on eight centuries of the regional Hanse tradition of craftsmanship and trade. Section 20.3 outlines a theoretical framework, inspired by neoclassical growth theory, for how sustainability, eco efficiency and the circular economy can function from the perspective of supply chain management. In Sects. 20.4, 20.5 and 20.6 these concepts are illustrated using practical cases of the production processes in three companies in the eastern region of the Netherlands. Section 20.7 concludes.

20.2 Economy in the Eastern Region of the Netherlands: Overview

Many cities on the banks of the river IJssel in the eastern part of The Netherlands played an active role in the German Hanse, the history of which can be traced back to 1241 when the first trade contract between Hamburg and Lübeck was signed. The

Hanse was a commercial federation of cities that dominated trade in the north of Western Europe for almost four centuries, and it still exists today. The river IJssel was an important water connection between the German hinterland and the Southern Sea, providing access to both the North Sea and the Baltic Sea. This geographical location stimulated 20 Dutch Hanse cities to act as intermediaries in the trade between the German and Baltic hinterland and the rest of the world (see Webers 2015).

In modern terms one could argue that the Dutch Hanse cities were intermediaries between the Rhineland model and the Anglo-Saxon model of trade. A characteristic of the Rhineland model, as opposed to the Anglo-Saxon model, is the use of informal contacts, i.e., mutual trust through long lasting trade relationships and intrinsic motivation to cooperate, which played a major role. This is reminiscent of the famous 'polder model', that reflects the Dutch culture of open dialogue, interactive consultation, deliberation and participation. In this way of organising transactions and obtaining compromises between various interests, the implementation costs of effectuating decisions remain low. In this cultural setting trust based informal contracts may replace costly formal contracts such as those used traditionally in the Anglo-Saxon world (see den Butter and Mosch 2003; den Butter 2011).

This Hanse history, coupled with a tradition of craftsmanship in the regional industry which includes family businesses with a long term focus on continuity and decision making, has prompted the ambition of the eastern Netherlands to become a circular economy hotspot in Europe. The innovation profile of the region comprises two priorities: smart and sustainable industries (smart manufacturing and materials, smart food production and biobased production) and concepts for a healthy life (healthy technology and delivery systems, healthy brain and personalised health and nutrition).

In the eastern Netherlands region three initiatives have been undertaken:

1. The foundation kiEMT, which was founded in 2005 and has 230 participants from industry, academic institutions and government, represents the innovation ecosystem EMT Valley (Energy and Environmental Technology Valley Eastern Netherlands). The EMT-sector comprises 1,400 companies, 30,000 jobs, 20,000 students; it has 2.2 billion Euros of added value, 73% of the companies are innovative and 76% are internationally oriented. EMT supports start-ups, cross overs and clusters, international collaboration (Nordrhein Westfalen) and has as its core activities: energy transition as well as the biobased and circular economy (Foundation kiEMT, VNO NCW 2015, 2016).
2. The Cleantech Region Stedendriehoek consists of seven municipalities (Apeldoorn, Brummen, Deventer, Epe, Lochem, Voorst, Zutphen) with a total of 400,000 inhabitants. The Strategic Board of the Cleantech Region chose cleantech as a major priority, setting forth the goal to be energy neutral in the year 2030.
3. Project Circles (Cluster for Innovation, Redesign of Value Chains and other Circular Lifestyle Economy Solutions) is a Triple Helix collaboration with 12 partners. In this project 65 business meetings are programmed with the support

of 146 companies. The plan is to make 35 business cases including the three circular economy pilot project cases reported in Sects. 20.4, 20.5 and 20.6.

20.3 Theoretical Framework

This section contains a stylized framework that illustrates the linkages between the concepts of eco efficiency, circular production and sustainability. The framework is used in the subsequent sections dealing with the cases to clarify which concepts are relevant and to what extent the firm in each case aims at achieving eco efficiency, circularity and sustainability.

We start with the traditional neo-classical production function from the theory of economic growth where volume y is produced with factor inputs labour (L), capital (K) and environment—or in practical cases energy—(E), which is included as an additional factor input to account for environmental issues in the analysis (see e.g. Hallegatte et al. 2012; den Butter and Hofkes 2001, 2006; Smulders and Withagen 2012):

$$y = af\left(bK, cL, dE\right)$$

Here a, b, c and d represent various forms of technical progress, induced for instance by innovation. The parameter a is neutral technical progress; an increase in a implies that the whole production process becomes more efficient such that a higher volume of production y is achieved at the same factor inputs K, L, and E. The parameter b represents capital saving technical progress, c is labour saving technical progress and d is environment (or energy) saving technical progress. In the case of environmental issues, it is especially this last form of technical progress that is relevant. When, for instance, d increases due to a new, innovative environment saving technology, the same volume of y can be produced using less E. The consequence of this increase in *eco efficiency*, d, is that the *environment or energy share in production*, E/y, decreases. In other words, an increase in eco efficiency leads to a lower impact on the environment. However, enhancing eco efficiency is not the only way to decrease the environmental impact of production. When in the production process the use of E is reduced due to substitution of K and/or L, E/y also deceases without a change in eco efficiency. Such substitution of factor inputs leading to a cleaner environment (or to less energy use) can be brought about by a relative change in the prices of the factor inputs, for instance by a tax shift from labour to environment/energy. Such a tax shift may, by the way, also imply that more effort is given to environment saving technical progress, d, and less to labour saving technical progress, c.

In order to include the concepts of circular economy and sustainability in the framework, we have to make a distinction between extractive use of the environment and renewable use of the environment (see den Butter and Hofkes 1995 for a related approach). To keep the discussion simple we can think of the use of fossil energy, E_f, and of renewable energy, E_r. Moreover, we must account for waste, W, caused by production and for the reuse of this waste in the production process, W_i, which is considered as an additional factor input. Now the framework reads:

$$y = a\,f\left(bK,\, cL,\, d_1E_f,\, d_2E_r,\, d_3W_i\right), \tag{20.1}$$

and

$$W = w\,y \tag{20.2}$$

Here the parameters d_1 and d_2 represent the fossil energy saving technical progress and the renewable energy saving technical progress, respectively. At the macro level these two forms of technical progress may, however, not be independent, as the use of renewable energy may require use of the fossil energy, or more broadly, of non-renewable resources in order to enable production of renewable energy through investing in, e.g., windmills, solar panels and related infrastructure. A shift from fossil energy to renewable energy can be brought about by price changes where fossil energy is made more expensive through taxation and renewable energy is made cheaper through direct subsidies or through subsidising innovations in the production and use (d_2) of renewable energy. However, in both cases there is the danger of the green paradox, as the owners of sources of fossil energy understand that prices of renewable energy will go down when subsidized technology progresses (Sinn 2015; van der Ploeg and Withagen 2015). For this reason fossil fuel firms will lower the price of fossil energy and spur demand, which may lead to more instead of less pollution.

The parameter w in the above framework represents the extent to which production of y brings about waste. In order to make production more sustainable and circular, w can be decreased by extending the life of products, through reuse and making products easier to repair and by restrictions on packing and waste disposal. Moreover, the use of waste in production can be made more efficient, which translates in the framework as an increase in parameter d_3. Examples of this include innovations in recycling technology and employing production processes which use recycled materials. It can be viewed as a special case of enhancing eco efficiency

The framework above allows for the following taxonomy of circularity and sustainability. W_i/W can be regarded as the *degree of circularity*. If $W_i/W = 1$, i.e., $W_i = W$, we have a *fully circular production process*, and a *full-fledged circular economy* at the macro level. *Full sustainability* is obtained when both $W_i = W$ and $E_f = 0$. Moreover, the framework indicates in what way and through which instruments and changes in the production chain a transition to these ideal states of circularity and sustainability can be achieved.

20.4 Case 1 Royal Auping Deventer

The mattress and bed manufacturer and retailer Royal Auping in Deventer (province of Overijssel) is a family business founded by Johannes Auping in 1888. The company started producing high quality and durable mattresses and beds for hospitals with an innovative bed base of woven steel in a metal frame instead of straw ('Auping's steel health mattress'). Nowadays it is the largest Dutch sleep comfort brand and manufacturer which serves the high quality segment of the market with the slogan '*Auping nights, Better days*'. It has 300 employees and 130 flex workers with an annual turnover of 73 million Euros. In the last few years Auping has invested over 35 million Euros in state of the art sustainable production facilities, product and process innovation and in retail and digital capabilities. Auping has marketing and sales subsidiaries in Germany, Belgium and Denmark, and worldwide the company serves over 15 other markets through exclusive distributors. The company strives to develop a 'good & healthy' product design; an ambition that does not only relate to the look and quality of their mattresses and beds, but also to their desire to bring about a transition to circularity in their production process. To this end a new factory using cost-efficient, demand driven production processes was developed and realized in Deventer from 2013 till summer 2015. The redesign of the production process in the new factory caused total consumption of electricity to decrease by 45% and use of natural gas by 90% while maintaining the same product output levels.

By striving towards being passionate, transparent, innovative and better every day the main object of the production process in the new factory is similar to the sustainability definition of the Brundtland commission: '… *to meet the needs of the present without compromising the ability of future generations to meet their own needs*'. The idea is both to reduce the use of the environment in production ('eco efficiency') and increase the positive impact on the environment through reducing waste and recycling, which the company labels as 'eco effectiveness'.

In 2010, Auping published its Cradle to Cradle Vision 2020 and became a certified Cradle to Cradle company in 2012. The bed model 'Essential' was launched with the Cradle to Cradle Silver Award as the first sustainable bed model worldwide. As a practical indicator for how far the company has advanced on the way to full circularity it has set up the Auping Take Back System for mattresses, which has the target of 90% recycling. Auping takes old mattresses back on delivery of new ones and ensures that they are recycled by an external company (RetourMatras) and that they do not disappear into an incinerator (which is what happens every year to an estimated 1.2 million mattresses in the Netherlands). The return system does not only take back the company's own mattresses, but also other brands, supplying 300,000 kilograms of raw, recycled materials. These materials are *inter alia* processed into cleaning cloths, judo mats or insulation. The return system was praised by the Deputy Minister of Infrastructure and Environment and the Minister of Economic Affairs in a letter to the Dutch Parliament, calling Auping an '*outstanding example of circular thinking*' (Dijkstra and Kamp 2016, p. 8).

Auping commits itself to the goal of a sustainable business model (Cradle to Cradle Gold) by 2020. By developing and producing manufactured goods that allow upcycling at the highest level (no waste but upcycle), full use of renewable energy (no fossil fuels used) and respect for natural systems (water). With respect to people as part of the people-planet-profit concept the aim is optimal working and social conditions for labour employment for all stakeholders in the supply chain. Considering four streams of the corporate circular strategy, namely (i) circular operations and eliminating waste, (ii) energy savings, (iii) use of renewable energy and (iv) radical transparency in the supply chain, Auping can be regarded as a frontrunner in the transition to both full circularity and sustainability.

The company is currently developing a Circular Economy Hub at their head office in Deventer. Auping has realised the importance of their Deventer roots in the Hanse tradition and made the production process LEAN and Cradle to Cradle in order to be more competitive and to become future-proof. The aim of making the production process fully circular also contributes to the profitability of the company and to its long run continuity. Moreover, Auping seeks cooperation with partners upstream in the supply chain that share their standards and values of socially responsible production. For example, instead of importing aluminium from natural resources in Vietnam, Auping uses recycled aluminium from local suppliers in the Netherlands. It appears that this does not only lead to lower use of natural resources and lower use of fossil fuels for transport, but also to more flexibility in the production process. In this way it even amounts to a cost reduction as well as less risk in the supply chain.

From the perspective of the model described in Sect. 20.3 Auping seeks to enhance circularity in its own production process by (1) extending the life of beds and mattresses, which involves a decline in w, (2) using recycled waste in the production, which implies substitution in (25.1) from E_f to E_r and W_i, and (3) innovating in ability to use recycled material in the production process, which is an increase in d_3. In order to contribute to circularity in the economy Auping is also selective in choosing its upstream suppliers in the supply chain.

20.5 Case 2 SolidPack Loenen

In 1630, paper manufacturer, Vincent Schoonman, who was born in Mulhausen (Germany), built the first paper mill in the small city of Eerbeek (province of Gelderland) because of the excellent water quality of the springs of the Veluwe Massif. Nowadays, most of the Dutch paper and board factories are concentrated in this area.

The total recycling rate in 2013 in the Netherlands was 70.5%, which is relatively high compared to the European average (65.3%). For glass recycling the rate was 78.8% that same year and for paper and board it reached as high as 88.8% thanks to national collection systems and waste logistics (data from Eurostat 2013). The main success factor is the separation of waste at source. The Dutch paper and board industry is well known for its efforts to reduce water and energy consumption in the last decades. Over the last century the production technologies used in the paper and board industry have become fully developed and mature and because of this they are relatively efficient. Yet, the development and introduction of new eco efficient technologies is expensive so that the risks in investing in these technologies are to be minimised. The branch organisation of the paper industry (VNP: Koninklijke Vereniging Nederlandse Papier en Karton Fabrieken) seeks to reduce the energy consumption in its entire production chain by 50% in 2020.

One of the innovative board companies is the privately owned SolidPack, founded in 1921 in Loenen with 178 employees (170 FTE) and a turnover of 55 million Euros (8–10% earnings before interest, taxes, depreciation, and amortization). SolidPack is an excellent example of a circular 24/7 company (365 days) avant la lettre, as it uses mainly recycled paper as basic raw material together with some virgin fibres and chemicals for the production of shelf and retail-ready packaging for worldwide applications. SolidPack produces high quality food-grade packaging solutions with high added value for specific applications. The solid board boxes cannot yet be reused but are recycled in the traditional way together with other board, corrugated, newsprint and magazines (for which the recycling rate in the Netherlands is very high). Whereas offices are becoming more digital and more paperless, the use of packaging by web shops is actually increasing. This poses a problem for the availability of recycled paper and board as input for the production. Therefore, SolidPack seeks to replace this raw material in the production process with inputs which are also non-fossil such as grass fibres ('grassbox') and other bio-based materials.

The company's circular strategy is based on the following priorities: zero waste, the reduction of water consumption and the reduction of its carbon footprint. The company searches to use alternative minerals for its own biological waste water treatment plant. In a 1.5 year pilot project SolidPack searches more specifically to use both phosphorus and nitrogen from municipal waste water instead of phosphorus from phosphoric acid and nitrogen from urea. To be successful these alternative mineral sources must be located in the neighbourhood of SolidPack. Moreover the company searches for new ways of processing 2000 tons of plastic (reject from waste paper defribration) into products with added value. However, for SolidPack the Achilles heel of sustainability is the use of fossil fuels and energy consumption. Therefore, the aim is to substitute fossil fuels with renewable energy such as bio-fermentation.

SolidPack collaborates successfully not only with the Water Authority Vallei and Veluwe in the field of industrial waste water treatment but also with the National Forestry Commission in the field of alternative raw material (grass). Company facilities (buildings, appliances and the waste water treatment plant) are shared with others to exploit economies of scale and to decrease the cost base. Through these kinds of new coalitions the various partners contribute to make the total value chain more circular.

In the province of Gelderland, ten paper and board companies with 1372 employees have an total annual energy consumption of 11 million Gigajoules (Quickscan

Gelderse papierindustrie. Duurzamer door Samenwerken, October 2010). SolidPack is currently exploring possibilities to reuse its steam and waste heat from the waste-water for municipal applications. In this way it also intends to support the Cleantech Region of Stedendriehoek to become energy neutral in 2030.

In terms of the model outlined in Sect. 20.3, SolidPack is substituting E_f for E_r in its production process and trying to enhance d_3, given the fact that the company already uses, with recycled paper and board, a considerable amount of 'waste' (W_i) as input. Moreover, efforts could be made to make the packaging more durable and reusable in order to decrease w, contributing to downstream circularity and sustainability. The plan to bring about a circular solution for the large amount of plastic as residual of the production process would also translate as a decrease of w in terms of the model.

20.6 Case 3 Interface Scherpenzeel

In 1956 the modular Heuga carpet tile was developed by Petrus Johannes van Heugten, who until then had produced saddle pads and felt hand warmers. For many years he experimented with waste material and through innovative product development he was able to use the waste material for the production of carpet tiles. In 1987 Heuga was taken over by the American company Interface. The story of Interface began in 1973, when founder Ray Anderson, who passed away in 2011, realised that at the time the office environment was changing dramatically and there was a need for flexible floor covering. Anderson's vision was: '…, *showing it's possible to create a better world, being restorative by the power of our influence…*' (see Anderson 1998, 2009). Since its takeover of Heuga, Interface has become the worldwide leader in modular flooring. With a market share of one third of the global carpet market, Interface leads the market in terms of design, durability and innovation. Worldwide Interface has 3500 employees in 110 countries (5 production sites) and an annual turnover of 900 million Euros. The Dutch site in Scherpenzeel (in the province of Gelderland) acts as the European headquarters for production and distribution. It has 315 employees, a turnover of 353 million Euros and net profits of 26 million Euros.

In 1994, Interface was one of the first companies to reverse its strategy towards sustainability. The company promises to be fully sustainable by 2020. The strategy is called Mission Zero® and it is challenging all parts of the business, throughout the supply chain. The aim is *'to become restorative by eliminating any negative impact our company may have on the environment by the year 2020'*. The company's sustainability strategy is based on learning from nature (biomimicry) (see Benyus 1998). This integrated approach following the way ecosystems function leads to circular business models, corporate social responsibility, green energy, sustainable production and clean and efficient transport. TacTiles® are an appealing example of biomimicry being put to use. TacTiles® are innovative adhesive-free installation system based on the feet of a gecko with zero VOC (Volatile Organic Compound) emissions. Interface aims not only to be fully sustainable by 2020 but even to be restorative, putting back more than the company takes out from the environment and the community. A good example of the inclusive circular business model is the Net-Works™ initiative, which collects discarded nylon fishing nets in the Philippines to be processed as a raw material for new carpet tiles. Thus, the Net-Works™ initiative is restorative because it supports social communities and makes a positive contribution to society.

For Interface co-innovation is a way of working that connects ideas, people and resources, removing barriers to progress and enabling to achieving goals together. In order to realise the goal to make the whole supply chain circular and sustainable Interface often plays the role of orchestrator of the supply chain.

December 2015 data from the company show that this strategy to make the production process fully circular and contribute to sustainability has already been very effective; Mission Zero® really works. Indicators of success include (compared to baseline 1996): zero waste to landfill, 98% reduction of direct GHG emissions, 95% renewable energy use, 50% reduction in transportation, 50% of the materials are recycled or bio-based and 50% reduction in product carbon footprint. In order to achieve these reductions Interface developed various circular business concepts and inclusive business models such as: buy with full service contract and return, leasing of carpet tiles, Tile ReUse, Tile Recycling (ReEntry 2.0), Tile ReEnergy, Tile Reclaim, Cool Tile (CoolCarpet®), Tile Take-back Guarantee, I Owe You, TacTiles® and in the future, subscription systems.

According to Interface the sustainability strategy is financing itself through the improvement of non-labour resource efficiency, investments in sustainable inputs and the commercialisation of competitive advantages. Customers become Interface ambassadors for life.

In relation to the modelling framework of Sect. 20.3 it seems that the strategy of Interface uses aspects of circularity mainly to reach the final aim of sustainability. The company realises that, in accordance with our taxonomy, circularity is a necessary albeit not a sufficient condition for sustainability. From that perspective, making not only its own production process but also the whole supply chain more circular is part of Interface's strategy. The use of recycled materials and efforts to reuse the tiles and prolong their lives by licensing, maintenance contracts as well as reclaim and take back programs can be seen as ways to reduce w and cause substitution from E_f to W_i in the production process. But the main part of the strategy is directed at substitution from E_f to E_r in production. Moreover, by making the employees of Interface intrinsically motivated to comply with the company's strategy and philosophy, the efficiency of labour (c) as factor input is enhanced.

20.7 Conclusion

The circular economy is a way to reduce the use of raw materials and significantly improve ecological values. The minimization of waste and fossil fuel use, prolonging the life of products and the use of waste as an input factor in the production process are all ways in which the economy can become circular. A transition to a circular economy, however, involves more than just less waste (Janssen and Stegeman 2016). This is even crucial for an economy that aims to be both circular and sustainable. To achieve a circular and sustainable economy we need to re-think the current business, organization and commercial models (van Arkel and Jonker 2012).

This chapter describes three cases of companies in the eastern region of the Netherlands which are frontrunners of the circular economy. They have indeed thoroughly changed their business strategy and the organization of the supply chain. The first two cases, Auping and SolidPack, focus on making the production process more circular by reducing waste and using recycled material instead of virgin material as input for production. The fact that sustainability of production is increased can be thought of as a side-effect of this focus. Moreover, in order to make the full supply chain more circular these companies take the role of orchestrators of the production chain. Control over the production chain is also part of the strategy of Interface. However, unlike the prior two companies, Interface focuses directly on sustainability, and circularity is seen as a means to comply with the Mission Zero® aim for full sustainability in 2020.

Making the production process and supply chain more circular and sustainable, and thereby making the economy more eco efficient, is also beneficial to welfare at the macro level of the economy. Thus, the government has a role to play in promoting the transition to circularity and sustainability. According to the Dutch Ministry of Infrastructure and Environment (2016) the government can shape the transition through five instruments or interventions (see also Janssen and Stegeman 2016). The first intervention is the removal of legislation that hinders the development of a circular economy where innovation is required. The second intervention is to implement smart market incentives. Market incentives increase the demand for circular products (such as bio-based and recycled materials) and encourage circular innovations and business models. The third intervention is to offer support and insight in investments in circular products and services. These have a different risk profile than traditional products and services, and a better understanding of the costs and benefits could stimulate further investment. The fourth intervention is organising knowledge spill-overs. The last intervention is to promote international partnerships and make arrangements that create the right conditions for the circular economy. It seems that the initiatives taken in the eastern region of the Netherlands, discussed in Sect. 20.2, mainly relate to these latter forms of intervention. In fact, the theory of public sector economics provides us with a main reason for government intervention: the positive externalities that the pioneering strategies of front running companies in circularity bring about should be internalised. This should be an underlying motive of the five interventions discussed above.

References

Anderson RC (1998) Mid-course correction, towards a sustainable enterprise: the interface model. Chelsea Green Publishing Company, White River Junction
Anderson RC (2009) Confessions of a radical industrialist. St Martin's Press, New York
Benyus J (1998) Biomimicry: innovation inspired by nature. Harper Collins Publishers, New York
Brundtland Commission (1987) Our common future. Oxford University Press, Oxford
den Butter FAG (2011) The industrial organisation of economic policy preparation in The Netherlands. In: Lentsch J, Weingart P (eds) The politics of scientific advice; institutional design for quality assurance. Cambridge University Press, Cambridge, pp 177–214
den Butter FAG, Hofkes MW (1995) Sustainable development with extractive and non-extractive use of the environment in production. Environ Resour Econ 6:341–358
den Butter FAG, Hofkes MW (2001) Endogenous technology and environmental quality in economic models. Int J Environ Technol Manag 1:32–44
den Butter FAG, Hofkes MW (2006) A neo-classical economics view on technological transitions. In: Olsthoorn X, Wieczorek AJ (eds) Understanding industrial transformation: views from different disciplines. Springer, Dordrecht, pp 141–162
den Butter FAG, Mosch RHJ (2003) The Dutch miracle: institutions, networks and trust. J Inst Theor Econ 159:362–391
Dijkstra SAM, Kamp HGJ (2016) Rijksbrede programma circulaire economie. Letter to the parliament, 14 September 2016
Dutch Ministry of Infrastructure and Environment (2016) Nederland circulair in 2050. The Hague, September 2016
European Commission (2015) Closing the loop—an EU action plan for the circular economy. Brussels, December 2015
Eurostat (2013) http://www.ec.europa.eu/eurostat
Foundation kiEMT, VNO NCW (2015) EFRO 2015—cluster application CIRCLES cluster for innovation redesign of value chains and other circular lifestyle economy solutions. Arnhem
Foundation kiEMT, VNO NCW (2016) Evaluatierapport regioproject circulaire economie. Arnhem, March 2016
Hallegatte S, Heal G, Fay M, Treguer D (2012) From growth to green growth—a framework. NBER Working Paper 17841
Janssen T, Stegeman H (2016) Circulaire economie vraagt om meer dan een ambitieus overheidsprogramma. Me Judice, 5 October 2016
Scottish Government (2016) Making things last, a circular economy strategy for Scotland. Edinburgh, February 2016
Sinn HW (2015) The green paradox: a supply-side view of the climate problem. Rev Environ Econ Policy 9(2):239–245
SITRA Studies 100 (2015) The opportunities of a circular economy for Finland. Helsinki, Oktober 2015
Smulders S, Withagen C (2012) Green growth. Lessons from growth theory. World Bank Policy Research Working Paper, WPS6230, Washington
United Nations (2015) New sustainable development agenda. UN.org, 25 September 2015
van Arkel G, Jonker J (2012) Sustainable innovation. Inspiring and accelerating, Working Paper, Nijmegen, June 2012
van der Ploeg F, Withagen C (2015) Global warming and the green paradox: a review of adverse effects of climate policy. Rev Environ Econ Policy 9(2):285–303
Webers HAAM (2015) Ambition document. The New Hanze: smart cities. Smart factories (Past results are a guarantee for the future!). Deventer, May 2015

Chapter 21
An Approach to Identify Resource Patterns on a Neighborhood Level

Magnus Österbring, Leonardo Rosado, Holger Wallbaum, and Paul Gontia

Abstract Resource stock and flow accounting has been increasingly applied on a country and city level. This paper elaborates on the possibilities to conduct a resource flow analysis even on a neighborhood/district level to identify, e.g., resource use related characteristics. Such resource patterns can be used to identify areas in different parts of a country or even larger geographical regions that are characterized by comparable types and amounts of resources entering the system boundaries. This identification might lead to archetypes that will help in deriving and applying policies, resource optimization strategies, etc. The approach presented in this paper stems from experiences made in the field of building stock modelling where archetypes are frequently used to characterize building stocks of a city or even a portfolio within a city. Furthermore, this approach is applied to develop energy and climate strategies for cities, renovation and maintenance strategies for real estate owners including investment planning. This paper reveals the shortcomings as well as the possibilities of resource pattern identification on a neighborhood/district level and closes with an outlook of necessary next steps to improve the quality of such an approach and increase the potential to use this approach as a strategic instrument.

Keywords Neighborhood • Building stock model • Material flow accounting • Material stock

21.1 Introduction

Top-down Material Flow accounting (MFA) methodologies have been widely used to account material flows in countries using the Eurostat Economy-Wide methodology (Eurostat 2001) and several adaptations to urban areas have been made under the urban metabolism research topic (Kennedy et al. 2011), but so far no one has

---

M. Österbring (✉) • L. Rosado • H. Wallbaum • P. Gontia
Department of Civil and Environmental Engineering, Chalmers University of Technology, SE 412 96 Gothenburg, Sweden
e-mail: magnus.osterbring@chalmers.se

H. Lehmann (ed.), *Factor X*, Eco-Efficiency in Industry and Science 32,
https://doi.org/10.1007/978-3-319-50079-9_21

attempted to conduct it at the neighborhood/district level (e.g. smallest administrative subdivision of a city) in a holistic way. Most of the Economy-Wide MFA adaptations to the urban level provide highly aggregated results for the material flows in material types (e.g. Niza et al. 2009; Barles 2009) and very few examples were able to allocate the flows to the needs of the specific economic sectors. Recently, Rosado et al. (2014) managed to develop an urban MFA model that bridges this gap and provides a detailed description for urban areas for 1000 product types, distributing them by more than 100 economic sectors and describing the flows in 28 material types, which covers the full techno-economic study in question.

The use of standardized statistical data such as international trade datasets, transport data and others provides also an opportunity to understand estimation errors and associated uncertainties in the results. In 2015 and 2016, two studies provided useful insights on these estimation errors (Schwab et al. 2016; Patrício et al. 2015), and despite increased error estimation from the country to the urban level, the uncertainty is still acceptable (Patrício et al. 2015).

In order to estimate material flows at the neighborhood level, top-down estimations need to be used with care due to the very likely high level of uncertainty. In that sense, the use of bottom-up techniques, such as the ones used for building stock modelling (BSM) can be useful to support the urban MFA techniques. Identifying types of districts based on drivers of consumption of resources might be a good compromise between the availability of urban MFA data at a higher level and the need to estimate neighborhood MFA.

21.2 Status and Development of Building Stock Modelling

To address the challenges of the stakeholders involved in the transformation of the building-stock, BSM have been used to calculate the energy demand of the existing building-stock (Ascione et al. 2013; Fabbri et al. 2012) as well as to evaluate its potential developments (Mata et al. 2013; Sartori et al. 2009). The bottom-up methods for describing and simulating the energy performance of building-stocks (Swan and Ugursal 2009; Kavgic et al. 2010) have evolved from being used on a country level to be applied at urban scale by incorporating geographical information systems (GIS). These bottom-up engineering models use a heat balance model to estimates the energy consumption for individual buildings. The buildings used as inputs to bottom-up models are defined by building properties such as geometry, U-values, climate data, indoor temperature and use of appliances. Thus, to apply a bottom-up engineering model requires detailed input data. As the scope of most research on building stocks are on a city or country level where data availability and computational time-constraints limits the use of detailed input data, representative buildings are often used to describe the building stock, where it is assumed that buildings with similar characteristics regarding year of construction, use of the building, type of heating system and building geometry will have similar values. Commonly, an archetype approach using representative theoretical buildings as described in

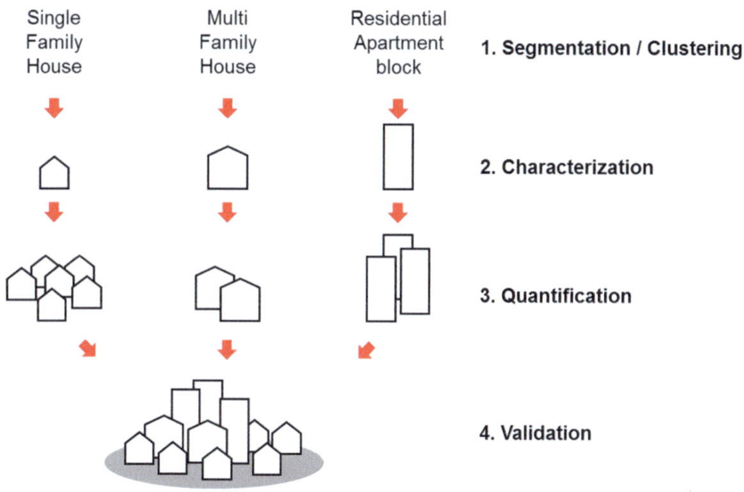

Fig. 21.1 Process of developing an archetype model (Source: reproduced from Mata 2013)

Fig. 21.1, often defined by construction year and the type or use of the building, is applied to represent all buildings with similar characteristics to allow for assessment of the entire stock.

In order to evaluate renovation potential of the existing building stock on an urban level, a local approach is needed to understand challenges and possibilities associated with the transformation of the building-stock. By incorporating building-specific information while considering the building in its setting, a more holistic view can be achieved. While the spatial resolution has increased and the energy models have become more advanced, the input used for these models have seen little development and are still largely based on using representative buildings which are used to scale results to the desired level of output. Such descriptions lose accuracy with increased spatial resolution and results are commonly only presented at aggregate levels for districts or entire cities and not on a building level (Reinhart and Davila 2016). The scaling of representative buildings to account for the energy performance of the entire stock is based on the assumption that buildings with a similar year of construction have a similar energy performance. The older the stock that the representative building is aiming to describe, the more problematic this assumption becomes as renovations have been applied to a varying degree and the energy performance may have changed significantly from its original state (Eriksson 1993; Aksoezen et al. 2015). Some models have gone further and have started incorporating building specific data, most commonly taking advantage of 3D city models based on LIDAR data (Steadman et al. 2014) or by analyzing differences in digital terrain models and digital elevation models (Mastrucci et al. 2015) as well as using building-specific data from energy performance certificates (EPC) to better describe the technical characteristics of individual buildings (Österbring et al. 2016). This development allows for the possibility of visualizing and communicating results on a building level as well as allowing results to be aggregated arbitrarily to suit communication with stakeholders that previously have not been reached. While the descriptions of buildings stocks are becoming more differentiated, the aspects of characterizing and classifying buildings as part of an archetype approach still holds value for other fields of research. Furthermore, aspects of visualization and communication that are enabled by using GIS in BSM can provide benefit when explaining the complex relationships of MFA in districts as well identifying districts with similar characteristics.

21.3 Lessons Learnt from Building Stock Modelling

Lessons learned from the use of archetypes in BSM should be considered before the methodology is applied in other fields. First, being a bottom-up approach, detailed input data is required which in turn brings up the questions of data availability, data access and data quality. Previous work for a differentiated bottom-up BSM for the city of Gothenburg (Österbring et al. 2016) showed that the EPC database and national property register could be retrieved for research purposes and linked to cadastral maps using GIS to provide a detailed description of the stock on a building level. While most technical characteristics could be directly linked to individual buildings, no database contains information on the U-value of the building components. To overcome this, an age-type classification was developed based on historical building regulations and a historic construction and architectural classification which was linked to individual buildings using information about year of construction and the economic extent of previous larger renovations.

In the process of gathering, processing and spatially matching the databases, questions regarding the quality of the EPC were highlighted. Issues with data quality in the EPC were due to both methodological and manual errors. The heated floor area given in the EPC proved to be problematic as it was not measured but rather derived from the living area in most cases where the suggested methodology was inadequate. Similarly, the most common manual error in the EPC was the number of stories, with 76 out 433 buildings in a sample set proved erroneous. Methods to overcome the methodological issues with deriving heated floor area from living area were developed (Mangold et al. 2015) and the correct number of stories had to be changed manually.

Second, with increased spatial resolution uncertainty becomes an issue. In energy performance calculations, it becomes difficult to capture the heterogeneity that naturally occurs within a building stock. One typical example is user behavior, which has a large impact on energy performance, is difficult to capture at this scale, and, while statistical calibration methods have been developed to quantify the uncertainties (Booth et al. 2012), the more common approach to overcome this issue is to aggregate results to a district or city level which reduces the heterogeneity. In this instance, the use of a differentiated description proves useful as aggregation can be done arbitrarily without having to conform to any imposed political or geographical division of city districts or zones.

Third, the use of GIS provides possibilities of communication and visualization of results as well as providing a way to consider local aspects by including census and socio-economic data. Spatiotemporal visualizations of BSM results can be used to support communication with urban planners, policy makers and property owners regarding the state of the building stock and coming renovation needs in an urban context as well to engage with the public. The use of local census and socio-economic data can provide further insight by considering a building in its setting. For example, local and socio-economic data in addition to the technical characteristics of the building-stock was used to visualize the socio-economic effects of renovation, rent increase and affordability (Mangold et al. 2016).

21.4 Material Stocks Accounting at Building Level

Accounting material stocks (MS) from the built environment also applies the same principals of the BSM bottom-up approach and uses typologies of buildings and infrastructures as representative units for the built environment. Bottom-up MS studies require access to two main different datasets: area and/or volume of the built environment components (buildings and infrastructure) and material intensity coefficients specific to each component. In terms of material composition of the built environment, there are very diverse and influencing factors like: local climate, historic and economic development, period of construction (which influences the material and construction technology), local resources availability (e.g. wood availability) and policies/regulations (e.g. in Zurich all public buildings need to apply 25% of recycled concrete aggregates). Therefore, the development of MS estimates relies on developing new Material Intensity databases, new methods (e.g. earth observation, image processing/machine learning) and data sources. Buildings have received a considerable amount of attention in recent studies (Kleemann et al. 2016; Ortlepp et al. 2015; Schebek et al. 2016) and cover information about area, type of use, components of the building, year of construction, spatial location, material categories types, reference quantity and database sources.

21.5 Resource Flow Accounting on a Neighborhood/District Level

In addition to accounting of stocks of resources on a neighborhood level, it is also crucial to provide an accounting of the flows, which will indicate the pressures put on resources to fulfil the needs of a specific part of a city.

There has been some discussion on the need to identify drivers of consumption on an urban level. The work by Kalmykova et al. (2015) shows that there are linkages between material consumption and the economy, as well as the lifestyle of the population. In that sense, understanding the drivers of consumption might be paramount to establishing reasonable estimation techniques for material flows. Among the variables that need to be included, it should be possible to identify socio-economic characteristics of the population and the characteristics of the economic activities placed in the neighborhoods/districts. This type of data is normally available for urban areas. However, work still needs to be done to identify the exact variables and indicators that need to be studied and what are the levels of correlation between them and the material flows.

The identification of the drivers of consumption will help in establishing typologies of neighborhoods that have the same types of material flows, hence allowing for their full accounting, in much the same way as the building stock modelling techniques. There are several examples in the literature on methodologies to identify typologies, among them some specific for the urban level (Schwarz 2010; Baum et al. 2005; Baum et al. 2007; Kladivo and Halás 2012; Creutzig et al. 2015), and a recent study for Gothenburg has preliminarily shown the existence of 4 main types of neighborhoods/districts that might have fundamentally different characteristics in terms of material consumption.

References

Aksoezen M et al (2015) Building age as an indicator for energy consumption. Energ Buildings 87:74–86

Ascione F et al (2013) Analysis and diagnosis of the energy performance of buildings and districts: methodology, validation and development of urban energy maps. Cities 35:270–283

Barles S (2009) Urban metabolism of Paris and its region. J Ind Ecol 13(6):898–913

Baum S et al (2005) Typologies of advantage and disadvantage: socio-economic outcomes in Australian metropolitan cities. Geogr Res 43(4):361–378

Baum S et al (2007) Considering regional socio-economic outcomes in non-metropolitan Australia: a typology building approach. Pap Reg Sci 86(2):261–286

Booth AT et al (2012) Handling uncertainty in housing stock models. Build Environ 48:35–47

Creutzig F et al (2015) Global typology of urban energy use and potentials for an urbanization mitigation wedge. P Natl Acad Sci USA 112(20):6283–6288

Eriksson B (1993) Energisparpotentialer i bostadsbeståndet: värmebalansmodell. In: Statens institut för byggnadsforskning. http://www.lth.se/fileadmin/byggnadsmaterial/BFR-publ/BFR_skriftutg-1985.pdf. Accessed Mar 2017

Eurostat (2001) Economy-wide material flow accounts and derived indicators. A methodological guide. Statistical Office of the European Union. http://ec.europa.eu/eurostat/documents/1798247/6191533/3-Economy-wide-material-flow-accounts...-A-methodological-guide-2001-edition.pdf/. Accessed Mar 2017

Fabbri K et al (2012) Heritage buildings and energy performance: mapping with GIS tools. Energ Buildings 48:137–145

Kalmykova Y et al (2015) Resource consumption drivers and pathways to reduction: economy, policy and lifestyle impact on material flows at the national and urban level. J Clean Prod 132:70–80

Kavgic M et al (2010) A review of bottom-up building stock models for energy consumption in the residential sector. Build Environ 45(7):1683–1697
Kennedy C et al (2011) The study of urban metabolism and its applications to urban planning and design. Environ Pollut 159(8–9):1965–1973
Kladivo P, Halás M (2012) Quality of life in an urban environment: a typology of urban units of Olomouc. Quaest Geogr 31(2):49–60
Kleemann F et al (2016) GIS-based analysis of Vienna's material stock in buildings. J Ind Ecol. doi:10.1111/jiec.12446
Mangold M et al (2015) Handling data uncertainties when using Swedish energy performance certificate data to describe energy usage in the building stock. Energ Buildings 102:328–336
Mangold M et al (2016) Socio-economic impact of renovation and energy retrofitting of the Gothenburg building stock. Energ Buildings 123:41–49
Mastrucci A et al (2015) GIS-based life cycle assessment of urban building stocks retrofitting-a bottom-up framework applied to Luxembourg. In: Johannsen VK et al (eds) Proceedings of EnviroInfo and ICT for sustainability 2015, Copenhagen, Denmark, September 2015. ASCR: Advances in Computer Science Research. Atlantis Press
Mata É (2013) Modelling energy conservation and CO2 mitigation in the European building stock. PhD dissertation, Chalmers University
Mata É et al (2013) A modelling strategy for energy, carbon, and cost assessments of building stocks. Energ Buildings 56:100–108
Niza S et al (2009) Urban metabolism: methodological advances in urban material flow accounting based on the Lisbon case study. J Ind Ecol 13(3):384–405
Ortlepp R et al (2015) Material stocks in Germany's non-domestic buildings: a new quantification method. Build Res Inf 44(8):840–862
Österbring M et al (2016) A differentiated description of building-stocks for a georeferenced urban bottom-up building-stock model. Energ Buildings 120:78–84
Patrício J et al (2015) Uncertainty in material flow analysis indicators at different spatial levels. J Ind Ecol 19(5):837–852
Reinhart CF, Davila CC (2016) Urban building energy modeling–a review of a nascent field. Build Environ 97:196–202
Rosado L et al (2014) A material flow accounting case study of the Lisbon metropolitan area using the urban metabolism analyst model. J Ind Ecol 18(1):84–101
Sartori I et al (2009) Energy demand in the Norwegian building stock: scenarios on potential reduction. Energ Policy 37(5):1614–1627
Schebek L et al (2016) Material stocks of the non-residential building sector: the case of the Rhine-main area. Resour Conserv Recy. doi:10.1016/j.resconrec.2016.06.001
Schwab O et al (2016) Quantitative evaluation of data quality in regional material flow analysis. J Ind Ecol. doi:10.1111/jiec.12490
Schwarz N (2010) Urban form revisited-selecting indicators for characterising European cities. Landsc Urban Plan 96(1):29–47
Steadman P et al (2014) Energy and urban built form: an empirical and statistical approach. Build Res Inf 42(1):17–31
Swan LG, Ugursal VI (2009) Modeling of end-use energy consumption in the residential sector: a review of modeling techniques. Renew Sust Energ Rev 13(8):1819–1835

Chapter 22
Strategic Business Examples from Finland: The Growth of the Smartup Industry

Tuuli Kaskinen, Satu Lähteenoja, Mikael Sokero, and Iiris Suomela

Abstract The transition to a carbon neutral and resource smart society is perhaps the largest societal transformation of our time. There is an urgent need to develop scalable solutions, which help individuals lead sustainable lives and radically reduce ecological footprints. The promising news is that there is a whole new branch of startups, small and medium-sized enterprises (SME's) and also larger companies creating new solutions to these challenges. We call these 'smartups'. Recently, smartups have emerged to provide new products and services particularly in the areas of housing, mobility and food. These three essential sectors are where most of our household material footprint originates from as well as the majority of monthly expenses. Therefore, solutions are needed dissolve the link, even a dependency, between the majority of household consumption and the root causes of emissions. Smartups are able to solve this dependency by providing cheaper, better and more ecological solutions in the areas that we spend most of our money in the first place.

This article explains what smartups are, focusing on their business models and what is needed to make them successful. We introduce two promising smartups in the food sector from Finland and analyse the reasons for their potential and success. We find that globally emerging smartups will only be successful if their added value is based both on making individual lives easier, happier and more comfortable while simultaneously enabling the transition to an ecologically sound lifestyle. Smartups can disrupt markets and create new ones, but the growth of the industry is dependent on determined support. Thus, we suggest that one of the best ways to support the development of smart concepts is to create strategic partnerships between smartups and the existing large companies.

T. Kaskinen • S. Lähteenoja (✉) • M. Sokero
Demos Helsinki, Töölönkatu 11A, 00100 Helsinki, Finland
e-mail: tuuli.kaskinen@demoshelsinki.fi; satu.lahteenoja@demoshelsinki.fi; mikael.sokero@demoshelsinki.fi

I. Suomela
University of Tampere, Kalevantie 4, 33100 Tampere, Finland
e-mail: iiris.suomela@gmail.com

H. Lehmann (ed.), *Factor X*, Eco-Efficiency in Industry and Science 32,
https://doi.org/10.1007/978-3-319-50079-9_22

Keywords Business models • Sustainable consumption and production • Consumer cleantech • Smartup • Food • Sharing • Optimisation • Circular economy • Dematerialisation

22.1 Introduction

In the Anthropocene era we are faced with vast, overarching and pressing global challenges. Population growth, unsustainable lifestyles, climate change, biodiversity loss and increasing uncertainty are characteristic examples of systemic challenges. Inherent to these challenges is unpredictability: the basic assumptions of how our ecosystems and natural systems operate will inevitably change if certain limits of growth are exceeded. For this reason we need rapid and scalable solutions that reduce the consumption of natural resources.

Increasing resource scarcity and the unsustainable consumption of resources at a household level is closely connected to three sectors: housing, mobility and food. These three sectors of consumer activity comprise approximately one half of all global greenhouse gas emissions (Faber et al. 2012). Moreover, the majority of household expenses is also in these three sectors (Gerstberger and Yaneva 2013). The signatory countries of the Paris climate agreement have agreed on long term ambition and emission reduction strategies to globally peak the greenhouse gas emissions as soon as possible and to reach the maximum 2° temperature target within the next decades (Paris climate agreement 2015). It is therefore fair to say that there are huge business opportunities in products and services that decrease household carbon footprint and, at the same time, make our lives easier, cheaper and happier. In general, climate change and Paris climate agreement set the framework for a global demand for solutions, some occurring trends such as digitalization, global economy, movement of people and labour, enable the solutions to prosper. In particular, the most urgently needed solutions with scalable impact in the aforementioned three sectors are enabled by technological development and digitalization.

This article begins with an introduction to the startup industry in the field of sustainable lifestyles and consumption choices. The companies in this sector are developing globally scalable business models, which are often based on innovative solutions to specific problems or challenges. We call these companies 'smartups'. Next, we introduce two interesting innovation cases from Finland. The first case is pulled oats, a vegan protein product, which aims to solve a number of problems at the same time: it is easy to use, healthy and replaces meat or soya based protein products so it can be produced from local ingredients. The second case is ResQ Club, a digital service that connects consumers with restaurants just before closing time, allowing the restaurants to sell meals at a discounted price that would otherwise go to waste. The final section concludes and presents directions for future research.

22.2 What Are Smartups?

In the mid-1990s business administration researchers Gary Hamel and C.K. Prahalad (1996) analysed ways in which companies find new markets. The study led to the recommendation that companies should focus on searching for undiscovered markets. Later, the same thought was popularised by W. Chan Kim and Renée Mauborgne (2015), who coined the concept of the "blue sea strategy". The core idea behind the "blue sea strategy" is that companies stand to benefit from the growth opportunities associated with finding entirely new markets with very little competition.

This is exactly what smartups are doing today. They are startups or sometimes large companies aiming at freeing consumers from the excessive use of natural resources. Their business models are based on finding value-creation opportunities by responding to major challenges posed by global phenomena, such as climate change or resource scarcity. Smartups use enabling drivers such as digitalization and user-friendliness to introduce innovative changes at the core of the business and tackle unsustainable practices and structures at their source. This is essential for taking on the pressing challenges pertaining to a sustainable future (Short et al. 2014). However, the level of ambition of business model innovations must be high and focused on maximising societal and environmental benefits, not just economic gain.

A shift towards more sustainable business models may be carried out using different strategies depending on whether the company aims to merely protect itself through risk and cost reduction or to genuinely redesign its business model for sustainable development. This distinguishes a defensive strategy from a proactive one. Although all business model innovations that deliver sustainability are welcomed, proactive innovation strategies tend to be the most impactful as they integrate environmental and social objectives with the core business logic of the company (Short et al. 2014). The strategies of smartups are usually proactive since business model innovation and sustainability are at the core of their business logic.

In recent decades, high-growth companies have focused largely on industries, where people spend about 10% or less of their annual income such as communications and entertainment. It is a fraction of the roughly 60% consumed by housing, mobility and food combined (Gerstberger and Yaneva 2013). In the EU, for example, people spend more than a third of their income on housing, a tenth on mobility and nearly a fifth on food. This amounts to tens of thousands of Euros in a year per household — and the price tag on commodities such as gasoline and meat products keeps on increasing as resources such as fossil fuels become more scarce.

Thus, there is an increasing interest in cheaper, smarter and more efficient ways of living, eating and transportation. This is the demand that smartups address through innovative and sustainable solutions. Convenience and consumer-centred design are at the core of what makes smartups successful in market segments other than those occupied by environmentally conscious consumers. They operate where our time, capital and natural resources intersect to provide quicker, cheaper and more efficient solutions to people's everyday problems.

22.3 Four Business Models Where New Smartup Businesses Arise

In addition to the three main value creation hotspots in consumer cleantech (living, mobility and food), Ritola et al. (2014) studied hundreds of up and coming smartup companies. The study identified *four main ways* in which companies drive down consumer dependency on natural resources and create new business activity. These *business models for smartups* are: *sharing*, *optimisation*, *circular economy* and *dematerialisation*. For concrete examples see the Table 22.1.

- *Sharing*. Sharing increases the utilisation rate of physical resources by dividing and allocating their use more efficiently. Sharing is the most commonly used model to improve resource smartness and is typically connected to platforms that enable it.
- *Optimisation*. Optimisation refers to improving energy and resource efficiency through, e.g., data management, smart home and metering applications.
- *Circular economy*. This category refers to circular supplies that provide renewable, recyclable or biodegradable resource inputs as well as solutions to eliminate material leakage and maximize the economic value of product return flows.
- *Dematerialisation and smart substitution*. Companies in this category replace resource intensive practices with new solutions. Examples include teleconferencing, virtual reality applications and replacing energy intensive animal proteins with less resource-intense produce (e.g. insect and vegetable proteins).

22.4 Two Stories of Successful Smartups that Drive Lifestyle Changes and Create New Markets

In addition to leveraging natural resources in more efficient ways, smartup companies must be inherently smart to survive on the market, in other words: accessible, easy and attractive to use. According to a famous estimation by Hart et al. (1990), a new service or practice must be approximately five times more functional and better to take over its predecessor in our daily life - that is, to take over the market.

The demand for smart solutions can be seen in the strong drive of smartup companies to develop top-level usability. Moreover, the focus of smartup companies on new resource efficiency models means that the innovative products often help consumers cut down their personal expenses as well as their consumption of natural resources. The customer-centred approach of smartups also means that they are able to offer more convenient and quicker solutions. Next, we will describe and analyse two case examples of smartups in the food sector.

Table 22.1 Examples of business models for smartups

Sharing	Optimisation	Circular economy	Dematerialisation and smart substitution
Digital marketplace platforms	Data management	Upgrading	Cloud mobile services
Carsharing	Smart systems for growing food	Tuning	Fundraising platforms
Grocery delivery	Need based optimisation	Turnkey solutions	Product innovation
Sharing goods	Smart products	Recycling nutrients	Substitution of ingredients
Lending	Ingredient optimisation	Recycling technologies	Product as a service
Ridesharing	No waste cooking	Repairing	Virtual services
Swapping	Smart homes and smart cities	Technical innovation	Gamification
P2P task sharing	Gamification	Product innovation	Digital services
Outsourcing	Source based optimisation	Water cleaning solutions	Online food shopping
Bikesharing	Digital services	Products made from recycled materials	Technical product innovation
Digital service innovation	Smart vehicles	Leasing	Virtual services for product development
Crowdfunding	Smart routing	Modular consumer electronics	Energy source substitution
Urban farming	Smart metering	Waste management	Turnkey solutions
Co-creation platforms	Energy management	Return and payback incentives	Solar energy
Space use optimisation	Carbon management	Resources recovery	Decentralised energy
Space rentals	Home automation		Reusable packaging and shipping materials
Workspace sharing	Turnkey solutions		All in one app for mobility
Social cooking	Food waste reuse		Compact housing solutions
			Product life extension

22.4.1 Smartups Drive Lifestyle Changes – The Story of Pulled Oats

Lifestyle changes and new business opportunities are entwined. On the one hand, business opportunities arise from changes in lifestyle. On the other hand, lifestyle changes create new business opportunities by creating entirely new markets.

Tubiello et al. (2014) estimate that greenhouse gas emissions from agriculture, forestry and fisheries have nearly doubled over the past 50 years and could increase an additional 30% by 2050 unless greater mitigation efforts are made. The environmental impact of animal based products is especially severe. A total of 70% of all arable land is used for the production of meat and other animal based products (Fiala 2008). Furthermore, agriculture uses approximately 70% of the world's available freshwater (Worldwatch institute 2014).

However, significant technological development and smartup innovations are also happening in the food sector. Twenty years ago there was virtually no profit made out of vegetarian protein products. Within the last couple of years, the situation has changed entirely: vegetarian products are gaining ground as more and more people become aware of the burden that red meat, in particular, has on their health as well as the environment (EVU 2016).

Pulled oats was developed collaboratively by serial entrepreneur, Maija Itkonen, and food scientist, Reetta Kivelä, who holds a PhD in Food Technology from the University of Helsinki. By combining the entrepreneurial skills of Itkonen and the scientific knowledge of Kivelä and Zhongqing Jiang, a food scientist who joined the team in the very beginning, their start-up company, Gold and Green Foods, was able to develop a fully vegan protein product that combines a small material footprint and principally Nordic ingredients with high nutritional value and convenience.

The company came up with an entirely new method of processing oat, broad beans and peas into a product with a texture akin to pulled pork or chicken. The product can be eaten straight out of the packet as there is no need to cook pulled oats. It can also be bought pre-seasoned, meaning that very minimal effort is required to transform pulled oats into a meal. This has made pulled oats an attractive alternative to meat also for non-vegetarians who may not consider buying other vegan products.

Design thinking and a thorough analysis of consumer preferences played an important part in the development of the business model of Gold and Green Foods. Their aim was to develop the "perfect protein"—something that would not require consumers to make compromises in terms of sustainability, health, taste or cost, all of which market research has proven to be very important to consumers. Rather than targeting the market segment of environmentally conscious consumers, pulled oats proactively transformed the market and generated demand among consumers who may never have tried vegan protein products before.

According to Itkonen—now CEO of Gold and Green Foods—the demand for pulled oats has spilled over into the sales of other vegetarian products, too. Despite the doubling of production capacity in just 3 months, in early 2016 pulled oats was so popular that many stores ran out within 1–2 h of the arrival of the delivery. This has lead consumers to choose other vegetarian options more readily than before. Without the mass popularity of pulled oats these individuals may never have considered buying vegetarian options in place of animal-based products. The vegetarian trend has also been fuelled by storekeepers who designate a vegetarian section of the store, meaning that replacements for pulled oats are easy to find as soon as one finds out that the pulled oats section is empty.

Gold and Green Foods has thus furthered its environmental impact far beyond just getting people to buy pulled oats, which Itkonen finds very encouraging. Rather than being cautious of potential competitors, she is happy about the vast impact that pulled oats has had on transforming the market and people's lifestyles in a sustainable direction. In August 2016 they also secured a deal with Paulig, an international food industry giant with an annual turnover of 905 million Euros, who bought 51% of the shares of Gold and Green Foods. The deal will help Gold and Green Foods increase production capacity and head into international markets.

22.4.2 A Smartup that Rescues Food

Around 88 million tonnes of food are wasted annually in the EU, with associated costs estimated at 143 billion Euros. Two-thirds of the costs are associated with food waste from households, which amounts to around 98 billion Euros (Stenmark et al. 2016). Food waste is an issue of importance to global food security and good environmental governance and is directly linked with environmental, economic and social impacts. Different studies show that between a third and a half of the world food production is not consumed (Gustavsson et al. 2011; Bio Intelligence service 2010), leading to negative impacts throughout the food supply chain.

The societal and environmental impacts of food waste are severe, but there are also significant costs imposed on ordinary households. If the lifecycle of food products were to become more efficient at every stage from production to consumption, the cost of food could be reduced significantly.

ResQ Club is a smartup that matches restaurants that have leftover food with consumers willing to buy these leftover meals for a discount up to 70%. The ResQ Club mobile application shows all available meals on a map, making it easy for the user to pick the most convenient option. ResQ meals are bought through the application, meaning that the consumer can just walk to their chosen restaurant at any time to pick up their order. Both the restaurant and the consumer end up on the winning side: restaurants are able to sell food that would otherwise go to waste due to health and safety regulations, whereas consumers purchase restaurant food for a price almost as low as that of a home cooked meal.

According to CEO, Tuure Parkkinen, the business model of ResQ Club is based on connecting the right customer with the right restaurant at the right time and the right place. The success of the product is thus based on the efficient handling of data combined with a simple user interface that is intuitive and convenient to use for both the consumer and the restaurant. ResQ Club has managed to combine quality supply with cheap prices and usability, making their product hugely successful. As of November 2016, over 80,000 meals from over 200 restaurants have been sold since the launch of the app in January of the same year. This means that over 25 tonnes of food have been saved, cutting carbon dioxide emissions by an amount equal to that of driving 3 million kilometres. Additionally, several restaurants have been able to improve the quality, availability and freshness of their food as they know that they can sell the leftovers at a decent price. ResQ Club has also brought new customers to lesser known restaurants both for regular-priced meals and ResQ Club meals.

22.5 Conclusions

Smartups can play a significant role in creating new sustainable markets and promoting sustainable lifestyles. Therefore, their importance cannot be overlooked in efforts to mitigate climate change and in the transition to a global low carbon and resource smart society.

Smartups are often small companies that struggle with getting the necessary investments and resources to scale up. They are, nevertheless, faster, more flexible and more open to new innovations than large, traditional companies. That is why their role in opening new markets is important. However, large companies can also create smartup businesses and facilitate innovations. What matters is the strong, user-centred design process, which is illustrated by the case of pulled oats. The company succeeded in combining sustainable lifestyles with an easier and healthier food product. That is the recipe for success.

There are several ways to support the growth of smartup businesses. Demos Helsinki (2016) has established a Peloton Club Smartup Accelerator, which supports early stage teams to develop and test their ideas. From 2012 onwards, Peloton Club has accelerated almost 100 smartup teams and companies, creating a whole ecosystem around new sustainable service production.

This work has highlighted the importance of strategic partnerships between big and small companies. Small companies and startups often have lots of exciting, fresh ideas but lack the resources or knowhow for broad implementation. Large companies, on the other hand, often stick to established practices and are more resistant to change and slower in their moves. Through strategic partnerships, it might be possible to get the best benefits from both sides. More research is needed to identify optimal ways of organising strategic collaboration to support the growth of smartup businesses. Nevertheless, it seems obvious that the already occurring strategic collaboration between new and traditional companies, cities and academia, is critical to facilitate a deeper and wider societal change.

References

Bio Intelligence Service (2010) Preparatory study on food waste across EU 27. European Commission. doi:10.2779/85947

EVU (2016) European vegetarian union. http://v-label.eu. Accessed 15 Nov 2016

Faber J, Schroten A, Bles M, Sevenster M, Markowska A, Smit M, Rohde C, Dütschke E, Köhler J, Gigli M, Zimmermann K, Soboh R, Van 't Riet J (2012) Behavioural climate change mitigation options and their appropriate inclusion in quantitative longer term policy scenarios – main report. In: DG Climate action contract. CE Delft. Available via CE Delft. http://www.cedelft.eu/publicatie/behavioural_climate_change_mitigation_options_and_their_appropriate_inclusion_in_quantitative_longer_term_policy_scenarios/1290. Accessed 16 Dec 2016

Fiala N (2008) Meeting the demand: an estimation of potential future greenhouse gas emissions from meat production. Ecol Econ 67(3):412–419

Gerstberger C, Yaneva D (2013) Analysis of EU-27 household final consumption expenditure. In: Statistics in focus 2/2013. Eurostat. Available via Eurostat. http://ec.europa.eu/eurostat/statistics-explained/index.php/Archive:Household_consumption_expenditure_-_national_accounts. Accessed 17 Nov 2016

Gustavsson J, Cederberg C, Sonesson U, Van Otterdijk R, Meybeck A (2011) Global food losses and food waste – extent, causes and prevention. In: SAVE FOOD. FAO. Available via: FAO. http://www.fao.org/docrep/014/mb060e/mb060e.pdf. Accessed 15 Nov 2016

Hamel G, Prahalad CK (1996) Competing for the future. Harvard Business School Press, Bosto

Hart CW, Heskett JL, Sasser WE (1990) The profitable art of service recovery. Harv Bus Rev 68(4):148–156

Helsinki D (2016) Peloton club accelerator. http://pelotonclub.me. Accessed 01 Nov 2016

Kim WC, Mauborgne R (2005) Blue ocean strategy: from theory to practice. Calif Manag Rev 47(3):103–121

Ritola M, Annala M, Hulkkonen S, Lahtinen V, Lätti R, Noponen E, Mäkelä K, Mizera R, Neuvonen A, Hietaniemi J, Mokka R (2014) Cleantech takes over consumer markets. In: Study about the potential of Finnish consumer cleantech in global markets. Tekes. Available via Demos Helsinki. http://www.demoshelsinki.fi/en/julkaisut/cleantech-takes-over-consumer-markets/. Accessed 15 Nov 2016

Short SW, Bocken NMP, Barlow CY, Chertow MR (2014) From refining sugar to growing tomatoes. J Ind Ecol 18(5):603–618

Stenmark Å, Jensen C, Quested T, Moates G (2016) Estimates of European food waste levels. In: Reducing food waste through social innovation. FUSIONS. Available via FUSIONS. http://www.eu-fusions.org/phocadownload/Publications/Estimates%20of%20European%20food%20waste%20levels.pdf. Accessed 15 Nov 2016

Tubiello FN, Salvatore M, Cóndor Golec RD, Ferrara A, Rossi S, Biancalani R, Federici S, Jacobs H, Flammini A (2014) Agriculture, forestry and other land use emissions by sources and removals by sinks. In: 1990–2011 analysis. FAO statistics division. Available via FAO. http://www.fao.org/docrep/019/i3671e/i3671e.pdf. Accessed 15 Nov 2016

United Nations (2015) Paris agreement. UN framework convention on climate change, Paris. http://unfccc.int/files/essential_background/convention/application/pdf/english_paris_agreement.pdf. Accessed 15 Nov 2016

Worldwatch Institute (2014) Peak meat production strains land and water resources. Ecos 199:27–28

Chapter 23
Circular Flanders: Adaptive Policy for a Circular Economy

Sam Deckmyn

Abstract Sustainable, systemic changes in the use of both energy and raw materials are key to safeguarding a strong economy and a high standard of living in Flanders. The transition to a circular economy is contingent upon comprehensive change; it hinges not only on technological advances but also on systemic organisational, societal, financial and policy innovations.

Flanders has already taken the first steps towards policy changes that promote a circular economy, beginning with the adoption of the Flanders' Materials Programme in 2011. Six years down the road, starting in 2017, the programme will be relaunched as Circular Flanders, with an even stronger focus on an integrated approach to a circula r future.

Keywords Circular economy • Governance • Flanders • Transition management • Multi-stakeholder partnerships

23.1 Why: Limits to Linear

23.1.1 The End of the Line

Ever since the Industrial Revolution, our economies have basically forged their growth paths by transforming natural resources into products that are disposed of after use. It's a linear model based on the assumption of infinite, readily available resources that can be cheaply discarded. But linearity is reaching the end of the line.

As the global middle class grows by a projected three billion people over the next couple of decades (Dobbs et al. 2012), our linear Western lifestyle will prove unsustainable. China and India are already doubling their real per capita incomes at about ten times the pace England achieved during the Industrial Revolution, and at around 200 times the scale. The demand for resources will soar at a time when finding and

S. Deckmyn (✉)
Circular Flanders/OVAM, Stationsstraat 110, 2800 Mechelen, Belgium
e-mail: sam@vlaanderen-circulair.be

H. Lehmann (ed.), *Factor X*, Eco-Efficiency in Industry and Science 32,
https://doi.org/10.1007/978-3-319-50079-9_23

extracting additional raw materials is becoming an ever greater challenge, and it will be all but impossible to keep up with, despite technological advances in natural resources exploitation. As a direct result, raw materials prices are becoming volatile, and certain critical resources will most likely become so scarce as to be prohibitively expensive—not to mention the environmental impact of their extraction (Fischer-Kowalski and Swilling 2011) (Fig. 23.1).

At the end of the day we are grappling with a straightforward physical fact: the amount of natural resources on the planet is finite. Unlike energy, with a seemingly endless supply for new technology to take advantage of, the finite nature of raw materials on earth is a problem we cannot work our way around—unless alchemists magically succeed in turning iron into gold.

Clearly, we need to change tactics. A new approach to resources—one that counteracts leakages and disposables, closes loops and enables high-value reuse—is our only option for safeguarding a high standard of living in the long term. A shift to (preferably radically) new, sustainable economic activities that keep materials *in the loop* offers opportunities for Flanders, and for the EU as a whole. The circular economy represents at least part of the solution.

23.1.2 The Circular Economy as a Strategic Solution

Our current linear economies transform resources into products to be discarded and destroyed at the end of their lifecycle. The circular economy offers an alternative approach based on maximising product and resource reusability and minimising losses of value.

Using a host of strategies, the circular economy focuses on keeping resources in a continual economic loop, while retaining as much value as possible. As such, products are designed to be repaired, have high second-hand value, be upgradable and be easily disassembled and transformed into new products. In a circular economy, products start out as recycled or biobased materials and are recyclable or biodegradable at end-of-life.

However, the circular economy aims to go beyond recycling, which—while it definitely has its use—is often essentially a form of downcycling: the end product (e.g. insulation filler) holds less value than the product it started out as (e.g. clothes). The circular economy aims to preserve as much value as possible: clothes will first be repaired and resold; when repairing is no longer an option, the fabrics will be reused to make new garments and when reuse is no longer an option, the fibres will be spun into new threads.

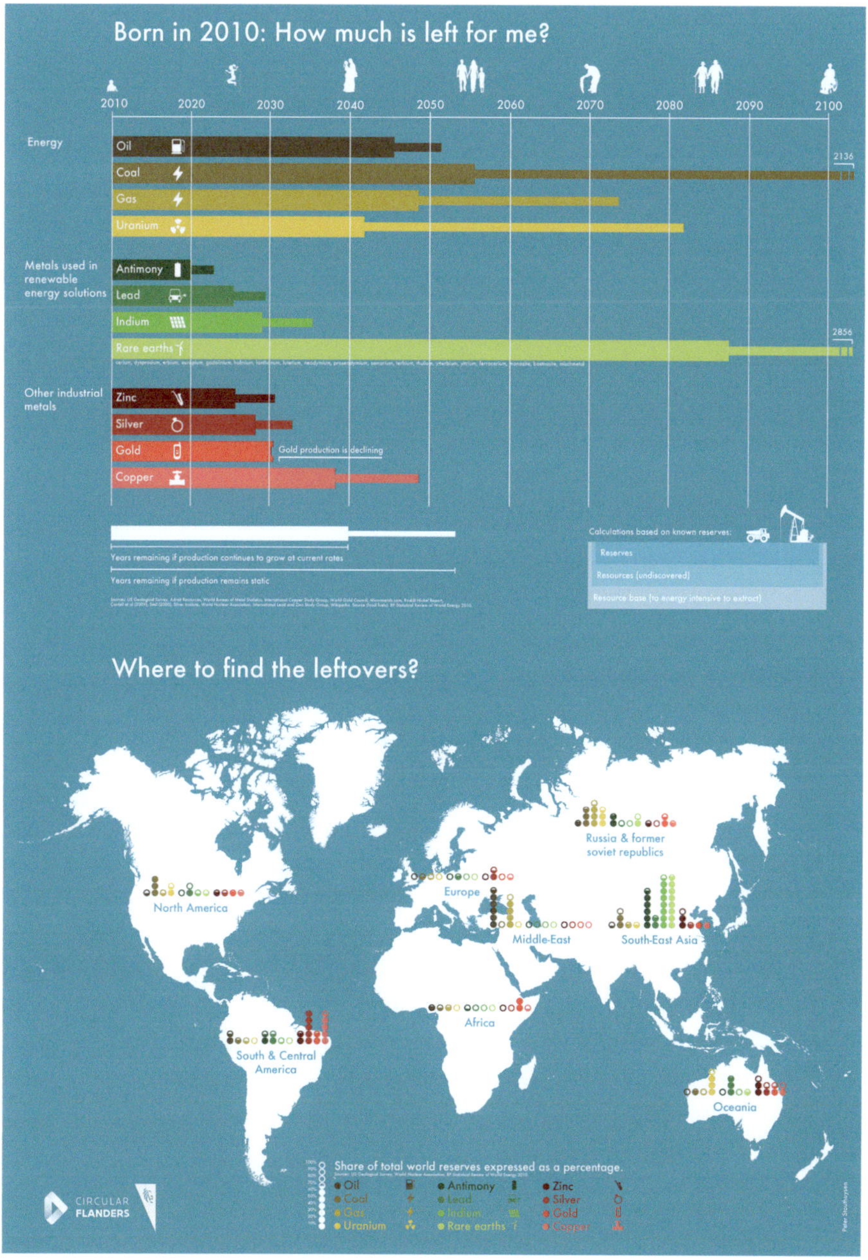

Fig. 23.1 How much is left? How long will several reserves of material resources last? (Source: Plan C)

23.1.3 Economic Gains

From a purely economic perspective, the advantages to the circular approach speak for themselves. Resource efficiency improvements throughout the value chain could decrease the demand for materials by an estimated 17 to 25% by 2030 (Meyer 2011). Optimising resource use could cut expenses for European businesses by 630 billion Euro (Ellen MacArthur Foundation 2012). EU GDP would grow by 3.9% because of eliminated costs for acquiring new materials and the creation of new products, services and value (Ellen MacArthur Foundation 2012). Furthermore, the circular economy would create 1.2 billion to 3 billion new jobs in Europe by 2030 (Mitchell and James 2015).

For Flanders, estimates show that investing in a circular economy could cut materials expenses by 2–3.5% of GDP and create 27.000 new jobs, ranging from high to low-skilled labor (Dubois and Christis 2014).

23.1.4 From Vision to Policy

Sustainable, systemic changes in the use of both energy and raw materials are key to safeguarding a strong economy and a high standard of living in Flanders. The transition to a circular economy is contingent upon comprehensive change; it hinges not only on technological advances but also on systemic organisational, societal, financial and policy innovations.

Flanders has already taken the first steps towards policy changes that promote a circular economy. It adopted the Flanders' Materials Programme in 2011. Six years down the road, starting in 2017, the programme will be relaunched as Circular Flanders, with an even stronger focus on an integrated approach to a circular future.

23.2 How: Circular Economy Policy

23.2.1 First Phase: Flanders' Materials Programme

23.2.1.1 A new Approach

In 2010, when Belgium held the EU presidency, the European Council conclusions on 'sustainable materials management and sustainable production and consumption' put the circular economy on EU member states' agendas (Council of the European Union 2010). The Public Waste Agency of Flanders (OVAM) played an instrumental role in developing these conclusions, using its vision of sustainable materials cycles and 30 years of experience in waste management.

OVAM brought its expertise to bear in 2011 when it founded the Flanders' Materials Programme[1] (FMP) with the purpose of providing Flanders with a future-proof economy where material cycles are closed.

The programme is an example of successful co-creation and co-ownership between business sectors, research institutions, government and environmental organisations. It is built on three pillars:

- A long-term vision: Plan C[2] is the circular economy hub in Flanders, created to encourage a change in mind-set from waste to resources and to accelerate the move towards a circular economy.
- Policy-relevant scientific research: SuMMa[3] (Policy Research Centre for Sustainable Materials Management) brings together a broad spectrum of researchers and investigates which economic, policy and social conditions need to be fulfilled in order to realise the transition towards a circular economy.
- Actions and projects in the field: Agenda 2020 is a list of 45 concrete projects with active partners and a clear time schedule.

23.2.1.2 Economic Clusters Paired with Enabling Functions

In order to develop the foundations for a circular economy by 2020, the FMP focuses on closing materials cycles among economic clusters and in providing enabling functions. Economic clusters are chosen for their potential for improvement from a primary resources and materials perspective and for the expertise in these domains present in Flanders. They include:

- sustainable materials management in construction;
- the bio-economy;
- sustainable chemistry and plastics;
- critical metals in a continuous cycle.

Enablers were chosen to break down obstacles encountered by every project, business case and innovation. They include:

- sustainable design;
- smart collaboration;
- smart investments;
- new materials and new material technologies
- better regulation.

Five priority actions have been identified for each cluster and each enabler. Agenda 2020 indicates what concrete steps need to be taken today to realise the ambition for 2020.

[1] www.flandersmaterialsprogramme.be

[2] www.plan-c.eu/en

[3] https://steunpuntsumma.be/english

23.2.1.3 Focus on Leverage Projects

The main focus of the programme is on six strategic leverage projects:

1. Transforming Flanders into a recycling hub for Europe through our seaports
2. New jobs in the cicular economy
3. Potential for a circular economy at several spatial scales (case region Genk)
4. Strengthening the metal recycling industry in Flanders
5. Innovative building concepts for energy- and material-conscious construction
6. Business model innovation

The leverage project 'Transforming Flanders into a recycling hub', for example, is about providing opportunities to import waste materials and complex end-of-life products for re-using, remanufacturing and recycling, as well as reinforcing the value chain for valuable metals and hard plastics.

Another leverage project, 'New jobs in the circular economy', is about developing the job potential of the circular economy by attuning education to provide the required knowledge and skills for new and different jobs related to circular economy opportunities.

23.2.1.4 Pursuing Catalytic Impact

Central to the whole agenda is reaching out to a small number of pioneering Small and medium-sized enterprises (SMEs) to encourage innovation in product design, business models and collaborative approaches across and between value chains.

Restricting intervention to a small number of companies was initially dictated by budget limitations. However, Plan C has now embraced its role as an 'activating' body that catalyses change by connecting and challenging the most innovative companies and enabling them to lead others, rather than trying to directly influence the large number of SMEs in Flanders. Its three core activities are: shaping a vision, activating a self-learning network around sustainable materials management within a circular economy and supporting transition experiments.

23.2.1.5 Facing Obstacles

The FMP worked to break down other barriers to its implementation beyond funding restrictions. An important enabling aspect of a circular system is public procurement, but procuring institutions often focus more on finding the lowest cost options rather than looking at total lifetime cost, which opens up opportunities to incorporate circular economic thinking. Low global resource prices, along with high labour costs, curb the profitability of remanufacturing. Another obstacle is a lack of knowledge and skills: a furniture remanufacturing plant was planned in Flanders, but it proved difficult to find workers sufficiently skilled to operate it.

23.2.1.6 Enhanced Cooperation

Successful cooperation between stakeholders, based on sharing responsibility for projects, is the keystone of the programme's approach to implementing its vision and overcoming barriers.

Of the 45 actions in the Agenda 2020 plan, 10 are run by OVAM, which focusses on encouraging action in larger organisations, 20 by industry associations such as FEBEM (Federation of Environmental Companies), Vlaamse Confederatie Bouw, (Flemish Construction Federation), essenscia Flanders (Federation for Chemistry and Life Sciences industries) and Agoria (Federation for the Technology Industry) and 15 by other organisations, including the Department of Economy, Science and Innovation, the Flemish Institute for Technological Research (VITO) and the Bond Beter Leefmilieu (Federation for a Better Environment).

23.2.1.7 Award-Winning First Step

The many projects set up and carried out under the umbrella of the Flanders' Materials Programme over the past couple of years (2012–2015) represent a first step towards a circular economy in Flanders.

The FMP has been quite successful so far, and these efforts were rewarded when OVAM won the Award for Circular Economy Government, Cities and Regions at the World Economic Forum in 2016 for the accomplishments of the programme.[4] A great merit of the FMP is the innovative partnership between government, industry, non-governmental organisations, universities and research centres, as well as its broad, cross-sectoral programme that creates fertile ground for the transition to a circular economy.

23.2.2 Second Phase: Circular Flanders

23.2.2.1 Keep Pushing

Despite appreciation for past achievements, there was no time for complacency. All in all, the total number of companies embracing sustainable materials management and the circular economy was still fairly small. There was also room for improvement in cross-sectoral collaboration and in interaction between the three pillars of the Flanders' Materials Programme.

At the same time, the circular economy had become more prominent at the European level. In December 2015, the European Commission launched its Circular Economy Package, consisting of the action plan 'Closing the loop' and six revised legislative proposals on waste (European Commission 2015).

[4] https://thecirculars.org/finalists

The time was ripe for a second step. The government of Flanders raised the bar with a vision for the future.

23.2.2.2 The Circular Economy Becomes a Transition Priority

In the spring of 2016, the government of Flanders adopted the cross-cutting policy paper 'Vision 2015. A long-term strategy for Flanders'. The document represents a commitment to a clear vision for the future and for the Flanders it wants to be in 2050: a social, open, resilient and international region where every individual counts that creates prosperity and well-being in smart, innovative and sustainable ways.

To attain this future objective, seven transition priorities were outlined that cut across policy areas and engage partners throughout business and society. '*Continuing the structural transition towards the circular economy*' is one of these seven transition priorities, thereby building on the previous experiences of the Flanders' Materials Programme (government of Flanders 2016).

23.2.2.3 Widening the Scope

From 2012 to 2015 our interpretation of the circular economy was mainly focused on closing materials loops, a focus that was evident in the Flanders' Materials Programme as well.

With its Vision 2050, the government of Flanders has now expressed its commitment to a broader scope of interpretation and has highlighted several key aspects it wants to target: materials, water, energy, land and food (government of Flanders 2016). A competitive bioeconomy in which biomass is sustainably produced and (residual) flows are (re)used in food, feed, materials, products and energy is also explicitly mentioned in Flanders' vision for 2050.

This broader interpretation of the circular economy indicates a paradigm shift in the policy of Flanders in which

- transversality is even more prominently featured. The transition priority 'circular economy' will be a joint responsibility of the Minister for Environment, the Minister for Economy and Innovation;
- even more emphasis is put on collaboration across industries and with a wide range of stakeholders;
- the scope of ambition is broadened to reach out to a significant number of stakeholders who were previously not or insufficiently involved. The scope is broadened to include materials, water, energy, land and food (Fig. 23.2).

A similar widening of scope is noticeable in the European Circular Economy Package: while closing resource loops is still considered the backbone of the circular economy, the Action Plan also stresses the importance of water reuse and bioeconomy.

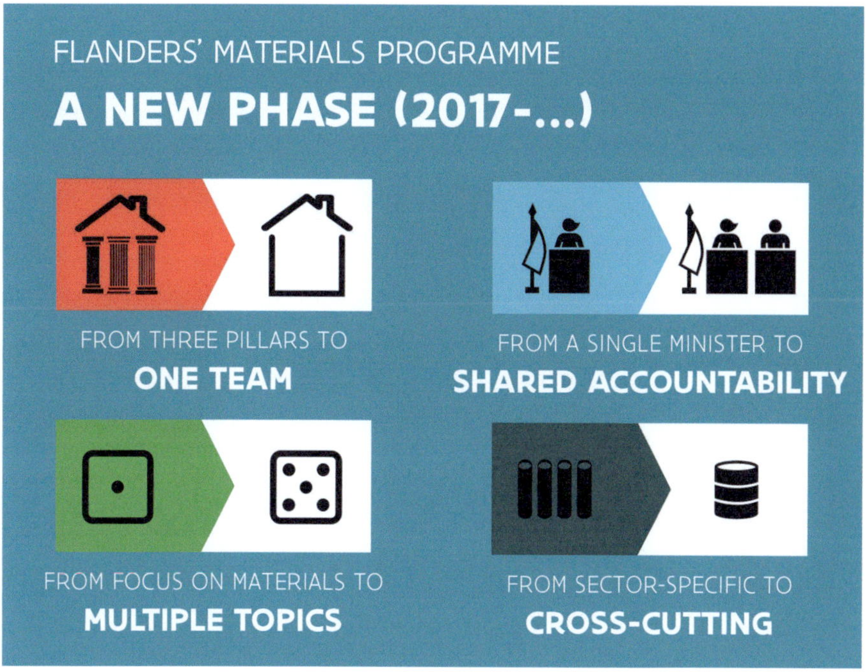

Fig. 23.2 Circular Flanders: the next phase (Source: Circular Flanders)

23.2.2.4 Joining Forces

At the same time the decision was made to merge the three pillars of the Flanders' Materials Programme (Plan C, SuMMa and Agenda 2020) within the Public Waste Agency of Flanders (OVAM). This operational team will be the designated 'delivery unit' for the circular economy transition priority.

A range of stakeholders will pool human and financial resources in the unit, which will be headed by a public-private steering group. It will provide a single, effective hub for the circular economy in Flanders.

The newly unified team will kick off on 1 January 2017, and it will initially focus on three cross-cutting themes, within which several projects will be initiated and facilitated:

- the circular city,
- circular business strategies
- and circular procurement.

In line with the broadened thematic scope outlined in the previous chapter, the team will also collaborate on projects related to water, materials, energy, land and food on a demand-driven basis.

23.3 What: A Sample of Our Projects

Flanders Recycling Hub
The collaborative research project 'Flanders recycling hub' was launched by the Flemish Institute for Logistics (VIL) and the Flanders' materials programme. It aims to develop a roadmap for putting Flanders on the map as an international recycling hub—a centre where materials are supplied, recuperated and exported again. Because of their extensive know-how, Flemish companies have already firmly established themselves as world leaders in the recycling industry. But it takes more than that to transform Flanders into a globally important recycling hub. Materials flows need to selected and attracted.

Additive Design Challenge
The Additive Design Challenge is a competition at the crossroads of circular economy and additive manufacturing—3D printing in all its forms. How often have you seen 3D printing applied to make a positive impact on society and the environment? That's what this challenge is about: the Additive Design Challenge invites thinkers, makers and doers to put 3D printing at work for people and the planet. Find out more online: www.additivedesignchallenge.com

New Jobs in the Circular Economy
A switch to a circular economy will generate new and different jobs. There is a need for further qualitative research into these 'new' and 'other' jobs. What jobs are we talking about in practice, and what competences and skills are required to develop them? Building on the qualitative research, we organise a job fair around the theme of 'materials' and existing training programmes are adapted or new ones are developed.

Masterclass Circular Entrepreneurship
It takes only four 1-day sessions to discover which opportunities the circular economy has in store for you as an entrepreneur. We check out trends in consumer behavior and policies, explore collaborations that could strengthen your enterprise and analyze how to financially cover various strategies.

Interdisciplinary Policy Research
The Policy Research Centre for Sustainable Materials Management investigates and fosters the role that policy can and should play in a transition towards a circular economy. The centre has published over 30 advisory papers in the past 3 years. Find out more online: www.steunpuntsumma.be/english

Close the Loop – The Ultimate Guide Towards a Circular Fashion Industry
In a circular fashion industry, designers, producers, retailers and consumers are challenged to take the whole lifecycle of a garment into account. Plan C and Flanders Fashion Institute built a web tool and exposition that guides you through the principles of this sustainable way of working. Learn more online: www.close-the-loop.be/en

The Circulator: The Circular Business Models Mixer
The Circulator is a web-based tool providing an overview of the most relevant business models in the context of the circular economy as well as relevant case studies. Users can choose and mix 36 different strategies and explore inspiring case studies. Learn more online: www.circulator.eu

Reburg: Imagineering a Circular City
To explore a circular society in the future we designed the city of Reburg, where the circular economy comes to life. The story of Reburg is told in four chapters—four circular futures. For each future, a Reburg citizen tells about his or her job and reveals related hot spots in the city. Online: www.reburg.world

Manufacturer Meets Maker
Businesses in the manufacturing industry join forces with makers, designers and experts, and head to a Solution Lab in search of circular, socially innovative solutions for today's challenges. View online: mmm.makers.be

References

Council of the European Union (2010) Council conclusions on sustainable materials management and sustainable production and consumption: key contribution to a resource-efficient Europe. https://www.consilium.europa.eu/uedocs/cms_data/docs/pressdata/en/envir/118642.pdf. Accessed 10 Jan 2016

Dobbs R, Oppenheim J, Thompson F (2012) Mobilizing for a resource revolution. McKinsey & Company. http://www.mckinsey.com/business-functions/sustainability-and-resource-productivity/our-insights/mobilizing-for-a-resource-revolution. Accessed 14 Dec 2016

Dubois M, Christis M (2014) Verkennende analyse van het economisch belang van afvalbeheer, recyclage en de circulaire economie voor Vlaanderen (Preliminary analysis of the economic importance of waste management, recycling and the circular economy in Flanders). Policy research centre on sustainable materials management Flanders. https://steunpuntsumma.be/nl/publicaties/summa-economisch-belang-8.pdf. Accessed 14 Dec 2016

Ellen MacArthur Foundation (2012) Towards the circular economy: economic and business rationale for an accelerated transition. https://www.ellenmacarthurfoundation.org/assets/downloads/publications/Ellen-MacArthur-Foundation-Towards-the-Circular-Economy-vol.1.pdf. Accessed 14 Dec 2016

European Commission (2015) Closing the loop – An EU action plan for the circular economy. http://eur-lex.europa.eu/resource.html?uri=cellar:8a8ef5e8-99a0-11e5-b3b7-01aa75ed71a1.0012.02/DOC_1&format=PDF. Accessed 14 Dec 2016

Fischer-Kowalski M, Swilling M (2011) Decoupling natural resource use and environmental impacts from economic growth. United Nations environment programme. http://www.unep.org/resourcepanel/decoupling/files/pdf/decoupling_report_english.pdf. Accessed 14 Dec 2016

Government of Flanders (2016) Visie 2050: een langetermijnstrategie voor Vlaanderen (Vision 2050: A long-term strategy for Flanders) (full report). https://www.vlaanderen.be/nl/vlaamse-regering/visie-2050#publicaties. Accessed 14 Dec 2016

Meyer B (2011) Macroeconomic modelling of sustainable development and the links between the economy and the environment. Gesellschaft für Wirtschaftliche Strukturforschung (for the European Commission). http://ec.europa.eu/environment/enveco/studies_modelling/pdf/report_macroeconomic.pdf. Accessed 14 Dec 2016

Mitchell P, James K (2015) Economic growth potential of more circular economies. WRAP. http://www.wrap.org.uk/sites/files/wrap/Economic%20growth%20potential%20of_more%20circular%20economies.pdf. Accessed 14 Dec 2016

Chapter 24
The 100 Companies Project Resource Efficient Practice Cases from Producing Industry

Mario Schmidt

Abstract Initiatives from industry and good corporate examples are required in politics to show how resource efficiency can be improved in industrial production processes. However, companies are usually reluctant to tell their success stories. In the highly industrialised state of Baden-Wuerttemberg, a project was carried out to present and analyse 100 concrete examples from companies. The aim was to overcome the barriers to communication about corporate resource efficiency and possible potentials for cost saving. To this end, it was necessary to consider the different objectives of the actors. It can be inferred that from the perspective of a company—i.e., an economic perspective—resource efficiency is perceived as a strategy to help to save costs.

Keywords Resource efficiency • Cleaner production

24.1 Introduction

The question of how natural resources and raw materials can be handled sparingly and efficiently has been a topic of dynamic discussions in the fields of environment and sustainability for many years, especially in the 1970s following the *Limits to Growth* report by the Club of Rome. However, the subject only attracted major public interest again ten years ago when raw material prices rose steeply on the world market and serious shortages occurred for some metals, including many rare earth elements (see Fig. 24.1). This situation was shaped by geopolitical and economic considerations and was by no means an indication of growing geological shortage, as had been prophesied for decades. Today, China holds a monopoly over many raw materials and in view of its enormous economic growth also has a high domestic demand. The companies that are dependent on raw materials have taken the tight supply situation very seriously in recent years. These include in particular the manufacturers of high-tech products which cannot be produced without technology

M. Schmidt (✉)
Institute for Industrial Ecology (INEC), Pforzheim University,
Tiefenbronner Str. 65, 75175 Pforzheim, Germany
e-mail: mario.schmidt@hs-pforzheim.de

H. Lehmann (ed.), *Factor X*, Eco-Efficiency in Industry and Science 32,
https://doi.org/10.1007/978-3-319-50079-9_24

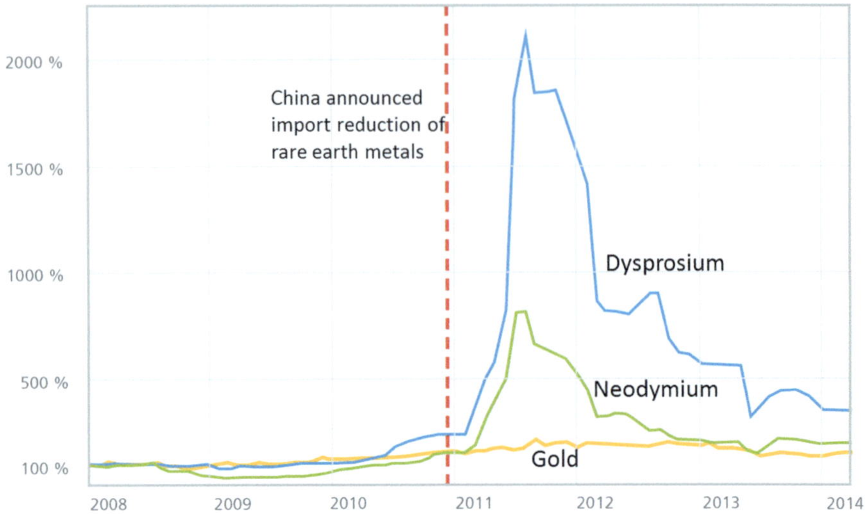

Fig. 24.1 Price development of the two rare earth elements neodymium and dysprosium by comparison with gold in per cent of the prices in January 2008 (Source: Schmidt et al. 2017)

metals such as rare earth elements, tantalum, cobalt, indium, germanium etc. For a short period the prices of some raw materials rose so considerably that certain technologies became less attractive—for instance, electric motors with permanent magnets for which large quantities of neodymium or dysprosium are needed.

Now, in the year 2016, the world market prices are low again and the economic pressure on companies is easing. It is, however, all the more important now to make it clear to manufacturers that resource efficiency is a permanent task. This includes the economical handling of raw materials in the production process, but above all targets the completion of material cycles in a closed loop system. It is helpful that resource efficiency has recently become one of the top issues in global politics. At the 2015 Summit in Elmau, Germany, G7 leaders made a statement on resource efficiency for the first time and initiated a series of further international activities at OECD and UN levels. They also called for initiatives from industry and good corporate examples. The European Union and individual countries have made resource efficiency a field of political action. In Germany, the Federal Government has already put forward a second programme on resource efficiency. Even the individual federal states in Germany are active and have discovered resource efficiency as an important sustainability issue.

24.2 Many Reasons for Resource Efficiency

The reasons for promoting resource efficiency are many and varied. At the corporate level, the focus is primarily an economic one. On the one hand, companies want to secure their supply situation with key raw materials. On the other hand, they want to reduce their costs. In general, the costs of materials have increased substantially in recent years. While in an average German industrial plant energy costs account for about 2%, material costs account for almost 45% and are more than twice as high as personnel costs. If the target is to lower costs, it is important to tackle the large cost blocks, in this case the material costs. Interestingly enough, considerable potentials for savings are apparent here too. Admittedly, attempts have always been made to reduce material costs. However, such attempts were understood primarily as a task for the procurement department, i.e., pertaining to the negotiation of low raw material prices. When it came to in-house potentials for saving, companies tended to look more at personnel or energy costs. It has been determined on the basis of various case studies that potentials for saving through material efficiency are distinctly higher and lie in the range of a few per cent of turnover. As many measures are organisational in nature and do not require any major investment, these savings also have a direct effect on profit margins. For this reason, material efficiency measures can leverage the profit situation of companies.

From an ecological standpoint, the careful and efficient handling of resources is called for. Here it is less the supposed geological scarcity that is crucial but the environmental impact of extraction. The extraction of raw materials involves considerable effects on the environment as it occurs in ecologically sensitive areas or in regions with low standards of protection and conservation. Around 10% of the world's primary energy consumption is used to extract raw materials. The share of raw material extraction in greenhouse gas emissions is correspondingly high. Reducing the use of raw materials thus additionally contributes to climate change mitigation.

Finally, the extraction of raw materials also carries social implications. Many raw materials are extracted in countries with low standards of labour protection, child labour, fair wages etc. In some countries the extraction of raw materials even finances civil wars or military conflicts. Resource efficiency is therefore called for on economic, ecological and social grounds and should be a central concern of any sustainability strategy.

For the actors, the reasons for implementing resource efficiency measures vary widely. Sustainability, environmental protection or social standards are important in society and the world of politics. However, for a company that operates in line with the rules of a market economy, these will only be of secondary importance. The most important argument is always profitability and hence the question of costs. This must be considered when it comes to implementation strategies. Sustainability and environmental consciousness have little power of conviction and drive only a few pioneering green companies.

24.3 Regional Commitment in a High-Tech Region

In Germany, the southern Federal State of Baden-Wuerttemberg is characterised by a high degree of industrialisation, a strong automotive industry and manufacturing industry of machinery and equipment. The manufacturing industry contributes 35% to the gross domestic product, while the figure for Germany as a whole is 23%. By way of comparison, in the UK the relevant figure is only 11%, and in the USA, 12%. At the same time Baden-Wuerttemberg is the first German state with a Prime Minister from the Green Party, which is particularly significant for sustainability. In March 2016 the first state strategy on resource efficiency was presented. This was preceded by a comprehensive stakeholder process with participants from research and academia, business and industry and civil society.

One major concern of the state government of Baden-Wuerttemberg is to motivate the companies in the manufacturing industry to produce in a more resource efficient way. However, this is not understood as a regulatory measure in which the use of raw materials is limited or controlled. Instead, the companies are encouraged to adopt innovative solutions and act on their own initiative. This lies in the tradition of the region, which is home to many high-tech companies, hidden champions and inventors. The cooperation between politics and business in the matter of resource efficiency therefore ranks very high.

The targets of the state strategy for resource efficiency are:

- Uncoupling economic growth from the consumption of resources while retaining and expanding the high share of manufacturing and preserving the economic structure of Baden-Wuerttemberg
- Supporting the goal of the National Sustainability Strategy—i.e., doubling raw material productivity between 1994 and 2020
- Developing Baden-Wuerttemberg to become the leading market for and leading supplier of resource efficiency technologies and thus one of the most resource-efficient regions
- Supplying safe raw material to industry through more efficient extraction of primary raw materials and increasing the share of secondary raw materials

For around 20 years now there have been repeated attempts with funded projects in Baden-Wuerttemberg to make production gentler on resources in the factories. One example here is the BEST programme, in which some 50 companies were funded. There have been many other funding programmes to promote resource efficiency both at a national level as well as in other German states. Consequently, a broad empirical data basis is available. It has been established that the savings potential for companies is substantial and leads to pay offs in both economic and ecological terms.

However, the problem that arises is that the positive experiences gained from such projects are not automatically perpetuated, i.e., they are not self-disseminating among the target group in the industry. This phenomenon has already been the subject of a number of research projects. One important explanation that has been discovered is that there is a lack of suitable, practical examples that are described publicly and that demonstrate the potential for savings. Companies that implement efficiency measures successfully are generally unwilling to talk about the economic returns, as they fear that customers in the B2B sector or competitors will quickly nullify their individual competitive advantage. That is why there is very little communication about particularly successful examples. The companies are quite reserved, even regarding non-monetary data, such as e.g. how much material or what level of CO_2 emission has been saved.

However, the commitment of industry is necessary to realise resource efficiency on a large scale. For this, impressive case examples are required. The leading industrial associations in Baden-Wuerttemberg recognised this and in 2013 together with the state government entered into an Alliance for Resource Efficiency. The goal of this Alliance was to convince over 100 companies to present examples of their operations publicly.

24.4 Approaches to Operational Resource Efficiency

For measures at the plant operational level, the crucial factor is what amount of primary raw materials and energy can be saved by resource efficiency and whether environmental impacts, e.g., greenhouse gas emissions, are reduced. The companies, on the other hand, as a rule evaluate efficiency measures solely on the basis of costs saved. Both viewpoints are legitimate as long as they serve the overarching goal. Figure 24.2 shows what importance an operational measure can have for the conservation of resources.

The most frequent starting point is to reduce residual materials and production wastes (No. 1 in Fig. 24.2). This reduction also leads to a lower material requirement, which has reverberations along the entire pre-chain. Residual materials that cannot be avoided can be recycled (No. 2). Even more important are the dematerialising (No. 3) and recycling of the products (No. 4) or the use of alternative materials (No. 5), e.g., from renewable sources.

It is notable that these measures generally also save energy along the value chain. By contrast, the reduction of in-house recycling flows admittedly does not contribute to conserving resources, but it does lead to a lower energy input and an increase in productivity (No. 6).

The opportunities for resource efficiency in a manufacturing company are diverse, but they depend on the nature of production and of the products. In contrast to the field of energy efficiency, material efficiency seldom displays groups of measures that can easily be transferred from one plant to another, such as, e.g., lighting, the use of compressed air or heat recovery. Consequently, each company needs to conduct specific analyses of its potential for saving on costs. Typical approaches are listed in Fig. 24.3.

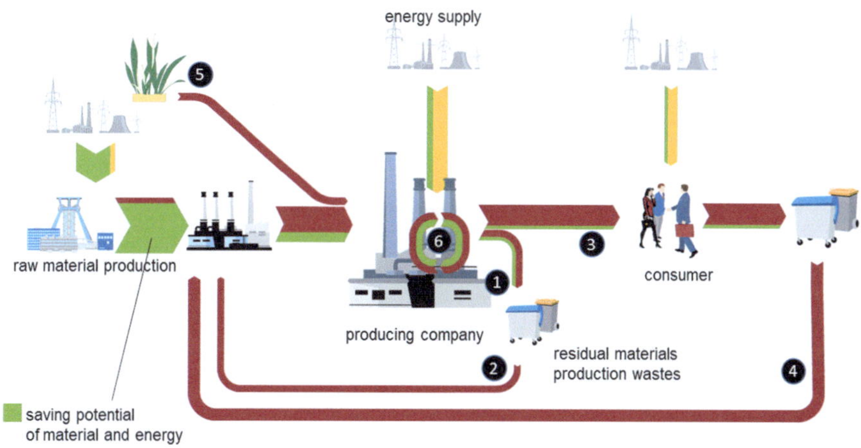

Fig. 24.2 Savings in the material and energy flows in the overall context of the product life cycle (Source: Schmidt et al. 2017)

Fig. 24.3 Overview of possible operational measures for resource efficiency (Source: Schmidt et al. 2017)

Legend

1. Delivery
- Transport damage
- Shrinkage

2. Store intake warehouse
- Incorrect arrangements

3. Product development
- Functionality
- Dimensioning
- Material selection
- Recycling capability

24.5 The 100 Companies Project

The 100 companies project sought case examples from manufacturing in which the companies successfully saved resources, either energy or materials, or distinctly improved their resource efficiency (e.g. through increasing output yields). The focus lay on material efficiency and production methods. Purely product-related measures were not taken into account.

Fig. 24.3 (continued)

4. Production planning
- Production optimising
- Batch sizes

5. Yield planning
- Offcuts

6. Individual process
- Energy and operating consumption
- Rejects
- Residual materials
- Productivity
- alternative technology

7. Set-up processes
- Start-up losses

8. Post-processing/re-working
- Energy and operating material consumption
- Rejects

9. Internal cycle management of product
- Energy consumption
- Operating material consumption

10. Auxiliary processes (Energy and operating materials)
- Efficiency
- alternative technologies
- alternative operating materials

11. Internal cycle management of operating materials
- Wastes
- Emissions
- Use of waste heat

12. Packing and packaging
- Selection of packing materials
- Suitability for re-use

13. Dispatch or intermediate store
- Over-production
- Storage damage
- Perishability

14. Distribution
- Breakage
- Shrinkage

15. Building management
- Lighting
- Heating/air conditioning
- IT
- Canteen and sanitary sector

The project started in autumn 2014 and was funded by the State of Baden-Wuerttemberg. It was implemented and provided with professional support by three partners: the Institute for Industrial Ecology of the University of Pforzheim (INEC), the State Environmental Technology Agency Baden-Wuerttemberg (UTBW) and the Institute of Human Factors and Technology Management (IAT) of the University of Stuttgart.

Companies had to apply actively to be included among the "100 companies". Application was open to manufacturing companies that had implemented measures at a location in Baden-Wuerttemberg and were willing to describe these publicly and quantify the savings effect. Application was possible via a comparatively simple form in which the company described the measure and crucially the calculated potential for savings.

The quantifying of savings potentials was mandatory for all applicants, but there were several possible ways of doing this, as experience has shown that companies have difficulties here. Physical quantities or energy saved in kilowatt hours per year or savings of certain raw materials in tons per year could be quantified. However, it was also possible to state potentials for cost savings (in EUR per year). As one essential goal of resource efficiency is also to reduce environmental impact, it was further possible to quantify savings in the form of CO_2 emissions avoided. In special cases the potentials for saving could be stated in percentage terms. However, figures always had to be plausible and transparent. The model character of the case example was crucial for the project.

The applications received were subjected to technical evaluation by the project partners. A jury consisting of one representative from each of the participating industrial associations, the state government and the project partners then made the final selection. By November 2016 around 120 applications had been received, from which the jury selected approximately 80 examples. The first 50 examples were presented at the Resource Efficiency Congress of the State of Baden-Wuerttemberg with over 700 participants in autumn 2016. The other examples will be published in 2017.

24.6 Examples from the Project

A few selected examples are presented briefly below. The complete list with detailed descriptions has been published in book form and will also be available online.

24.6.1 Fewer Residual Substances: Protektorwerk Florenz Maisch GmbH, Gaggenau

In interior finishing work in building construction, metal profiles are used for plastering edges and special structural details. These metal profiles normally have perforations so that the plaster adheres better. The profiles are made from a metal strip into which holes are punched. Consequently, large quantities of punching wastes are produced.

The firm Protektorwerk set itself the goal of producing perforations without waste. For this purpose the sheet metal is now no longer punched, but instead cut using a special technique and then pulled apart as in Japanese origami folding techniques (see Fig. 24.4). In this way no waste is generated, but the metal profiles satisfy their function equally well. The company developed the production machinery for this operation itself.

Galvanised steel is used for the sheet metal and with the new method no punching wastes result anymore. In this way it was possible to save 20% of steel used.

24.6.2 Lighter Products: Duravit AG, Hornberg

The firm Duravit AG is based in the Black Forest and produces ceramic objects for bathrooms, e.g., washbasins. Ceramic is heavy and has to be fired. In other words, the process also requires a high energy input. The wall thickness of washbasins is 12–14 mm. This wall thickness is necessary in order to guarantee the strength of the product. That is why complicated forms go hand in hand with a very high weight. To date, it was not even possible to realise some product geometries at all.

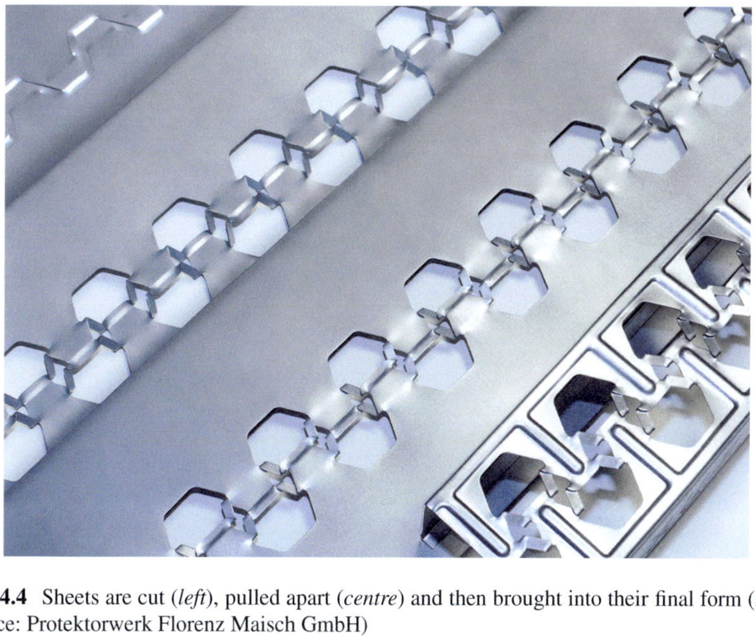

Fig. 24.4 Sheets are cut (*left*), pulled apart (*centre*) and then brought into their final form (*right*) (Source: Protektorwerk Florenz Maisch GmbH)

Duravit AG has now developed a new material that allows strength at lower wall thicknesses. A washbasin that was previously 11 mm thick and weighed 10.3 kg can now be produced with a thickness of 7 mm and a weight of 7.4 kg. This saves not only raw materials, but also energy for the firing process. Furthermore, cost savings and lower CO_2 emissions are achieved.

An interesting aspect of this example is that it is now also possible to make products with a new design (see Fig. 24.5). In particular thin-walled washbasins with delicate processing suddenly offer new competitive advantages for the company in addition to the other cost-related benefits.

24.6.3 Re-use of Machinery: Robert Bosch GmbH, Stuttgart

The large Bosch corporation operates 25 manufacturing plants worldwide. It repeatedly occurs that production plants and machinery are no longer needed because production has been converted or because a product line is discontinued. The goal of the company was to re-use these machines, to coordinate such re-use worldwide and thus to reduce new investment requirements.

For this purpose Bosch set up a Re-Use-Team that coordinates this further use. The machines have to be upgraded in order to bring them into line with the latest state of the art and to satisfy current energy standards.

Fig. 24.5 The innovative material DuraCeram® allows lighter and more delicate structures (Source: Duravit GmbH)

To date, 1565 machines have been re-used again in this way since 2009. This is equivalent to EUR 260 million in savings on new investments. At the same time 6000 t steel, 500 t aluminium, 141 t copper and 282 t plastic were saved. The amount of steel saved alone avoided over 35,000 t CO_2 emissions.

24.6.4 Paint Separation: Daimler AG Gaggenau

When automotive parts are painted, one wasteful result is large amounts of overspray. An important function in painting facilities is to collect the overspray, i.e., the paint, which does not land on the object during the coating operation. This is generally done through wet separation; that is, the overspray is absorbed by water. The water then subsequently has to be separated from the paint in an expensive process and the paint sludge has to be disposed of as waste.

One alternative to this is a dry method in which the overspray is caught by plastic fins (see Fig. 24.6). After drying of the paint, the fins can be dismounted and cleaned

Fig. 24.6 View from the dry separation with the fins (*bottom*) in the painting booth (Source: Daimler AG)

using specially designed brushing machines. This saves water at the site, and also avoids 9 t/a hazardous waste. At the same time it is possible to save electric power and natural gas, which leads to reductions in CO_2 emissions of 160 t p.a.

24.7 Conclusion

Resource efficiency at plant operational level ranks high among policy implementation strategies. This is where the technical innovation potential that can achieve the desired reduction in use of resources is based. However, addressing and motivating the actors in the industrial sector is a difficult task, as the companies act and react in accordance with different rules than those prevailing in the world of politics or academia. For companies, the focus is primarily on economic arguments. Moreover, at the same time competition prevents them from communicating innovative ideas or good case examples frankly and to a broad base. This makes it more difficult to disseminate best practice cases. So far, very little has been published about these impediments and about the success factors of positive examples.

If we want to integrate companies and industry as important partners in resource efficiency strategies, the dissemination of innovative ideas and technologies must be made attractive for manufacturing companies and be conducive to their competitiveness concerns. This has not been achieved to date. Communicating best practice cases and promoting participating companies through media and PR outlets is one possible way of facilitating resource efficiency among the actors in industry.

Acknowledgments Financial support of the Ministry of Environment, Climate and Energy of the State of Baden-Wuerttemberg (FKZ L75 14008-10)) is gratefully acknowledged. The project was carried out by INEC Pforzheim (www.hs-pforzheim.de/INEC), UTBW (www.umwelttechnik-bw.de) and IAT (www.iat.uni-stuttgart.de).

Reference

Facts, Graphics and Examples Are Cited from

Schmidt M, Spieth H, Bauer J, Haubach C (2017) 100 Betriebe für Ressourceneffizienz. Band 1. Praxisbeispiele aus der produzierenden Wirtschaft. Springer, Heidelberg

Chapter 25
Lifestyle Material Footprint of Finnish Households – Insights, Targets, Transitions

Michael Lettenmeier

Abstract This chapter presents the results of and reflects on several projects and studies that analysed the material footprints of Finnish households on a micro level and developed options for a transition towards sustainable household consumption. We started 10 years ago by establishing a database and methodology for assessing the material footprint of household consumption. The first projects thus concentrated on analysing Lifestyle Material Footprints and understanding the facts behind them. A sustainable annual target level of eight tonnes per person was introduced in a later study. On the basis of this, measures for a transition towards the target were developed and implemented in experiments with households and other actors in recent projects. Currently, a joint effort between municipal, private and educational actors seeks to challenge households to systematically reduce their material footprints. The chapter concludes that the Lifestyle Material Footprint is suitable for providing relevant insights into household consumption, illustrating sustainability targets and facilitating the transition towards sustainable lifestyles.

Keywords Household • Consumption • Lifestyle • Material footprint • Natural resources • Sustainability

25.1 Introduction

Household consumption is a relevant factor in terms of natural resource use. On a global level, the role of household consumption is increasing because affluence and population are growing faster than technological efficiency (Lorek and Spangenberg 2014). In the European Union approximately 55% of the final use of resources is due to households, a share that exceeds public consumption and capital formation

M. Lettenmeier (✉)
Department of Design, NODUS Research Group on Sustainable Design, Aalto University, Hämeentie 135 C, FIN-00560 Helsinki, Finland

Research Group Sustainable Production and Consumption, Wuppertal Institut, Döppersberg 19, D-42103 Wuppertal, Germany
e-mail: michael.lettenmeier@aalto.fi

H. Lehmann (ed.), *Factor X*, Eco-Efficiency in Industry and Science 32,
https://doi.org/10.1007/978-3-319-50079-9_25

(Watson et al. 2013). Already in the 1990s authors stated that in order to achieve sustainability a considerable reduction of resource intensity has to take place in Western economies (Schmidt-Bleek 1993) as well as Western lifestyles (Heiskanen and Pantzar 1997).

This chapter reports and reflects on several studies on the Lifestyle Material Footprint (LMF) of Finnish households and the potential transition to a more sustainable consumption level. The projects started by analyzing LMFs to develop a LMF sustainable consumption target of eight tonnes and then used this target to implement the transition in household consumption.

25.2 Materials and Methods

The principal method utilised in this chapter is based on a micro-level LMF calculation for households. The LMF as a methodology is itself based on the MIPS concept (material input per unit of service, Schmidt-Bleek 1993). It encompasses the natural resource categories of abiotic materials, biotic materials and erosion in agri- and silviculture and includes both materials used by the human economy and unused extraction (see Schmidt-Bleek et al. 1998, Ritthoff et al. 2002, Aachener Stiftung 2011). Lettenmeier et al. (2009) proposed the term *material footprint* as a synonym for the *ecological backpack* in order to extend the footprint metaphor to material use. On the micro-level, this has become state of the art (see Liedtke et al. 2014). The *micro-level* material footprint as used in this chapter includes the same resources as the two macro-level indicators: TMC (total material consumption) and TMR (total material requirement). On the *macro-level*, however, Wiedmann et al. (2015) used the term *material footprint* as a synonym for RMC (raw-material consumption), which has since become common practice on macro-level (see e.g. Chap. 4 of this book). The RMC includes the use of materials throughout the value-chain but excludes unused extraction as the data here is not yet available or systemised sufficiently (see also Chap. 4). Therefore, while methodologies for assessing the RMC of nations have developed faster and further than TMC accounting so far, the problem with leaving unused extraction out of the calculations has been recognised and should be tackled further (e.g. Dittrich et al. 2012, Lettenmeier and Heikkilä 2015).

As a micro-level indicator, MIPS was initially used for calculating the natural resource use of products and companies (e.g. Ritthoff et al. 2002). Before the year 2006, households were covered mainly as one sector in the macroeconomic TMR and TMC calculations (e.g. Adriaanse et al. 1997). However, these calculations do not provide insight in the resource use of specific households and the factors that influence resource use (Teubler et al. 2016). In addition, their results can differ from micro-level calculations (e.g. Lähteenoja et al. 2007) because of different allocation procedures—for example, in relation to infrastructure (for details, see Lettenmeier et al. 2014).

Table 25.1 gives an overview on the projects that analysed the LMF of Finnish households. The series of projects covered by this chapter started in 2006 by estab-

Table 25.1 Overview of projects (own compilation based on the references mentioned and internal project data)

Project number	Project name	Duration	Main focus of project	Main methods applied	Number of participating/ applied households	Main sources of further information
(1)	FIN-MIPS household	04/2006–12/2008	Database establishment	MIPS calculation of goods and activities	27/90	Kotakorpi et al. (2008)
			Insight in (LMF of) different households	Consumption surveys (6 weeks) on paper forms		
				LMF calculation		
				Focus group discussions		
(2)	Back to basics	10/2010–02/2014	Insight in (LMF of) low-income single households	Consumption survey (2 weeks) on paper form	18 (/170)	Hirvilammi et al. (2013) and Lettenmeier et al. (2012)
			Relation of social and ecological sustainability	In-depth interviews		
				LMF calculation		
(3)	SPREAD	01/2012–12/2012	Future visions and scenarios for sustainable lifestyles	Web-based consumption survey, in-depth interviews and LMF calculation	14/50 (fi), 38 (DE + ES + HU)	Kuittinen et al. (2013), Leppänen et al. (2012), Neuvonen et al. (2014) and Mont et al. (2014)
				Scenario-building (delphi and back-casting)		

(continued)

Table 25.1 (continued)

Project number	Project name	Duration	Main focus of project	Main methods applied	Number of participating/ applied households	Main sources of further information
(4)	Future household	02/2014–01/2015	Transition to sustainable lifestyles on household level	Consumption survey (3 weeks) on excel forms	5/46	Laakso and Lettenmeier (2016) and Lettenmeier et al. (2015)
			Developing roadmaps for reducing household's LMF	LMF and CF calculation		
				Roadmapping (co-creation, back-casting)		
			Testing solutions	Experiments and their evaluation		
				Several interviews		
(5)	Low-carbon April	01/2016–05/2016	Developing and testing low-carbon and low-resource options for households	Consumption survey (3 weeks) on excel forms	5/7	Vähähiilinen (2016)
				LMF and CF calculation		
				Light roadmapping		
				Experiments and their evaluation		
	Average Finn		Calculation for projects (1) and (2) and others	LMF calculation on micro-level based on statistics etc.		Lähteenoja et al. (2007) and Lettenmeier et al. (2014)

lishing a database and methodology for assessing the LMFs of households. The database and methodology were used in all the projects described here with some adjustments over the course of each project. In each project, the LMF of a number of households was analysed on the basis of consumption surveys. The households were then chosen out of a larger number of applicants. The main goal of the final choice was to get a sample of widely varying households to allow for insights in terms of size, affluence, living conditions, etc. An exception to this procedure was project 2 where a random sample of 170 low-income single households out of the database of the Finnish social security institution KELA was invited by paper mail and all 18 respondents were accepted to participate (Hirvilammi et al. 2013).

While LMF calculation was a basic part of all projects, the other methods applied varied according to the projects' foci. Project 1 had to establish a database for LMF calculation by means of a MIPS calculation for numerous household-related goods and activities. After this, the project had to develop an acquisition method for household data that was both sufficient in terms of data quality and feasible in terms of the amount of effort demanded of the households. The consumption components were monitored or reported on during a period of 6 weeks and included housing, mobility and leisure time activities, nutrition, packaging and wastes, tourism and summer houses as well as household goods and appliances. Project 2 was motivated by project 1 but the focus was instead at the interface of ecological and social sustainability. For data collection, the project combined in-depth interviews with questionnaires so that data collection was reliable but required less time and effort on the part of the households. Single households were chosen for the study as this is the most common household size in Finland and because individual material footprints are more unambiguous to calculate and compare. The project was also the first one that applied the concept of a sustainable level of LMF partly initiated in project 3 and finally published by Lettenmeier et al. (2014). Project 3 covered sustainable future lifestyles on a very broad basis and used the LMF as a means for visualizing and cross-checking the sustainability of different lifestyles and future options. Compared to the other projects, the LMF assessment in this project used a web survey that provided less detailed results, probably on the low side in terms of LMF level but still sufficient level for considering and discussing future lifestyle options. Project 4 focused on the active transition process at the household level and the supporting systems around the household for reducing footprints to a more sustainable level. It covered the whole transition cycle (see Laakso and Lettenmeier 2016) from problem assessment (LMF calculation) to future visioning (roadmaps), experimenting and considerations for mainstreaming. Project 5 focused in a pragmatic way on the local feasibility of making lifestyles sustainable and improving the prerequisites for doing so. Both projects 4 and 5 were based on a detailed 3 weeks long consumption survey that included the consumption components: housing, mobility, leisure time activities, tourism (incl. Summer houses), nutrition, and household goods and appliances. Both projects also included an assessment of the households' carbon footprints (CF).

25.3 Insights – Lifestyle Material Footprints of Finnish Households

Project 1 (FIN-MIPS Household) assessed the natural resource consumption of 27 households from three regions in Southern Finland. The households had a total of 78 members and the household size varied from one to nine persons. With a natural resource consumption level of 39 tonnes per person in a year, the average LMF of the participating households was close to the Finnish average. However, differences between the individual households ranged from 13 to 118 tonnes, respectively, i.e., a factor of nine between the households with the lowest and highest levels of consumption. In most cases, mobility, housing, nutrition and tourism were the most relevant consumption components of the participating households (Kotakorpi et al. 2008).

As a part of the project ‘Back to basics: Consumption and basic income security’, the LMF and the quality of life of 18 single households living on basic social security was analysed. The income of the participants ranged from 400 to 1200 Euros per month, which is low compared to the average Finnish income of approximately 3000 Euros. The material footprint of the participating single households ranged from 7 to 35 tonnes per year, which means a factor five of maximum difference. The households’ consumption patterns were rather similar and comprised primarily of housing, nutrition and daily mobility but varied highly in leisure time activities and travelling (Hirvilammi et al. 2013; Lettenmeier et al. 2012).

Figure 25.1 shows that the LMFs and the material footprints of the three most relevant consumption components of the participants in projects 3, 4 and 5 show similar patterns as in project 1. Conversely, the low-income participants in project 2 show smaller footprints in nearly every aspect. Nutrition and housing had a higher material footprint than mobility for most participants in project 2. This underlines the prominent role of basic needs for low-income households (Lettenmeier et al. 2012).

Out of the 69 households that participated in the studies mentioned, only one person in project 2 had a LMF within the ecologically sustainable level of eight tonnes per person per year. However, during the study this person did not have a home but lived instead with friends on a temporary basis. Thus, only electricity and heat consumption in the friends’ apartment and no built infrastructure was allocated to that participant’s housing footprint. As this example shows, in the present Finnish society an ecologically and socially sustainable lifestyle can hardly be achieved simultaneously. The fact that all other low-income households exceeded the sustainable level of eight tonnes shows that sustainable lifestyles cannot solely be achieved by reducing consumption and that material structures, e.g., the physical infrastructure provided by society and the products and services supplied by producers, greatly determine the environmental impacts of people’s lifestyles (e.g. Hirvilammi et al. 2013).

Projects 4 and 5 also assessed the carbon footprint (CF) of households in addition to the LMF. The results show that in general, both footprint methods provide similar results. In general, mobility shows a higher share in the LMF than in the CF of house-

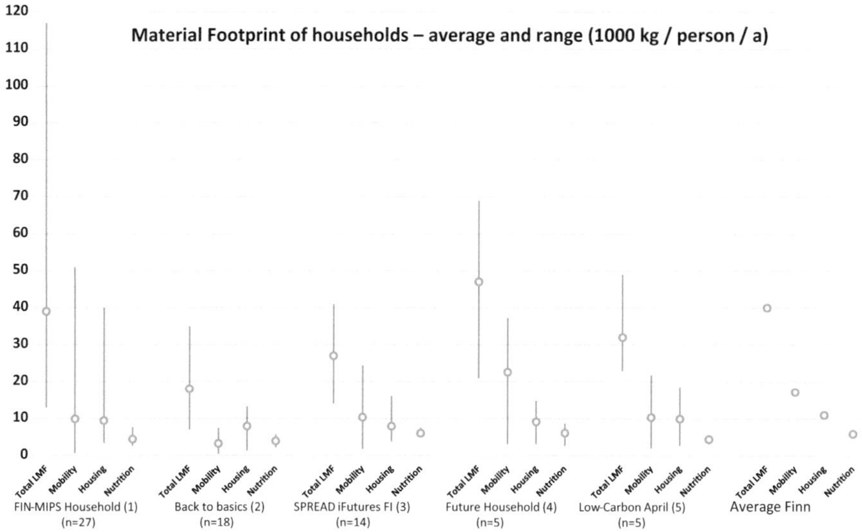

Fig. 25.1 Overview of LMF results (Sources: Kotakorpi et al. 2008; Lettenmeier et al. 2012, 2015; Verbree et al. 2013; Vähähiilinen 2016)

holds, while nutrition usually has a slightly lower share in the LMF. This is mostly due to the fact that infrastructure plays a more visible role in the LMF of households than with most other indicators. Nevertheless, both the LMF and CF of households provide rather similar results—with a few relevant exceptions. Oil heating and air travel greatly affect the CF but do not play a special role in the LMF because burning oil and kerosene releases large amounts of CO_2 but their production is not especially material-intensive and air traffic requires relatively little infrastructure (e.g. Kotakorpi et al. 2008; Lähteenoja et al. 2006). On the other hand, the use of electric cars decreases the CF of car-driving whereas it increases material footprints because electric cars require the same amount of infrastructure and the motive system of electric cars is more material-intensive than that of conventional cars. Additionally, electric power itself may also be material-intensive (e.g. Frieske et al. 2015).

25.4 Targets: The Eight Tonnes Resource Cap as a Vision and Benchmark for Future Lifestyles

During project 3 (SPREAD), the need arose to determine a specific sustainable level of LMF and what it could consist of. On the basis of macro-level considerations (Bringezu 2009), a sustainable LMF level of eight tonnes per person per year by 2050 was suggested (Lettenmeier et al. 2014). The purpose of the eight tonnes vision is to illustrate and support the understanding of future sustainable lifestyles. Although future lifestyles will be diverse and vary widely between different

Table 25.2 Present average Finnish material footprints and future requirements in the different consumption components

Consumption component	Current material footprint		Sustainable material footprint		Change required	
	kg/(person·a)	Share	kg/(person·a)	Share	%	Factor
Nutrition	5900	15%	3000	38%	−49%	2.0
Housing	10,800	27%	1600	20%	−85%	6.8
Household goods	3000	7%	500	6%	−83%	6.0
Mobility	17,300	43%	2000	25%	−88%	8.7
Leisure activities	2000	5%	500	6%	−75%	4.0
Other purposes	1400	3%	400	5%	−71%	3.5
Total	**40,400**	**100%**	**8000**	**100%**	**−80%**	**5.1**

Source: Lettenmeier et al. (2014)

households (Leppänen et al. 2012; Kuittinen et al. 2013), the eight tonnes vision can serve as a benchmark illustrating the amount and quality of goods, services and activities that fit into a sustainable lifestyle, accessible to all humans on Earth. Table 25.2 shows one example of the LMF reduction in different consumption components to achieve a sustainable LMF on a household level.

Lähteenoja et al. (2013) point out that a "lot of imagination is needed to understand how the shift from the current overconsumption can be turned into sustainable lifestyles for all. On the one hand, we need a deeper understanding on how to scale up current promising practices. On the other hand, we need to know how far these practices will take us towards sustainable living for all, and what kinds of new creative solutions, currently unknown or 'unthinkable', might bring us there." Based on various weak signals and promising examples, the eight tonnes concept is a suggestion for a possible sustainable lifestyle, and thus helps to clarify and envision the shift from overconsumption to sustainability.

Although the calculations and database of the eight tonnes concept are mainly based on the data used for LMF calculations of Finnish households, the magnitude of the target itself is relevant for other countries as well. The TMC target (10 tonnes per capita per year), which the eight tonnes suggestion for household consumption is based on, has been reinforced by recent research on sustainable resource use (Bringezu 2015). The challenge in terms of lifestyle transition will be similar for other Western industrialized countries because similar levels of TMC have been reported for many Western countries (Bringezu et al. 2009).

The eight tonnes benchmark can also be used for developing sustainable solutions for future markets. Lettenmeier (2017) presents an 'Evaluation Framework for Design for One Planet' that is based on the eight tonnes concept. The framework is intended as both an evaluation tool and a source of inspiration to promote design solutions supporting one-planet lifestyles. The World Business Council for Sustainable Development (WBCSD) has considered options and pathways for achieving sustainable lifestyles in the huge emerging markets of India, Brazil and China. With average LMFs of 8.4, 11.4 and 15.2 tonnes per person, respectively, these countries are currently much closer to the sustainable level of LMF than

Western countries although projections show that on a business as usual basis their average LMFs are likely to increase (WBCSD 2016a, b, c). The present average LMF values of these countries show that currently large groups of people must have considerably lower footprints than the sustainable level. Thus, if dematerialized goods and services are developed systematically and soon, the living standard of many people could be improved immensely without a need for exceeding the sustainable LMF level.

25.5 Transitions: Households on the Road to Sustainability

The European Commission's Framework Programme 7 (FP7) project 'SPREAD Sustainable Lifestyles 2050' (project 3 in Table 25.1 and Fig. 25.1) consisted of a European social platform investigating ways of more sustainable everyday living at the household level. In the project, four future scenarios for European sustainable lifestyles in 2050 were developed that described links between emerging lifestyle patterns and infrastructural development and were based on the prerequisite of a LMF of eight tonnes per capita per year. The scenarios demonstrate that there are several ways and alternatives to sustainable lifestyles and societies. By using a backcasting methodology, the project tried to understand how changes in societies emerge and transform, and how lifestyle-level experiments could serve as bottom-up actions for macro-level transitions (Leppänen et al. 2012; Lähteenoja et al. 2013; Neuvonen et al. 2014; Mont et al. 2014). One part of the project called 'iFuture' investigated how new ideas could reduce peoples' material footprints in practice. The material footprints of participants from four European countries were calculated and, in addition, deep interviews were conducted on the participants' life situations, everyday choices as consumers, own experiences, values and futures. On the basis of the participants' lifestyle profiles and material footprints, sustainable future projections were developed and discussed, which again showed the diversity of future sustainable lifestyles (Kuittinen et al. 2013).

While project 3 already largely focused on visioning future lifestyles, the projects 'Future household' (4) and 'Low-carbon April' (5) also experimented with future lifestyles and considered options for mainstreaming. In project 4, households developed roadmaps for reducing their LMF until 2030 by 37–61%, which is about half way towards the eight tonnes target for 2050. During the experiment phase, households were able to reduce their LMFs by 26–54%. Although a part of the experiment was based on simulated services not yet existing on a regular basis in the region, the results showed that relevant reductions in LMF can be achieved even in the short term. In addition, households reported an increase in quality of life during the experiments because of, e.g., better mobility planning, home delivery of groceries or decreased living space. In an interview round several months later the project households conveyed that several options tested or developed in the project were still going on, e.g., ride-sharing, giving up the second car, planning co-housing in the city centre and increasing vegetable-based food. On the other hand, some changes in life situations had also increased resource use, such as measures that required a new car. This implies the need for changes in supply structures that go beyond individual behaviour changes and temporary experiments in order to facilitate sustainable resource use.

Project 5 was similar to project 4, but due to a much smaller budget, the solutions tested had to be on the market in one way or another. This facilitated a pragmatic collaboration between the organizing local authority and local companies for supplying solutions. Also the households' target-setting was completed more quickly and with an easier process compared to the roadmapping process of project 4. On the basis of the options tested during the project and considered for the near future, the participating households were able to drop their material footprints by 5–49% and their carbon footprints by 8–77%.

25.6 Outlook: The Eight Tonnes Concept Spreading

The eight tonnes resource cap suggestion for households has gained interest both in and outside of Finland. The WBCSD activities for developing sustainable market solutions were described in section 26.4. In Germany, an online resource calculator[1] gives its users a LMF target for 2030 that is based on the eight tonnes target for 2050. The calculator has already attracted tens of thousands of users. One project on the LMFs of 16 German households used a consumption survey adopted from project 1 and calculated the LMFs with a mostly Germany-specific, partly more accurate database (Greiff et al. 2017). The project showed similar results as the Finnish projects mentioned, with a tendency for lower footprints for mobility and higher ones for household goods. These differences can be explained by different consumption patterns and different databases.

In the city of Lahti in Southern Finland, a joint effort of municipal, commercial and educational actors has started to challenge households to gradually but systematically reduce their material footprints. The challenge is called 'Tonni lähti', which in English has a meaning similar to *one tonne less*. The idea of the challenge is that by reducing one tonne of LMF per year, the average Finn would achieve the eight tonnes sustainable LMF by 2050. A website[2] provides ideas and background for the challenge, including a link to the recently developed resource calculator.[3] The calculator was initiated by the Tonni lähti consortium. It corresponds to the German resource calculator, with a footprint database adapted to Finland but otherwise with the same questions on the user's consumption and socioeconomic background. Thus, the data gathered by the Finnish calculator can be compared to the associated German data once it has been used widely enough. In addition, we have started to consider how to combine business interests and models with the calculator in order

[1] www.ressourcen-rechner.de

[2] www.tonnilahti.fi

[3] www.niisku.lamk.fi/~matlaskuri

to promote one-planet businesses on the one hand and on the other to increase the challenge's significance and reach by incentivising businesses to promote it.

25.7 Discussion and Conclusions

On the basis of the experiences and conclusions from the projects described in this article, a LMF assessment of households appears feasible and reasonable. It can provide relevant insights that can help to direct the transition efforts of both households and other actors. While macro-level calculations can provide data on the resource use from household consumption, micro-level data as provided by the projects described here can highlight additional aspects, e.g., on the strong links between infrastructure and household consumption (Kotakorpi et al. 2008; Hirvilammi et al. 2013). Although the general correlation between income and footprints is widely known, micro-level studies can provide additional insights into questions of social practices, social status and the like (Kotakorpi et al. 2008; Laakso and Lettenmeier 2016; Teubler et al. 2016). The projects described in this chapter have also shown the inter- and transdisciplinary dimensions related to LMF calculation and reduction, which has links to, for instance, the social sciences, politics, business and design.

As a holistic indicator, the LMF is well suited for covering the multifaced nature of household consumption. Overall, LMF results correspond well to CF results but the LMF illuminates a greater range of influencing factors than a CF analysis. Therefore, in cases with a low material footprint but high additional footprints, it may be necessary to reduce specific environmental impacts (e.g. climate change with oil-heating or air travel). Conversely, solutions with high material footprints despite low CFs (e.g. electric cars) may actually be insufficient for reducing overall environmental pressure.

Although there were large differences across households in both the level and composition of LMFs, out of all the households studied in all five projects only a homeless person achieved a LMF below the sustainable level of eight tonnes. This shows that present Finnish (and Western) lifestyles are far from sustainable. In present-day Finland, a decent lifestyle is hardly achievable within the ecological limits of natural resource use.

Yet, there are huge potentials for reducing LMFs. Considerable reductions are possible even in the short term (Laakso and Lettenmeier 2016) and even with increasing quality of life. A variety of solutions for decreasing LMFs already exists (e.g. Lettenmeier et al. 2014; Lettenmeier et al. 2015; Vähähiilinen 2016) and with a concerted effort by companies and authorities, remarkable reductions in LMFs can be achieved in the following years. From this perspective, achieving a sustainable LMF by 2050 appears possible although it will require notable improvements in consumption, production and infrastructure. There is every reason to believe that future sustainable lifestyles will be as diverse as present lifestyles (Leppänen et al. 2012; Kuittinen et al. 2013).

Acknowledgements I would like to express my acknowledgements to all colleagues, co-authors and last but not least the households involved in the projects. Without your contributions the results and outcomes presented in this chapter would never have been achieved. Hopefully our projects, methodologies and results can inspire scientists and practicians, politicians and servants, producers and consumers, each of us in our different roles, to promote sustainable lifestyles for all humans on Earth.

References

Aachener Stiftung Kathy Beys (2011) Factsheet measuring resource extraction. Sustainable resource management needs to consider both used and unused extraction. Aachener Stiftung Kathy Beys. https://www.aachener-stiftung.de/fileadmin/ASKB/Media/AcStift_factsheet_1102.pdf

Adriaanse A, Bringezu S, Hammond A, Moriguchi Y, Rodenburg E, Rogich D, Schütz H (1997) Resource flows: the material basis of industrial economies. World Resources Institute, Washington, DC

Bringezu S (2009) Visions of a sustainable resource use. In: Bringezu S, Bleischwitz R (eds) Sustainable resource management. Global trends, visions and policies. Greenleaf, Sheffield, pp 155–215

Bringezu S (2015) Possible target corridor for sustainable use of global material resources. Resources 4(1):25–54

Bringezu S, Schütz H, Saurat M, Moll S, Acosta-Fernández J, Steger S (2009) Europe's resource use. Basic trends, global and sectoral patterns and environmental and socioeconomic impacts. In: Bringezu S, Bleischwitz R (eds) Sustainable resource management. Global trends, visions and policies. Greenleaf, Sheffield, pp 52–154

Dittrich M, Giljum S, Lutter S, Polzin C (2012) Green economies around the world. Implications of resource use for development and the environment. Sustainable Europe Research Institute, Vienna

Frieske B, Klötzke M, Kreyenberg D, Bienge K, Hillebrand P, Hüging H, Koska T, Monscheidt J, Ritthoff M, Soukup O, Tenbergen, J (2015) Begleitforschung zu Technologien, Perspektiven und Ökobilanzen der Elektromobilität: STROMbegleitung; Abschlussbericht des Verbundvorhabens. DLR, Wuppertal Institut. https://epub.wupperinst.org/files/5966/5966_STROMbegleitung.pdf

Greiff K, Teubler J, Baedeker C, Liedtke C, Rohn H (2017) Material and carbon footprint of household activities. In: Keyson DV, Guerra-Santin O, Lockton D (eds) Living labs. Springer, New York, pp 259–275

Heiskanen E, Pantzar M (1997) Toward sustainable consumption: two new perspectives. J Consum Policy 20(4):409–442

Hirvilammi T, Laakso S, Lettenmeier M, Lähteenoja S (2013) Studying well-being and its environmental impacts: a case study of minimum income receivers in Finland. J Hum Dev Capabilit 14(1):134–154

Kotakorpi E, Lähteenoja S, Lettenmeier M (2008) Household MIPS–natural resource consumption of Finnish households and its reduction. The Finnish environment, vol 43en|2008. https://helda.helsinki.fi/handle/10138/38369

Kuittinen O, Mokka R, Neuvonen A, Orjasniemi M, Ritola M, Wikholm M (2013) iFuture—the diversity of sustainable lifestyles. Report D7.3 People's Forum Workshop Summaries of the Project SPREAD Sustainable Lifestyles 2050. Demos Helsinki, Collaborating Centre on Sustainable Consumption and Production (CSCP). http://sustainable-lifestyles.eu/fileadmin/images/content/D7.3_iFuture_report.pdf

Laakso S, Lettenmeier M (2016) Household-level transition methodology towards sustainable material footprints. J Clean Prod 132:184–191

Lähteenoja S, Lettenmeier M, Saari A (2006) Transport MIPS–the natural resource consumption of the Finnish transport system. Finn Environ 820. https://helda.helsinki.fi/bitstream/handle/10138/40369/FE_820en.pdf?sequence=3

Lähteenoja S, Lettenmeier M, Kauppinen T, Luoto K, Moisio T, Salo M, Tamminen P, Veuro, S (2007) Natural resource consumption caused by Finnish households. In: Proceedings of the Nordic Consumer Policy Research Conference Helsinki, pp 1406–1420. https://helda.helsinki.fi/handle/10138/152254

Lähteenoja S, Neuvonen A, Groezinger R (2013) From global champions to local loops: sustainable lifestyles in Europe 2050. Paper presented to the World Resources Forum, Davos, 6–9 October. www.demoshelsinki.fi/julkaisut/from-global-champions-to-local-loops-sustainable-lifestyles-in-europe-2050/

Leppänen J, Neuvonen A, Ritola M, Ahola I, Hirvonen S, Hyötyläinen M, Kaskinen T, Kauppinen T, Kuittinen O, Kärki K, Lettenmeier M, Mokka R (2012) Scenarios for sustainable lifestyles 2050: from global champions to local loops. Report D4.1 future scenarios for new European social models with visualisations of the project SPREAD sustainable lifestyles 2050. Demos Helsinki, Collaborating Centre on Sustainable Consumption and Production (CSCP). http://sustainable-lifestyles.eu/fileadmin/images/content/D4.1_FourFutureScenarios.pdf

Lettenmeier M (2017) Design for one planet – a framework for evaluating design solutions. Submitted manuscript under review

Lettenmeier M, Heikkilä I (2015) Decoupling of economic growth and environmental impacts in Germany. Paper presented to the World Resources Forum, Davos, 11–14 October

Lettenmeier M, Rohn H, Liedtke C, Schmidt-Bleek F (2009) Resource productivity in 7 steps, how to develop eco-innovative products and services and improve their material footprint. Wuppertal Spezial 41. https://epub.wupperinst.org/files/3384/WS41.pdf

Lettenmeier M, Hirvilammi T, Laakso S, Lähteenoja S, Aalto K (2012) Material footprint of low-income households in Finland—consequences for the sustainability debate. Sustainability 4(7):1426–1447

Lettenmeier M, Liedtke C, Rohn H (2014) Eight tons of material footprint—suggestion for a resource cap for household consumption in Finland. Resources 3(3):488–515

Lettenmeier M, Wikholm M, Salovaara JJ, Laakso S, Toivio V (2015) Tulevaisuuden kotitalous – Resurssiviisaan arjen tavat ja palvelut. Tulevaisuuden kotitalous -hankkeen loppuraportti (Future household – the habits and services of a resource-wise everyday, final report of the future household project, in Finnish). SITRA. http://www.sitra.fi/sites/default/files/u753/tulevaisuuden_kotitalous_loppuraportti.pdf

Liedtke C, Bienge K, Wiesen K, Teubler J, Greiff K, Lettenmeier M, Rohn H (2014) Resource use in the production and consumption system—the MIPS approach. Resources 3(3):544–574

Lorek S, Spangenberg JH (2014) Sustainable consumption within a sustainable economy–beyond green growth and green economies. J Clean Prod 63:33–44

Mont O, Neuvonen A, Lähteenoja S (2014) Sustainable lifestyles 2050: stakeholder visions, emerging practices and future research. J Clean Prod 63:24–32

Neuvonen A, Kaskinen T, Leppänen J, Lähteenoja S, Mokka R, Ritola M (2014) Low-carbon futures and sustainable lifestyles: a backcasting scenario approach. Futures 58:66–76

Ritthoff M, Rohn H, Liedtke C (2002) Calculating MIPS: resource productivity of products and services. Wuppertal Spezial 27. https://epub.wupperinst.org/frontdoor/index/index/docId/1577

Schmidt-Bleek F (1993) Wieviel Umwelt braucht der Mensch? MIPS – das Maß für ökologisches Wirtschaften. Birkhäuser

Schmidt-Bleek F, Bringezu S, Hinterberger F, Liedtke C, Spangenberg JH, Stiller H, Welfens MJ (1998) MAIA: Einführung in die Material-Intensitäts-Analyse nach dem MIPS-Konzept. Birkhäuser

Teubler J, Buhl J, Lettenmeier M, Greiff K (2016) A Household's precious burden – typical patterns of household equipment and its resource use in Germany. Submitted manuscript under review

Vähähiilinen huhtikuu (2016) Vähähiilinen huhtikuu, loppuraportti (Low-carbon April, final report, in Finnish). City of Joensuu. http://www.joensuu.fi/vahahiilinen-huhtikuu
Verbree R, Groezinger R, Hicks C, Kuhndt M, Kuittinen O (2013) Sustainable lifestyles–from today's realities to future solutions. SPREAD project internal paper, Demos Helsinki, Collaborating Centre on Sustainable Consumption and Production (CSCP)
Watson D, Acosta Fernandez J, Wittmer D, Pedersen OG (2013) Environmental pressures from European consumption and production: a study in integrated environmental and economic analysis. EEA Technical Report 2013, vol 2. Publications Office of the European Union. http://www.eea.europa.eu/publications/environmental-pressures-from-european-consumption
WBCSD (2016a) Sustainable lifestyles report Brazil. World Business Council for Sustainable Development. http://wbcsdpublications.org/project/sustainable-lifestyles-report-brazil/
WBCSD (2016b) Sustainable lifestyles report China. World Business Council for Sustainable Development. http://wbcsdpublications.org/project/sustainable-lifestyles-report-china/
WBCSD (2016c) Sustainable lifestyles report India. World Business Council for Sustainable Development. http://wbcsdpublications.org/project/sustainable-lifestyles-report-india/
Wiedmann TO, Schandl H, Lenzen M, Moran D, Suh S, West J, Kanemoto K (2015) The material footprint of nations. Proc Natl Acad Sci 112(20):6271–6276

Chapter 26
Construction 4.0: The LifeCycle Tower and Digitalised Timber Construction

Hubert Rhomberg

Abstract The purpose of this paper is to rethink the world of construction anew and to transfer it to the future with modern methods and technologies. To this end, we present the LifeCycle Tower (LCT)-construction system that was developed at the end of 2010 as part of a multi-annual, cooperative and interdisciplinary research project lasting over several years. It combines two problem solving approaches that significantly contribute to a CO_2-neutral building sector: the utilisation of wood as a building material and the regional realisable modular construction system. Additionally, a comprehensive energy supply concept was developed that helps to reduce energy demand to a minimum and is at the same time in tune with the needs of the users. Rhomberg is also changing the development and planning process with his disruptive approach: Turning and anchor point of the new building will be an interdisciplinary, digital, living platform – a kind of cybernetic table for everything involving the topics "planning, erecting and operation" of buildings. With these developments, construction, as we know it, is completely modified and turned on its head.

Keywords Bauen 4.0 • Internet of things • 3th revolution • BIM • LifeCycle tower • Hybrid slabs • Tall buildings • Illwerke Zentrum Montafon • Prefabrication • Carbon storage technology • Less tech • Dismantling • LCT • From the ego to the lego principle • Education • Cree by Rhomberg • Platform technology • Digital twin • Digitization

26.1 Introduction

The LifeCycle Tower construction system was developed as the result of a two-stage cooperative and multi-disciplinary research project that spanned over several years. It combines two approaches which contribute significantly to a CO_2-neutral building sector: the use of timber as a construction material and a regionally available prefabricated

H. Rhomberg (✉)
Cree GmbH and Rhomberg Holding GmbH, Mariahilfstr. 29, A-6900 Bregenz, Austria
e-mail: hubert.rhomberg@creebyrhomberg.com

H. Lehmann (ed.), *Factor X*, Eco-Efficiency in Industry and Science 32,
https://doi.org/10.1007/978-3-319-50079-9_26

Fig. 26.1 The LifeCycle Tower ONE in Dornbirn (Source: Norman A. Müller)

construction system. The practicality of the system has now been demonstrated with the LifeCycle Tower ONE as a prototype and with several client projects (Fig. 26.1).

Compared to conventional building methods, the modular building system and prefabrication contribute significantly to a reduction in life-cycle costs. Timber is not only an ideal construction material, but is also totally renewable, CO_2-neutral and represents careful use of resources. In addition, we have developed a comprehensive energy supply concept, which reduces energy consumption to a minimum whilst also satisfying user requirements. Furthermore, we are currently transferring our system to a digital platform, which will turn the building process we are familiar with on its head.

26.2 The Status Quo

The construction arm of Rhomberg operates in an industry that consumes 40% of all resources and is responsible for about 40% of the waste generated worldwide and for 60% of all transport movements (UNEP 2009, p. 5). These statistics are a clear indicator that there is an urgent need for action in the construction sector. This is not made easier by the fact that the industry is mostly conservative in its approach, and innovations are only slowly and reluctantly accepted. Furthermore, almost all building projects are essentially producing prototypes, that is to say that each building is newly designed from scratch. In practice, it is usually not possible to simply replicate existing design details, which means that experience gained from similar projects is lost. And – in most cases – the question as to what will happen to the building at the end of its service life is not even asked, not to mention answered. It is no surprise then that current construction methods are costly, use a lot of resources and are not very sustainable. This is bad news – particularly when considering that the availability of resources will be dramatically reduced in the near future. Last but not least, another challenge for the construction industry is the amount of red tape and the many regulations (such as those for obtaining housing subsidies), which can hinder the implementation of more sustainable methods. In Austria, about 6000 standards regulate the methods used to erect buildings; this increases the construction costs by up to 15%.

Another important aspect to consider is the trend towards urbanisation. While in 1800 about 25% of the population lived in cities, the proportion of city dwellers in 2008 was 50% for the first time in the history of mankind (UNFPA 2014). Already today, over three-quarters of worldwide energy consumption is attributed to cities. It is expected that by 2050, over 75% of the world population will live in so-called megacities, i.e. cities with more than one million residents (UNFPA 2014). One consequence of this rapid growth in urbanisation is the scarcity of space, which in many cases can only be resolved by constructing multi-storey buildings. Up to now, urban construction methods have focused on traditional, conventionally produced prototypes that involve long construction periods and complex construction management. That means that currently, new buildings are mostly built in conventional solid construction (reinforced concrete skeleton construction).

26.3 Method: "LifeCycle Tower" Timber Hybrid Construction System

In response to the aforementioned challenges, Rhomberg set itself the task of developing a construction method to build in urban settings that is suitable for digitalisation and satisfies the criteria for long-term sustainable construction and living. With support funding from the Austrian Federal Ministry for Transport, Innovation and Technology, we set up a research project to address these issues and – over a period of several years – we developed the LifeCycle Tower solution, which is based on the principles mentioned above, i.e. "building with timber" and "system construction".

We were further inspired in our efforts by Prof. Schmidt-Bleek, who demands "de-materialisation" by a factor of 10 and – with his concept of "ecological backpack" – has developed an indicator for determining how sustainable a product, or building, is (Schmidt-Bleek 2004). The ecological backpack is a measure of the amount of resources consumed in the manufacture, use and disposal of products. From that we concluded that we should replace those materials that have a very large backpack with other, more resource-efficient materials. The solution is wood, one of the oldest building materials there is. It is a renewable raw material that is available in many countries all over the globe. It is also a CO_2 sink and therefore plays an important role in balancing the world's climate; when used as a building material, it can reduce the total weight of a building by about 50%. Additional advantages are its high strength, good thermal insulation and the fact that it is 100% recyclable. Furthermore, modern timber construction offers a wide range of possibilities in terms of construction and conceptual design.

26.3.1 An Ambitious Goal: Building High-Rise with Timber

When timber is to be used in cities, it is not only important to build large volumes, but also to be able to construct high-rise buildings. What was considered impossible until a few years ago has now become reality – a resource-efficient solution for timber-based multi-storey buildings up to 30 floors or 100 m in height. With our subsidiary Cree GmbH (Creative Resource & Energy Efficiency), which was founded in 2010, we have developed a fully marketable and world-wide usable timber hybrid system for large buildings which can be individually designed and takes very little time to construct – the LifeCycle Tower (LCT) (Fig. 26.2).

Green building standards can be achieved in many ways. Many design solutions rely on sophisticated services installations to achieve significant energy savings and hence sustainability. However, in our opinion, that is not enough. If we really want to take a step towards a low-carbon construction industry, we have to consider the product from the cradle to the cradle, and we have to select materials accordingly early on during the design process. In the Cree system, we are conscious of the benefits of urban mining; we know the type and quantity of materials installed in the

Fig. 26.2 The LCT ONE was constructed in 8 days (Source: Cree GmbH)

building and we know how these materials can be put to new use again at the end of the building's life cycle. What is new – and hitherto unique – in our approach is that timber is used without being encapsulated or treated in load-bearing elements, i.e. no cladding or lining is applied. Using this "unencapsulated" structure is another way of saving resources. In this construction, timber remains the surface material and can therefore be experienced by the users; timber also has a positive and healthy effect on the interior climate.

26.3.2 Maximum Flexibility Based on Modular System Construction

Up to now, construction in urban settings has used conventionally produced prototypes with complex construction processes. This entails high construction costs, long erection times and significant design and construction risks. With the Cree system, we opt for a systems approach and an industrialised prefabrication process which has been used for decades in other industries, e.g. the automotive and shipbuilding industry.

The fully detailed timber hybrid construction elements are prefabricated based on drawings and can be used universally – in office buildings, hotels or housing, for restaurants or retail facilities. Modular system construction presents several other advantages:

- Both changes of use and renovation are considerably simplified;
- The facades can be configured to suit various requirements and desires, making it possible for each LCT to be custom-designed;
- Since no load-bearing partition walls are required, there is greater flexibility in the sub-division of spaces;
- The energy supply (heat, electricity) is highly efficient and can easily be adapted to the local situation; the focus is on the use of renewable energy sources (following standards such as Plus Energy, Passivhaus or Low Energy).

Fig. 26.3 The LCT ONE (Source: Architekten Hermann Kaufmann)

26.3.3 Efficient Use of Resources on a Large Scale

With the construction of the first projects in Vorarlberg and beyond, Cree has demonstrated that the system functions and is fully compliant with regulatory requirements, including fire protection – a key issue in timber construction.

LCT ONE, the eight-storey office building which was built in Dornbirn in 2012, is the prototype which is used by Cree GmbH and other tenants as company headquarters (Fig. 26.3).

It also accommodates the LifeCycle Hub, which is in a way a exhibition for items of the future, presenting sustainable solutions for the construction industry. This was followed in 2012 and 2013 by the new hydro power competence centre of Vorarlberger Illwerke AG in Vandans, Montafon, which was built using the Cree system and provides over 10,000 m^2 of gross floor area – one of the largest and most sustainable timber hybrid buildings in the world (Fig. 26.4).

LCT technology was also used for extending the offices of plant manufacturer Wagnertec in Nüziders. Since then, the system has been used for the first time in a combined commercial and residential building in Memmingen (Bavaria). We will continue this success story and revolutionise the construction process with the LCT NEXT, which will shortly be constructed in Bregenz.

The Cree system demonstrates that the goals stipulated by the EU in 2011 in the resource efficiency program can be achieved, and that de-materialisation by a factor of 2 is already possible in the construction industry. This clever system can be used worldwide as a universal model and thereby make a big contribution to the transition

Fig. 26.4 The Illwerke Zentrum Montafon (Source: Norman A. Müller)

to a low carbon economy. The distribution of the system will stimulate economic growth which is intelligent (innovative, and based on knowledge), sustainable (compatible with the environment and hence more sustainable in the long run) and integrative (new, regional "green jobs" strengthen social and territorial cohesion).

In short, I am convinced that multi-storey timber construction provides the solution in the quest for a sustainable resource-conserving building method, particularly in an urban context. But that is only part of the story.

26.4 Bauen 4.0

The LCT construction system devised by Cree GmbH is only the beginning of a completely new and much more sustainable method of erecting buildings. Throughout the world we are experiencing a cultural, ecological and economic process of change in which working and living can only be managed using innovative technologies and ecological concepts. When it comes to planning and designing change in the human environment, architecture and the construction industry face a challenge. To do this it is not sufficient to simply modernise existing processes and to digitalise the classic construction process – something that today, BIM (Building Information Modelling) is much too often understood to be. Instead, we must completely rethink construction ("Bauen" in German) as we know it today. The objective is to create a digital platform where all relevant information on building projects, regulatory requirements, building materials, building elements and stakeholders in the building process are compiled, so that this data is available, user-friendly and developed by all those involved in the process.

26.4.1 Digital Twin

This will make it possible, for example, to design a digital twin of the planned building before it is erected – in other words "building before building". All simulations regarding the load-bearing structure, thermal insulation, fire protection, building automation, etc. can be incorporated in the form of such a digital twin. However, this means that all electric components need an appropriate IP address to enable them to be selected, operated and maintained from a cloud. This is a prerequisite for establishing the digital twin in the first step and, in the second step, for making the building suitable for the Internet of Things (IoT)[1], rendering it into a smart building. This means that all processes relating to prefabrication, logistics, assembly and also of course dismantling, in addition to preventive and predictive measures for the use phase of the building, can be defined and derived from the model. This is an easy way of saving on electric wiring and cables by wireless connectivity, and hence of conserving valuable resources. This evolution makes the traditional design process in support of the construction, sometimes also in project rooms, obsolete. The digitalisation of the building process and the associated development have an impact on all parties involved. For example, architects and specialist engineers will in future take on the role of experts in detailed design. Submissions and building applications

[1] The notion of the "Internet of Things" is that the (personal) computer will gradually disappear and will be replaced by "intelligent objects". Instead of being the object of human attention – as is the case now – the "Internet of Things" is intended to seamlessly support people in their activities. The computers embedded in things will become ever smaller, and are supposed to support people without causing any distraction or even being noticed. (Source: wikipedia.org, 24 October 2016).

in the conventional written form and 2D drawings will progressively be replaced by digital models, thus enabling municipalities to take a step towards the digital city.

Of course, such development has a corresponding impact on all current stakeholders, not only the architects and specialist engineers. Everybody is involved, the tradesmen, authorities, service providers, contractors and customers – and above all, the manufacturing industries. For example, in future the choice of tradesman will be determined by which supplier can make his products available in the required Building Information Modelling (BIM) quality early on in the design phase. Shifting work processes from the building site to industrial production facilities can only succeed by radically reducing the existing interfaces and using appropriate semi-finished and finished products. This shift requires tradesmen to group together so that they can cooperatively produce modular elements. From this moment onward, many components of the services installations and the fit-out will be pre-assembled in components of the load-bearing structure (e.g. ceiling or column) and the external walls, providing a simple plug connection for these installations once the building shell has been assembled. With the help of this design method and pre-assembly of elements, it will be possible to apply quality and safety management methods such as FMEA[2] as used in industrial production and shipbuilding.

However, the digital evolution cannot take place until it is no longer necessary to design each building from scratch and, instead, the plans of previously built projects are actually available for download on a platform. Once this has been established the foundations will be laid for a "sharing community" which will aim to make its plans available for use by everybody else, although its planned buildings will be adapted in accordance with the respective social, cultural and urban requirements. Everything is permitted, within the boundaries of compliance with certain rules and the use of specified building components. This is a prerequisite to generate the size of components (semi finished goods) needed by industry, to preserve resources and keep costs down while simultaneously improving quality. It follows that such an approach requires different tendering procedures such as design & build contracts[3], or even design/build/operate contracts, in order to generate the maximum cascaded use of the digital twin. Once this step has been accomplished, it will be possible to duplicate and simulate the functions of 3D building models through to 7D[4] facility management information in the digital twin.

[2] Failure Mode and Effects Analysis.

[3] Collaboration of design and build teams reduces interfaces during the complete building process and enable an optimization of resource input, process quality and timing. In combination with operate the complete lifecycle of a building can be optimized with know-how and experience of all participants.

[4] BIM is much more than 3 dimension modelling. It includes also 4D (scheduling), 5D (estimating), 6D (sustainability) and 7D (facility management applications).

26.4.2 Learning from Others

Other industries, especially the automotive industry, are clearly ahead of construction here; items such as the chassis, engine and body are standardised and modular, and can therefore be manufactured cheaply and – most importantly – resource efficiently. In the automotive industry, some components are used even in different brands. For example, technical components produced by VW are also used in Audi, Seat and Skoda cars. Nevertheless, it is possible to individualise the car and adapt it to specific requirements using different fit-out components, colours and a choice of finishing material – the so-called modular transverse matrix. Using a "car configurator", the customer can configure his/her desired model and order it by pressing a button. In the near future, purchasing a house will follow a similar process – the client will configure the building at the computer, walk through it with 3D glasses in order to plan details and furnishings, and then order the house with a mouse-click (see Fig. 26.5). We are not there yet however. But it's only a question of time.

This time will come quicker than we think. The reason for this is also the logarithmic nature of human thinking: in order to anticipate the world in one year's time, we have to look back into the past – not one year back, but ten. The steps necessary to engender progress are becoming increasingly shorter due to digitalisation and

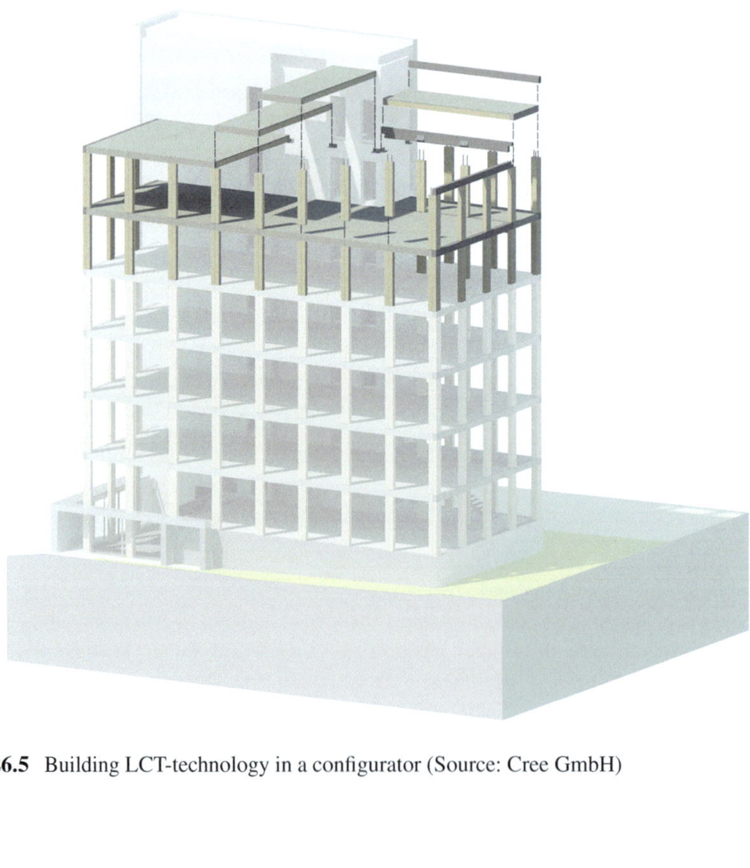

Fig. 26.5 Building LCT-technology in a configurator (Source: Cree GmbH)

technical development. Therefore the time for "Bauen 4.0" is now – the timber hybrid system construction method has established the prerequisites and ideal conditions for the next step. Some significant drivers around the world are sharing and open innovation in order to scale the revolution.

The focal point of the new method of building will be a lively, interdisciplinary digital platform. It could be described as a cybernetic table for all data relating to the design, construction and operation of buildings. A place to find all the information, contacts, products, forms and case studies from the world of systematic timber-hybrid construction. And also the option of erecting "digital twins". This platform will be subject to continuous development, and therefore continuous improvement; every newly developed building element, every successfully completed project, every new producer, architect or designer will enrich the platform and fill it with knowledge, thereby enhancing the knowledge of all those involved in the cybernetic table. This also means that in the future, knowledge will no longer be owned by a select group, but rather collectively owned for the benefit of the construction process and all the parties involved.

This is also the reason why we do not want to use our Cree system in isolation. In fact, we hope that many similarly creative spirits and pioneers will join us. Therefore we offer building contractors and architects the opportunity to use the innovative LCT system with us, and to drive its development. In order to place this offer on an equal footing and to make it possible for users to concentrate on their core competences as architects or (structural) engineers, we support our partners with items such as controlling, benchmarking, the organisation of key account management and our configuration software, CREEator. In this way we want to make the principle of a "sharing community" mentioned above a reality, and want to play our part in a "sharing economy" – and share our success.

The platform will be subdivided into various project rooms. There will be areas dedicated to different political administrations, which will contain statutory regulations and approval guidelines. There will be virtual meeting rooms for architects and engineers where they can set up meetings and even start virtual companies[5]. There will be product shops which will contain information on available construction elements – including all information on materials, costs, their ecological footprint and the contact details of their manufacturers. There will also be evaluation lists based on feedback from customer assessments, for example for the best projects (which will also include information on the elements and manufacturers involved), for the best timber architects, the best structural engineers and the best fire safety experts. All this knowledge, all the experiences and all the contacts on the platform are available to all users –without limitation in time and space.

We will say goodbye to the conventional construction process and the design process that supports construction will no longer exist. It will simply become superfluous. Instead, the architects will face a more demanding task in terms of urban design and social and cultural values. Today, shipbuilding and design in the

[5] Virtual companies are necessary to simulate all processes for digital fabrics according to Industrial 4.0.

automotive industry already work that way – the design engineers of a yacht or a cruise ship no longer have any influence on their plans once they have submitted them to the shipyard. And why should they? At the shipyard the ship will simply proceed along the defined and hundred-fold tried-and-tested process. The functions of architects and structural engineers will change. New players will come to the table – from IT, from community management, and from other areas that are currently quite separated from the industry. They will see the opportunities in this systematic design process, and use them. We might call it the "Uber" principle of the construction industry – the rules will be redefined. We want to, and will be, part of that. We owe this to our company, its employees, society at large and our specific trading location. But above all we owe it to the environment and future generations. All in the spirit of sustainability!

References

Schmidt-Bleek F (ed) (2004) The ecological backpack: An economy for a future with a future. [Original title: Der ökologische Rucksack: Wirtschaft für eine Zukunft mit Zukunft]. Stuttgart [et al], Hirzel

United Nations Environment Programme Sustainable Buildings & Climate Initiative (UNEP SBCI) (2009) Buildings and climate change, summary for decision-makers. www.unep.org/sbci/pdfs/sbci-bccsummary.pdf. Accessed May 2014

United Nations Population Fund (UNFPA) (2014) www.unfpa.org/pds/urbanization.htm. Accessed May 2014

Chapter 27
Protect Resources, Strengthen the Economy: Good Examples for Resource Efficiency in Industry and Handicraft Businesses

Peter Jahns

Abstract A lack of knowledge about internal business flows, lack of a cross-company perspective, dependence on customer requirements and the lack of time, personnel and financial resources, especially in small and medium-sized enterprises, mean that the advantages of the resource-efficient economy are not sufficiently exploited. Laws and regulations alone can not provide a remedy, and instead entrepreneurial ambition must be aroused. The experiences and successes of the work of the Duisburg Effizinez-Agentur NRW show that this is possible. Concrete and individually tailored support offers as well as the intensive cooperation of all participants often lead to a start-up in entrepreneurial activity. The public authorities can provide decisive support here, as the NRW Ministry of the Environment has shown with the establishment of the Effizienz-Agentur NRW. In the following we will outline how this work is carried out and explain the results by illustrating some examples.

Keywords Effizienz-Agentur NRW • Resource efficiency • Saving potential • Industrial companies • Handicraft business • Resource productivity

27.1 Introduction – The Effizienz-Agentur NRW

In the German state of North Rhine-Westphalia, the state-financed Effizienz-Agentur NRW (EFA) is the centre of excellence for resource efficiency. The agency supports the manufacturing industry to enhance productivity by reducing the use of natural resources. Analysing technical and organisational processes should reveal a company's potential for saving costs and resources. It was set up in 1998 at the

P. Jahns (✉)
Effizienz-Agentur NRW, Dr.-Hammacher-Str. 49 47119, Duisburg, Germany
e-mail: pja@efanrw.de

H. Lehmann (ed.), *Factor X*, Eco-Efficiency in Industry and Science 32,
https://doi.org/10.1007/978-3-319-50079-9_27

initiative of the Environmental Ministry of the State of North Rhine-Westphalia and works by order of the Ministry.

With the Effizienz-Agentur NRW the state of North Rhine-Westphalia provides the about 9,000 industrial companies and the about 18,000 craft enterprises with custom fit support in the identifying savings potential and the implementation of the best solution. As an independent and neutral institution, it has established a wide consulting range for the improvement of resource efficiency.

The agency is a business partner for small and medium-sized businesses as well as for large industrial corporations. Its range of services encompasses consultation on resource efficiency, financing activities and putting on events and training courses. As an innovative business partner, EFA launches and guides more than 250 projects per year in and with companies from industry and handicraft businesses in North Rhine-Westphalia (NRW). With its experience from NRW, EFA is also a business partner in demand on the national and international level for networks and consultation services.

The identification of saving potential is the first step on the road to efficient production processes and products. Nevertheless, only after implementation is material and energy consumption reduced and the environment and our climate protected. Therefore, EFA focuses its work on finding practical solutions that are easy to implement and have high probability to be put in practice by the companies. The following successful resource efficiency projects from North Rhine-Westphalia initiated and established by EFA gives an overview of the wide range of possibilities to save resources and strengthen the economy.

27.2 Good Examples for Resource Efficiency in Industry and Handicraft Businesses

27.2.1 Aluminium Norf GmbH (2014a)

By means of innovative heat treatment, Alunorf is able to decrease its energy demand by approx. 31,000 MWh per annum. EFA's financing experts assist the build up of the innovative measure.

Initial Situation

Aluminium Norf GmbH (Alunorf) is one of the largest aluminium smelting plants and rolling mills worldwide. The semi-finished and primary products, which have been made in Neuss since 1965, are used for – among other things –food packages and vehicle components. According to customer requirements, the plant with approx 2,200 employees produces aluminium strips in different material thicknesses, diameters and strip lengths. In the cold rolling process the strips heat up to 190 °C. In order to achieve the desired metallurgical properties, the rolled strips are heat-treated in an annealing furnace at 480 °C at the end of the process. The heat already introduced into the strips through the rolling process was not able to be used for the heat treatment by the system which existed until 2011. The strips needed to cool down below 60 °C prior to the heat treatment in order to provide process-reliable conditions for the existing time and temperature programs for the entire furnace chamber.

Measures and Advantages

In 2011, Alunorf implemented an innovative furnace concept consisting of an energy-efficient furnace group with five units in which the thermal state of every single strip can be controlled online, in order to utilize the residual heat of the strips for the heat treatment. The furnaces have a 4-single-zone control with an online process control which allows the individual annealing of four single coils. This makes it possible to record and control the coil temperatures during the entire annealing process online. For the first time, hot coils can directly be heat-treated by means of this new technology. Thus, the residual heat from the rolling process can be utilized in the annealing furnaces. Furthermore, Alunorf uses the hot furnace exhaust gases to preheat the protective gases utilized in the furnace chamber which also saves energy. Also, the internal logistics were able to be improved leading to shorter processing and machining times. The innovative heat treatment facility amounts to overall energy savings of 45% a year compared with the old facility, an equivalent of approx. 31,000 MWh annually. In this way, a CO_2 equivalent of about 8,500 tons at an annual production of 180,000 tons of aluminium strips can be avoided. Due to the innovative approach and the achieved savings, this implemented measure is considered a best practice for metal-working companies which operate multi-stage and interlinked processes with integrated heat treatment processes.

The Way to Financing and Implementation

In the run-up to the implementation, Alunorf benefitted from the financing advice provided by EFA in July 2010. After careful consideration of the planned measure, the company created a project sketch for the environmental innovation programme of the Federal Ministry for the Environment, Nature Conservation, Building and Nuclear Safety with the assistance of EFA. The project was subsidised with funds from the environmental innovation programme of the Federal Ministry for the Environment, Nature Conservation, Construction and Nuclear Safety worth € 1.5 million. After the approval of the subsidy in January 2011, EFA was assigned with the creation of the final report as well as the coordination of the measuring programme in August 2012. In December 2013 the project was finished. Alunorf invested approximately € 7.5 million altogether in the new heat treatment (Table 27.1).

Table 27.1 Resource effects at a glance

Energy savings	Approx. 31,000 MWh/a
CO_2 equivalents	Approx. 8,500t/a

27.2.2 Brauerei und Wirtschaft "Im Füchschen" Peter König e.K (2012a)

New purification process reduces waste water and chemicals consumption through a combination of state-of-the-art technology with traditional beer brewing.

Initial Situation

Since 1848, the brewery "Im Füchschen"has been producing Düsseldorf's most loved drink. The company of 104 employees brews 32,500 hl of so called "Altbier" (a special, dark type of German beer) annually, 70% of which is filled in kegs and 30% is bottled. In 2009 the brewery wanted to reduce the loss of beer during the production process as well as the organic waste water contamination caused by yeast and beer residues. The problem, however, was that due to the confined spatial conditions which are typical of the historic city of Düsseldorf, there was no way to install a conventional waste water treatment facility. In this situation the brewery contacted the Effizienz-Agentur NRW via their bank, the Stadtsparkasse Düsseldorf. The EFA specialists focussed on existing efficiency potentials in the production process.

An initial assessment of the potential was carried out followed by a detailed analysis. On the basis of the results, an innovative concept for an integrated purification process by means of one single CIP (Cleaning in Place) plant inside a craft brewery was developed for the first time and eventually implemented in 2011.

Measures and Advantages

The new CIP plant is adapted to the spatial conditions of the craft brewery. Due to the confined spatial conditions, an external suction line was developed to make sure that containers for return water and other liquids can be operated independently from the installation site of the CIP plant. With an inductive quantity and flow rate control, all purification functions and steps such as the dosage of the detergents are controlled automatically today. The system controls whether the programmed amounts of detergents are pumped through the system with the required pressure and flow rate. This means turning away from the traditional time and conductivity-controlled CIP systems. Lower mixed phases of water and beer lead to a clear reduction of the waste water load. Furthermore, the spreading of alkaline solutions and acids during the phase separation could be reduced, with the consequence that detergents are saved and the waste water is less contaminated. The brewery "Im Füchschen" invested € 150,000 in total and saves approximately € 20,500 a year.

The Way to Financing and Implementation

As a part of the project financing, the Effizienz-Agentur NRW supported the private brewery. After the EFA advice, the company applied for subsidies from the investment programme called "Abwasser NRW" (Wastewater NRW) (Table 27.2).

Table 27.2 Savings of the CIP plant as opposed to the manual purification

Equipment	Amount
Water/waste water	3,304 m^3/p.a.
Alkaline solution	7,920 kg/p.a.
Acid	4,000 kg/p.a.

27.2.3 IDEALSPATEN – BREDT GmbH & Co. KG (2014b)

Innovative timed production line reduces material loss. This company of 80 employees produces spades in a "timed" way and ensures not only a reduction of material consumption, but also to offers competitive "Made in Germany" products in the medium and low price segment.

Initial Situation

The company IDEALSPATEN has been producing high quality hand tools such as shovels, spades, pitchforks and pickaxes for the construction and gardening industries on their Herdecke site for more than 100 years. In order to enter the medium and low price segment of DIY stores, the company initially imported spade blades from Southeast Asia which were finally processed in Germany. But due to increased costs and existing quality problems regarding the purchased spade blades, IDEALSPATEN decided in 2010 to re-establish their own production for this segment in Germany. In order to be able to produce within all price segments competitively, the company developed a material and energy efficient concept which was industrially used in this way for the very first time.

Measures and Advantages

The core of the new timed production is the interconnection of all the process steps in one line, from uncoiling to the punch and bending process all the way up to the powder coating of spade blades. Thus, the cost-intensive internal logistics between the individual steps are no longer required which leads to a considerable reduction of the processing time. In order to reduce the material and the energy consumption in production, IDEALSPATEN has additionally enhanced the resource efficiency of the individual process steps, e.g., an optimized punch processor of the board reduces steel scrap significantly and the heat treatment is done through accurate induction. Also, process steps such as quenching and hardening of the spade blanks are combined in one step in the timed production line. A measuring programme carried out by EFA prior to commissioning the plant proved that the anticipated savings were even exceeded. For example, the company reduced the required amount of steel by 22.7% which corresponds to savings of approximately 60 tons of steel a year. In conjunction with the energy savings during the heat treatment, the CO_2 emissions resulting from the spade production also went down by 26%. The timed production serves as a best practice model; the process can be transferred to all punching-bending shops where the produced parts must undergo heat treatment.

Table 27.3 Resource effects at a glance

Reduction of material losses	Approx. 60 t/a
Reduction of CO_2 emissions	Approx. 115 t/a

The Way to Financing and Implementation
In 2010 IDEALSPATEN requested the financing advice of the Effizienz-Agentur NRW, and a project sketch for the environmental innovation programme of the Federal Ministry for the Environment, Nature Conservation, Building and Nuclear Safety was drawn up. After the approval of grants from the environmental innovation programme worth approximately € 241,000, EFA was assigned with the transaction of the grants at the beginning of 2011. In October 2013 the project was completed. IDEALSPATEN invested approximately € 1.5 million in total in the new production line (Table 27.3).

27.2.4 Walzwerke Einsal GmbH (2012b)

Innovative drawing bench increases flexibility and reduces resource consumption. A combination of inductive material heating, continuous coating and a special drawing process reduces the steel consumption by 124 tons a year.

Initial Situation
The company Walzwerke Einsal produces, among other things, cold drawn profiles in different shapes such as rectangular, square bar, hexagonal and round, as well as a variety of special profiles. In order to meet the customers' high requirements regarding accuracy of fit, the drawing process in the field of cold metal forming is of great significance. Prior to the cold drawing process, a surface coating had to be applied on the profiles as a preparatory measure. The coating agents were permanently kept in large dip tanks at a liquid temperature of 70 °C. In addition to that, the profile endings were pre-machined (they have to be milled) to ensure the pushing process into the drawing device. This process causes substantial amounts of material loss. In order to reduce the consumption of resources, Einsal decided to introduce a flexible and resource-saving new drawing line. For the first time, a combination of inductive material heating, a continuous coating line and a drawing bench with a specifically adjusted drawing tool was used.

Measures and Advantages
After pickling, the hot-rolled raw profiles are put ion the machine at the material inlet and fed to an inductive heating device. Regarding their dimensions, the used inductors are adjusted to the respective profile dimensions. Thus, the used heating power is reduced to a minimum. Directly after that, the pre-heated profile goes into the coating chamber where it receives an even and thin drawing coating. In this way, the surfaces of the work peaces are improved and coating agents saved. For the most part, the excess coating agents are returned to the production cycle. Afterwards, the profile is pressed into the drawing tool by means of the so-called pushing process.

Table 27.4 Resource effects at a glance

Omission of coating tank	Avoidance of pre-machining steps	Energy consumption drawing process
Primary energy during the coating process approx. 140,000 m³/a natural gas	Material savings 124 t/a	Old (natural gas, electricity) 482.17 kWh/t
Coating agent (lime and salt) 3,504 kg/a (90%)	Energy savings 59,824 kWh/a electricity	New (electricity) 120.97 kWh/t
Rinse water of the coating tank 2,688 m³/a (99%)		

Immediately after that, the drawing die takes over the drawing part which is, from now on, drawn through the tool at a constant speed. The implemented solution of drawing bench inlet and drawing bench outlet as well as the used drawing slide ensure a minimum of material loss at the profile beginning. The new drawing facility enhanced productivity and increased the process variations considerably. Due to the comission of the coating tank, air and noise pollution could be reduced significantly.

The Way to Financing and Implementation

In 2009 EFA carried out provided financial advice prior to the implementation. As a result, the company applied for subsidies from the environmental innovation programme of the Federal Ministry for the Environment in October 2009. After the approval of the project by the KfW Bank (Kreditanstalt für Wiederaufbau) in December 2009, EFA was assigned with a measuring programme. The results of the project, which was vcompleted in October 2011, were recorded in a final report created by all parties involved. The costs for the measures amounted to approx. € 2.5 million. The project was funded with subsidies worth € 750,0000 from the environmental innovation programme of the Federal Ministry for the Environment, Nature Conservation, Building and Nuclear Safety (Table 27.4).

27.2.5 P.F. FREUND & CIE. GmbH (2014c)

Modular and ergonomic – new pinch roller saves resources and health. Thanks to its modular structure, the innovative pinch roller requires up to eight percent less components, is almost completely recyclable and saves the roof tiler's arm due to its ergonomic design.

Initial Situation

When it comes to hand tools, the medium-sized company FREUND is a worldwide leader in the field of slate and brick processing. Roof tilers, carpenters and plumbers are their biggest customers. It was the company's target to improve a reliable tiler's tool, the pinch roller, in a way that it increases quality, facilitates both production and handling, and is made from environmentally-friendly and recyclable materials. During the development process of the new tool, FREUND utilized the Effizienz-Agentur NRW's method so-called JUMP-Tool JUMP-Tool communicates method competence regarding environmentally compatible product design and development. The Effizienz-Agentur NRW supported the conception of the new pinch roller.

The Resource Efficient Product
The characteristic feature of the new generation of pinch rollers is the improved ergonomic design of the tool. The changed arrangement of the handle and the roller ensures a considerably more efficient power transmission and thus a relief for the user. Due to the new product design, eight percent fewer components for the tool in total are required, while at the same time the number of working steps for the production and assembly processes can be reduced by 22%. Furthermore, the material change of the tool head from steel to stainless steel and aluminium makes a galvanizing process unnecessary. By means of the improved tool head, all single parts can be replaced easily. This was previously possible for only 60% of the parts which is why this option was rarely used by the customer. FREUND will offer an extensive repair service for this new product generation. A special feature of the pinch roller is a handle made of Arboform® – also known as "liquid wood" ("Flüssigholz"). This composite which is made of renewable raw materials is 100% recyclable. The ergonomic handle is supposed to be used for all types of devices at FREUND in the mid-term and replace the 49 handle variations which have been produced so far. The positive resource effects of the new tool are convincing; FREUND was awarded the "Efficiency Prize of North Rhine-Westphalia" (Effizienz-Preis NRW) for the development of the pinch roller.

27.3 Conclusion

Material costs have been constantly falling in the last few years in Germany. Resource productivity is rising but not to a degree which will effect a trend reversal. We need a decoupling from economic growth and resource consumption due to the climate change and the finite nature of resources.

Resource efficiency is not merely an environmental-politically necessity but also a clear economic competitive edge – particulary for a heavy industry oriented state like North Rhine-Westphalia.

But the potential for the increase in resource efficiency is not always easy to discover in daily operations of a business. Frequently, an outside view of production and products by experienced experts can help.

The examples described above show that through targeted measures for the increase in resource efficiency in production and products, companies can meet growing costs, reduce environmental burden, improve their competitiveness and enter new business fields. The identification of savings potential is the first step on the road to efficient production and products, but only with implementation is material and energy consumption can be reduced: With the Effizienz-Agentur NRW, the state of North Rhine-Westphalia provides industrial companies and tradesman with concrete support in the identification and implementation of efficiency potential.

References

Effizienz-Agentur NRW (2012a) New purification process reduces waste water amount and chemical consumption www.ressourceneffizienz.de/fileadmin/user_upload/Englische_Flyer___Loseblaetter/EFA_Loseblatt_Fuechschen_engl_Web.pdf

Effizienz-Agentur NRW (2012b) Innovative drawing bench increases flexibility and reduces resource consumption www.ressourceneffizienz.de/fileadmin/user_upload/Englische_Flyer___Loseblaetter/efa_Loseblatt_EINSAL_engl_Web.pdf

Effizienz-Agentur NRW (2014a) Innovative heat treatment enhances energy efficiency www.ressourceneffizienz.de/fileadmin/user_upload/Englische_Flyer___Loseblaetter/EFA_Loseblatt_Alunorf_engl_Web.pdf

Effizienz-Agentur NRW (2014b) Innovative timed production line reduces material loss www.ressourceneffizienz.de/fileadmin/user_upload/Englische_Flyer___Loseblaetter/EFA_Loseblatt_Idealspaten_engl_Web.pdf

Effizienz-Agentur NRW (2014c) Modular and ergonomic – new pinch roller saves resources and health www.ressourceneffizienz.de/fileadmin/user_upload/Dokumente_2015/EFA_Loseblatt_Freund_engl_2015_WEB.pdf

Chapter 28
Chemical Leasing: A Business Model to Drive Resource Efficiency in the Supply Chain

Reinhard Joas, Veronika Abraham, and Anke Joas

Abstract Chemicals are indispensable in our daily life and most economic sectors depend on chemical supplies for the manufacturing of their goods. However, the use and production of chemicals is a resource intensive business that faces growing global demand. An efficient use of resources by the chemical industry and downstream chemical users can therefore significantly reduce the overall resource consumption in the economy and contribute to sustainable development. Thus, increasing the sustainable use of chemicals supports the achievement of international political objectives such as the Agenda 2030 Sustainable Development Goals.

In order to achieve the sustainability in the use of chemicals, innovative and service-oriented business models are needed that provide incentives for companies to reduce their chemical consumption. Chemical Leasing offers such an approach. In this business model, the functions performed by the chemical serve as the unit of payment and chemical suppliers and users work together to optimize chemical use in fulfilling the function. Due to the service-based approach, chemical consumption is a cost factor for the business partners, which they aim to minimize. This induces economic benefits that are shared among the business partners. Moreover, environmental and health benefits are generated since fewer chemicals are applied, stored, handled and produced, and fewer emissions are released into environmental compartments. In the upstream supply chain, reduced chemical demand results in tremendous savings of raw materials and energy for the production of the chemicals. Hence, large resource and cost saving potentials are seen with the application of Chemical Leasing in a variety of branches.

Keywords Chemical leasing • Resource efficiency • Innovative business model • Sustainable chemistry • Environmental and economic benefits • Emission reduction • Cost savings

---

R. Joas (✉) • V. Abraham • A. Joas
BiPRO GmbH, Grauertstraße 12, 81545, Munich, Germany
e-mail: reinhard.joas@bipro.de

H. Lehmann (ed.), *Factor X*, Eco-Efficiency in Industry and Science 32,
https://doi.org/10.1007/978-3-319-50079-9_28

28.1 Introduction and Background

The challenge of finite resources and the simultaneously growing demand for products and services worldwide requires a reduction of raw material consumption and emissions to the environment as well as the sustainable use of resources. In achieving these goals, an increase in resource efficiency across all economic sectors is inevitable.

The chemical industry is an important pillar of the global economy and provides goods and services to almost every other economic sector, some of the most important ones being construction, electronics, household, paper and packaging, the automotive industry (Cefic and EPCA 2004) as well as food and beverage, agriculture and the textile industry. The chemical industry has grown tremendously throughout the last six decades, in particular during the last 15 years, in which the increasing demand in developing countries and countries with economies in transition spurred an increase in global sales of over 2.5 times. This trend is expected to continue in years to come (e.g. UNEP 2013).

Despite numerous efforts and achievements with regard to the improvement of resource efficiency, the chemical industry continues to consume significant amounts of energy and raw materials to fulfil the needs of the supply chain. In Europe, Germany is the leading producer of chemical goods and ranks fourth among the world's largest chemical producers after China, the United States and Japan (VCI 2014a, b). The German chemical sector increased its resource efficiency by 35% between 1994 and 2007 and reduced its overall energy demand by 21% between 1990 and 2010. Nevertheless, the chemical industry was responsible for about a quarter of the manufacturing industry's energy demand in 2012 while contributing around 11% to the turnover (based on IG BCE 2011; Statista 2015; VDI ZRE 2014). Thus, this industry sector can contribute considerably to the increased resource efficiency in the economy, especially when cooperating with supply chain partners and increasing the efficient consumption of chemicals.

Several governmental and private initiatives or activities already promote the reduction of resource demand in the chemical industry and call for a shift towards a sustainable chemical industry. Examples include the Resource Efficiency and Cleaner Production (RECP) Programme, a joint initiative of the United Nations Environment Programme (UNEP) and the United Nations Industrial Development Organization (UNIDO), the industry initiative Responsible Care® as well as regional activities such as SusChem or Chemical Regions for Resource Efficiency (R4R) and the activities of individual enterprises and associations. Furthermore, resource and energy efficiency as well as sustainable production and consumption in different economic sectors are fostered by, e.g., the European Circular Economy Strategy, the G7 Alliance on Resource Efficiency, SWITCH-Asia and UNEP's resource efficiency program.

However, conventional business models along the chemical supply chain predominantly aim at increasing consumption and thus hamper the uptake of sustainable practices. Therefore, alternative business models need to be found and established that promote the efficient use of resources and give priority to sustainable products and services. Chemical Leasing is one such approach.

28.2 The Chemical Leasing Business Model

Chemical Leasing is an innovative business model that aims at the efficient and sustainable use of resources and may significantly contribute to the more sustainable use of chemicals. The business model promotes both environmental and economic aspects of chemical use along the supply chain through increased resource efficiency.

The Austrian Ministry of Agriculture, Forestry, Environment and Water Management (BMLFUW) has promoted Chemical Leasing since 2002. Since 2007, the German Ministry for the Environment, Nature Conservation, Building and Nuclear Safety (BMUB) and the German Environment Agency (UBA) have also supported the business model. The Swiss government joined in 2014. On the international level, UNIDO's Global Chemical Leasing Programme has continuously promoted and advanced the business model since 2004.

The official UNIDO definition of Chemical Leasing is presented in the text box. Within a Chemical Leasing business model, suppliers and users of chemicals cooperate more closely and aim at reduced chemical consumption by maintaining high quality standards of the performed service or function.

Chemical Leasing is a service-oriented business model that shifts the focus from increasing sales volume of chemicals towards a value-added approach.

The producer mainly sells the functions performed by the chemical and functional units are the main basis for payment.

Within Chemical Leasing business models the responsibility of the producer and service provider is extended and may include management of the entire life cycle.

Chemical Leasing strives for a win-win situation. It aims at increasing the efficient use of chemicals while reducing the risks of chemicals and protecting human health. It improves the economic and environmental performance of participating companies and enhances their access to new markets.

Key elements of successful Chemical Leasing business models are proper benefit sharing, high quality standards and mutual trust between participating companies (UNIDO 2016).

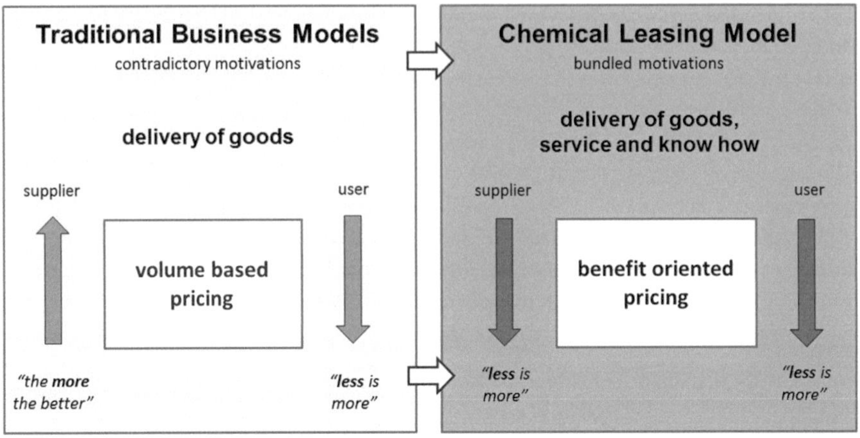

Fig. 28.1 Major differences between traditional business models and Chemical Leasing in terms of the motivations of the business partners and the basis of payment (own graph)

The implementation of Chemical Leasing thus leads to the common motivation to reduce chemical demand for a certain function. The major differences between traditional business models and Chemical Leasing are depicted in Fig. 28.1. Most often, the cooperation leads to long-term partnerships among the involved companies and new technologies and/or substances are introduced in the process.

Throughout the last decade, numerous companies of different branches have successfully introduced Chemical Leasing, and many of them are still applying the innovative business model. All of the applications have been able to realise several of the benefits listed in Table 28.1.

Chemical Leasing can be applied whenever chemicals are needed to fulfil a certain function or service. Some examples for the implementation of Chemical Leasing include the following processes and industries:

- Industrial cleaning and degreasing of metal parts in the metal-processing industry
- Bonding of boxes in the packaging industry
- Cleaning of bottles, pipes and vessels in the beverage industry
- Lubrication of conveyor belts in the beverage industry
- Application of agrochemicals
- Corrosion and surface protection in the automotive and electric appliances industry
- Cleaning of hotel rooms in the hospitality sector

In order to maintain high standards and to ensure a high quality of all Chemical Leasing applications, five sustainability criteria have been introduced, which must be fulfilled:

1. Reduction of adverse impacts for environment, health, energy and resource consumption caused by chemicals and their application and production processes

Table 28.1 Overview of the major benefits generated by the implementation of the Chemical Leasing business model

Selection of benefits induced by Chemical Leasing
Reduction of chemical consumption
Reduction of energy demand (directly in the application and indirectly in the supply chain)
Decrease of emissions
Improved handling and storage of chemicals
Improvement of the overall environment and health performance
Generation of economic benefits for the involved business partners
Planning security and long-term business relations
New customers and sales opportunities
Development and market launch of innovations
Process optimisation and improved process understanding

2. Improved handling and storage of chemicals to prevent and minimize risks
3. No substitution of chemicals by substances with a higher risk
4. Economic and social benefits are generated; a contract should contain the objective of continuous improvements and should enable a fair and transparent sharing of the benefits between the partners
5. Monitoring of the improvements needs to be possible

If these five sustainability criteria are met, companies are eligible for the Global Chemical Leasing Award that is presented by UNIDO in order to acknowledge successful Chemical Leasing applications and to increase the awareness about the business model.

28.3 Chemical Leasing's Contribution to the Policy Framework on Chemicals and Potentials for the Business Model

To reduce the resource consumption in the chemical industry as well as in the sectors along the supply chain despite the growing demand for products and materials, improved processes and technologies need to be found that use chemicals more efficiently, improve recycling strategies and focus on fulfilling the function of the chemical products. Furthermore, adverse effects from chemicals on human health and the environment need to be reduced, which generates a growing need for safer chemicals and sound chemicals management. In particular, very hazardous chemicals should be substituted wherever possible or, if no substitute is found, their production, handling and use should be carried out as safely as possible in order to reduce environmental and health burdens.

The need to solve these challenges is underpinned by the policy framework for chemicals. Various international treaties and conventions, multinational and national regulations and directives as well as a range of voluntary initiatives and activities exist to foster increased resource efficiency, safety and sound management in the chemicals industry. The broadest of these are several United Nations (UN) conferences and their declarations, which call for a greener and more sustainable chemical industry – either directly or indirectly by demanding sustainable development overall. One of the recent and most important initiatives is the 2030 Agenda for Sustainable Development adopted by the UN in 2015. Many of its Sustainable Development Goals directly target the chemical industry, e.g., target 12.4, which reaffirms the 2020 goal for chemicals that was first formulated in the outcome document of the World Summit for Sustainable Development in Johannesburg in 2002. Target 12.4 of the 2030 Agenda states: "By 2020, achieve the environmentally sound management of chemicals and all wastes throughout their life cycle, in accordance with agreed international frameworks, and significantly reduce their release to air, water and soil in order to minimize their adverse impacts on human health and the environment". This 2020 goal is also the core of the Strategic Approach to International Chemicals Management (SAICM), a unique voluntary multi-stakeholder approach aimed at achieving sound management of chemicals worldwide.

These international activities are complemented by national or regional chemical regulations as well as by political strategies and action programs, such as the European regulation REACH (Registration, Evaluation, Authorisation and restriction of Chemicals) or the European Commission's Circular Economy Strategy. Furthermore, there is an increasing awareness among consumers for more sustainable products as well as a range of industry activities that promote the sustainable and safe production and use of chemicals.

It is in the interest of suppliers and users of chemicals to react to these trends and developments in order to stay competitive in the market. Chemical Leasing can help companies in doing so, as this business model promotes the efficient use of chemicals, reduces harmful emissions, leads to reduced resource consumption and adverse impacts, demands save handling and storage of chemicals and simultaneously provides economic benefits for the involved business partners. Moreover, Chemical Leasing can support companies in complying with regulatory requirements.

The service-oriented business model builds upon the notion that the service provided by product and not its pure ownership is the chief aim of chemical consumption (Jakl and Schwager 2008). Due to the improved cooperation among suppliers and users, processes are optimised, safer alternatives may be introduced and experiences are shared in order to further improve the process and reduce the demand for chemicals. Besides the direct benefits in its application, Chemical Leasing also leads to indirect benefits in the upstream supply chain, where high amounts of energy are required for many chemical production processes. Since Chemical Leasing reduces the amount of chemicals needed for a function or service, fewer chemicals need to be produced on the supplier side. Thus, the energy demand in the production process decreases. Moreover, fewer raw materials are required for the production of the chemical itself. The decreased energy and raw material demand thus reduces overall production costs significantly.

Within a research project conducted on behalf of the German UBA, the potentials for Chemical Leasing in Germany have been analysed (UBA 2014). From an economic perspective, the potential for Chemical Leasing refers to the reduction of costs for the user and increased profits for the manufacturer or distributor. In both cases application of the business model results in positive effects on competitiveness. It is estimated that between 46 and 100 million Euros could be saved in Germany if Chemical Leasing would be applied broadly for industries such as solvents, paints and varnishes, adhesives and disinfectants and detergents. Cost reductions also occur due to reduced auxiliaries, such as water, or due to a lower energy demand.

Chemical Leasing also supports increased resource efficiency through the optimisation of the application or production process. This is a result of the cooperation and exchange of know-how between the business partners that aim at reducing chemical consumption by maintaining defined high quality standards for the result. It is estimated that in Germany between 25 and 50 kt of chemical such as solvents, paints and disinfectants can be saved and that their respective waste generation can be reduced tremendously.

The following case study provides an example of how Chemical Leasing is applied and how this reduces the resource demand in the chemical application.

28.4 Case Study: Conveyor Lubrication in the Beverage Industry

The following case study presents some of the highlights achieved during a Chemical Leasing application in the beverage industry – one representative sector for the application of the business model. Two companies have introduced the Chemical Leasing business model in order to improve and stabilise the lubrication process and reduce chemicals consumption during the bottling process of mineral water. The user is Knjaz Miloš a.d., Serbia's largest producer of mineral water and beverages. The bottling company obtains a lubricant from its supplier Ecolab Inc. that is used for the conveyor, which transports the bottles across the production facility during the bottling process. This process should be as stable and reliable as possible to avoid unnecessary stops and the associated costs during production.

Chemical Leasing typically improves the business partnership and cooperation among companies. In this example, supplier and user joined forces with the National Cleaner Production Centre, which helped to implement the Chemical Leasing model. The cooperation among the business partners and the exchange of experiences and ideas led to a new approach for lubrication that is both safer and more efficient. Due to the new and improved technology, the lubricant was substituted with an alternative substance. In contrast to the old lubricant, the substitute is not hazardous and does not require water for the lubrication of the conveyor belts. New nozzles and a new dosing system complemented the new approach.

Several advantages occurred due to the new technology and the substitution. First of all, the process became safer as the new lubricant does not contain hazardous substances and because no water is applied for the lubrication process in the production hall, which reduces the risk of accidents occurring from slippery floors.

Besides these safety improvements, the new lubrication process also created environmental benefits since the resource demand in the process has been significantly reduced: 30% less lubricant is required compared to the former process. This generates additional (not quantified) savings in resources in the upstream supply chain, where fewer chemicals need to be produced, in turn reducing the raw material and energy demand. Furthermore, the new lubricant does not require water, thus eliminating the entire water demand from the process. Consequently, no wastewater is generated. Before the introduction of Chemical Leasing, the wastewater required special treatment at the production site using chemicals. This process became redundant; the water demand and the chemicals for wastewater treatment have been reduced by 100% in this production step.

The unit of payment is no longer based on the litre of chemical product but on the operating hours of the conveyor belt. The companies have been able to generate economic advantages due to reduced costs for chemicals, water, water treatment and less downtime of the production process. Overall, savings of 39% were generated and additional benefits occurred for the supplier, such as becoming the sole supplier and having less costs for transportation and storage of the chemicals. The partners are still using this business model and have since expanded its application to more lines of production.

Similar examples exist in Germany, Uganda and Croatia and have been successfully running for several years. In some cases, the Chemical Leasing model has been extended to other parts of the production, e.g., to the cleaning of surfaces and bottles.

28.5 Conclusion

Chemical Leasing is an innovative, service-oriented business model that provides environmental and economic benefits to the involved business partners, thus resulting in a win-win situation for the chemical supplier and user as well as the environment. It contributes to the reduction of resource demand by optimising the use of chemicals in production processes through focusing on the function or service of the applied chemicals. The business partners cooperate and share knowledge about the process in order to optimise the chemical use, which in addition facilitates stable and long-term business relations. The unit of payment is based on the function or service of the chemical, and suppliers and users aim at fulfilling this service with the least amount of product possible in order to reduce their costs. The economic benefits are shared among the business partners.

The reduced demand for chemicals coincides with reduced energy or water consumption in the application and induces significant energy and raw materials savings in the upstream supply chain as fewer chemicals need to be produced. The cooperation may also lead to process innovations, new technology or the substitution of chemicals with lower risk alternatives. Consequently, Chemical Leasing also contributes to the improved health and safety performance of companies in the chemical supply chain and helps them to respond to initiatives, regulations and strategies on sound chemicals management, sustainable chemistry, resource efficiency as well as sustainable development.

References

Cefic, EPCA (2004) Supply chain excellence in the European chemical industry: results of the EPCA-Cefic supply chain excellence think tank sessions

IG BCE (2011) Informationen zur Industriepolitik: Ressourceneffizienz in der Industrie aus Sicht der IG BCE

Jakl T, Schwager P (2008) Chemical leasing goes global. Selling services instead of barrels: a win-win business model for environment nd industry

Statista (2015) Anteil der Chemieindustrie am Energiegesamtverbrauch des Verarbeitenden Gewerbes in Deutschland nach Energieträger in den Jahren 2010 bis 2012 http://de.statista.com/statistik/daten/studie/241328/umfrage/anteil-der-chemieindustrie-am-energiegesamtverbrauch-des-verarbeitenden-gewerbes/

UBA (2014) Resource efficient businesses in practice by applying the alternative business model Chemical Leasing

UNEP (2013) Global chemicals outlook: towards sound management of chemicals

UNIDO (2016) Chemical leasing http://chemicalleasing.org/

United Nations (2015) Transforming our world: the 2030 agenda for sustainable development. A/RES/70/1

VCI (2014a) Branchenporträt zur deutschen chemischen Industrie 2014. https://www.vci.de/Die-Branche/WirtschaftMarktinformationen/Berichte-und-Analysen/Seiten/Branchenportraet-deutsche-chemisch-pharmazeutische-Industrie.aspx

VCI (2014b) Branchenporträt der deutschen chemisch-pharmazeutischen Industrie. Foliensatz zum Branchenporträt 2014. https://www.vci.de/Die-Branche/WirtschaftMarktinformationen/Berichte-und-Analysen/Seiten/Branchenportraet-deutsche-chemisch-pharmazeutische-Industrie.aspx#

VDI ZRE (2014) Analyse von Ressourceneffizienzpotenzialen in KMU der chemischen Industrie

Chapter 29
Resource Efficiency for the Manufacturing Industries – A Holistic Approach

Werner Maass, Christof Oberender, and Martin Vogt

Abstract Resource efficiency is a crucial factor for the competitiveness of the manufacturing industry in Germany. It provides a significant potential for cost reduction and helps companies to become more independent from the supply and cost risks related to the international raw material markets. Resource efficient manufacturing is also a key element for achieving the international and national goal for climate protection. The Federal German Climate Protection Plan demands an industrial CO_2 reduction of 50% by 2030 (see German Ministry for the Environment, Nature Conservation, Building and Nuclear Safety BMUB. Klimaschutzplan 2050. http://www.bmub.bund.de/fileadmin/Daten_BMU/Download_PDF/Klimaschutz/klimaschutzplan_2050_bf.pdf. Accessed 3 Jan 2017, 2016a). This can only be achieved by the continuous implementation of more resource efficient technologies and processes. To this end, national and international norms and standards can contribute significantly to the dissemination of best available technologies.

To our knowledge, the new VDI Standard 4800, published in 2016, is the first technical standard for defining, evaluating and calculating resource efficiency (see VDI 4800 Part 1, Ressourceneffizienz – Methodische Grundlagen, Prinzipien und Strategien. Beuth Verlag GmbH, Berlin, 2016). The standard contains 37 strategies and measures to increase resource efficiency in companies. They address the improvement of in-house company procedures, adapting or changing technologies in the production processes or new approaches in the product development process.

The sector and technology-specific services offered by the VDI Centre for Resource Efficiency (VDI ZRE) address this complexity in all its aspects. In addition to supplying extensive information and examples, companies are offered cross-regional informative events and qualification courses for successful material and energy efficiency strategies. VDI ZRE furthermore cooperates closely with German state and regional agencies and authorities, Chambers of Commerce, Chambers of Trade and many related associations.

Keywords Resource efficiency • Processing industry • VDI standard 4800 • Material efficiency • Instrument

---

W. Maass (✉) • C. Oberender • M. Vogt
VDI Zentrum Ressourceneffizienz GmbH (VDI ZRE),
Bertolt-Brecht-Platz 3, 10117 Berlin, Germany
e-mail: maass@vdi.de

H. Lehmann (ed.), *Factor X*, Eco-Efficiency in Industry and Science 32,
https://doi.org/10.1007/978-3-319-50079-9_29

29.1 Preface

Upon introducing the German Resource Efficiency Programme (ProgRess) in February 2012, Germany became one of the first countries to define guiding concepts and commit itself to specific courses of action to protect their natural resources (see BMUB 2012). The programme's continuation, the German Resource Efficiency Programme II (ProgRess II), was enacted by the Federal Cabinet on 2 March 2016 (see BMUB 2016b). The main objective of ProgRess II is to decouple economic growth from material and energy consumption. The four guiding principles are: (1) joining ecological necessities with economic opportunities, innovation support and social responsibility; (2) viewing global responsibility as a key focus of the German national resource policy; (3) gradually making economic and production practices in Germany less dependent on primary resources, developing and expanding closed cycle management and (4) securing sustainable resource use for the long term by guiding society towards quality growth (see BMUB 2016b). ProgRess and ProgRess II thus emphasise both the economic and ecological benefits of resource efficiency.

The focus on both aspects is especially significant in Germany, where the industrial sector contributes 26% (see Statistisches Bundesamt 2015) to the gross national product (GNP), which is significantly higher than the European average value of 19%.

If one analyses the cost structure of the manufacturing industry in Germany, material costs come out on top. The share of material costs has increased from 36% in 1993 to 43% in 2014. In contrast, energy costs within the same period have remained stable at 2%, while personnel costs decreased from 27% to 18% (Fig. 29.1).[1]

The increase of material efficiency is therefore a powerful instrument to reduce costs and increase competitiveness of the German industry. According to the evaluation results of the *Bundeministerium für Bildung und Forschung/Federal Ministry of Education and Research* (BMBF) funding programme r^2, the supported projects in metal recycling and manufacturing, chemical and lamination, ceramics and innovative construction materials industries attained material cost reductions valued at 3349 million euros (see Woidasky et al. 2013).[2] A study by the VDI Centre for Resource Efficiency estimates potential material efficiency cost reductions in selected fields of the metal processing industry in the range of 763 million to 2346 million euros, corresponding to material savings in the range of 2 to 6% p.a. (see VDI ZRE 2013). Both studies also show a very high potential for CO_2 reductions due to resource efficiency measures.

Resource efficiency is therefore a crucial strategy for increasing the competitiveness of the German industry by decreasing costs and reducing the dependency on

[1]Own illustration on the basis of the Statistical Yearbooks of the Federal Sta-tistical Office of Germany (values for 1993–2009) and the yearly Cost Structure Statistic Publications (values for 2010–2014) (see Statistisches Bundesamt 2016).

[2]Note: the authors apply the term *differential costs*, as the complexity of material-related cost reduction prohibits calculating all parameters for their analysis. A common procedure, differential cost analyses are applied to economic analyses to cope with complex parameters.

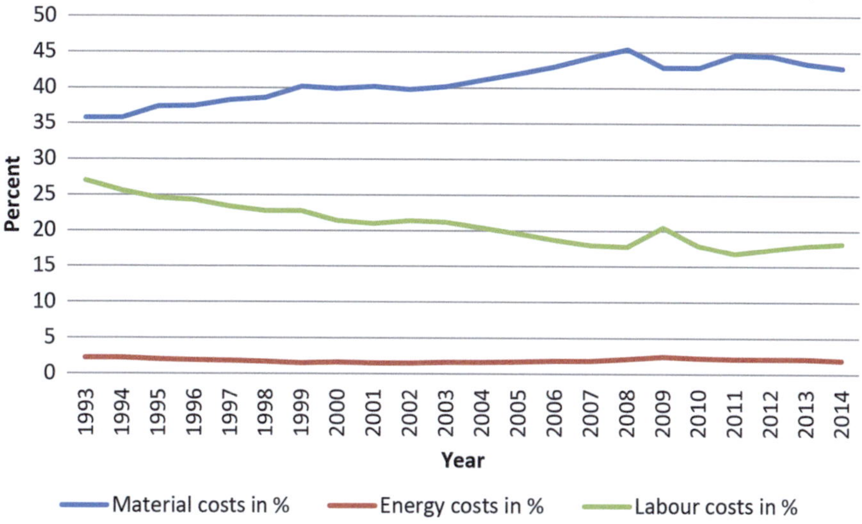

Fig. 29.1 Cost distributions in the manufacturing industry since 1993 (Destatis)

the external raw market. At the same time it contributes to the goals formulated in the national strategies for climate protection.

In order to foster the innovation and dissemination of new and more resource efficient processes and technologies in the German manufacturing industry, the VDI Centre for Resource Efficiency (VDI ZRE) was founded in 2009. It is a national competence centre facilitating the dissemination of resource efficient technologies and processes in the German manufacturing industry. Its specific focus lies in supporting small and medium-sized enterprises (SME). This work is carried out on behalf of the German Federal Ministry of the Environment, Nature Conservation, Building and Nuclear Safety (BMUB) and funded by the National Climate Initiative. The focus of this article is on two activities of VDI ZRE: (1) the development of norms and standards in the framework of the VDI standards and (2) the development of various tools to identify potentials for resource efficiency in companies.

29.2 VDI Standards for Measuring and Evaluating Resource Efficiency

As early as 2009, the Association of German Engineers (VDI) saw the need for developing standards addressing the topic of resource efficiency.[3] The goal was to develop methods to define and measure the resource efficiency of products,

[3] Today, approximately 200 Standards based on the latest technical develop-ments are produced by the VDI's technical divisions per year. That way the VDI has systematically built up a set of techni-

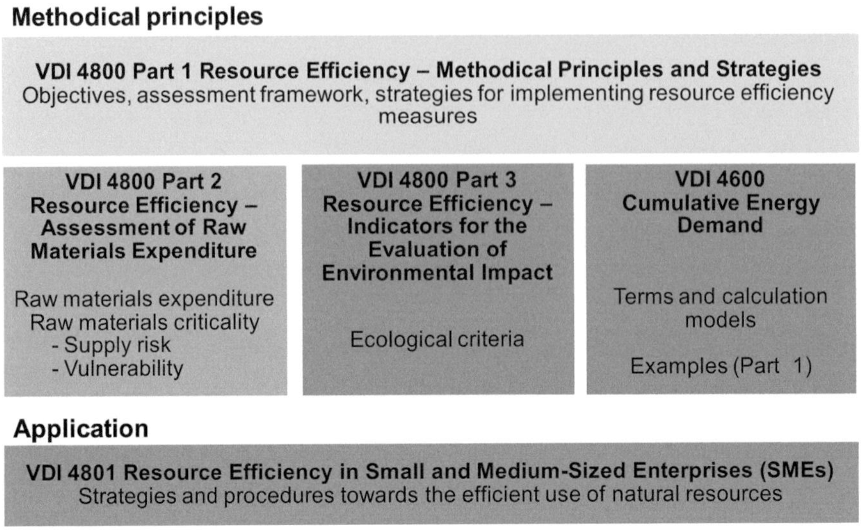

Fig. 29.2 Structure of the VDI Series of Standards about resource efficiency (VDI 4800 Part 1)

processes and services. To this end, working groups for the VDI Series of Standards were established in 2011. The structure of this series is shown in Fig. 29.2.

As a first result of this work, the standard VDI 4800 Part 1 was published as a white paper in 2016. The Standards VDI 4800 Part 2 and VDI 4801 are expected to be published as white papers in 2017. They address methods for the evaluation of raw material demand and strategies and procedures for resource efficiency in SME respectively.

VDI 4800 Part 1 addresses basic methods, principles and strategies. It also provides a definition of resource efficiency as the relation between the benefit associated with a product and the corresponding use of natural resources.

$$\text{resource efficiency} = \frac{\text{benefit}}{\text{need / effort}}$$

The term "natural resources" encompasses the categories shown in Fig. 29.3.

The resource efficiency of a product can only be determined by an assessment of the entire life cycle of a product. A methodology is provided in the VDI Standard 4800 Part 1.

The Standard also provides a list of qualitative strategies and measures to increase resource efficiency. They are divided into product-related and process-related strategies. The list gives an overview and a classification in terms of relatedness to the product or production process, influential parties in the company, life cycle

---

cal regulations, which today contains more than 2000 valid VDI Standards extensively covering the broad field of technology. Today's topics range from securing loads on road vehicles to testing of optical fibres up to biomimetics and monitoring the consequences of genetically modified organisms.

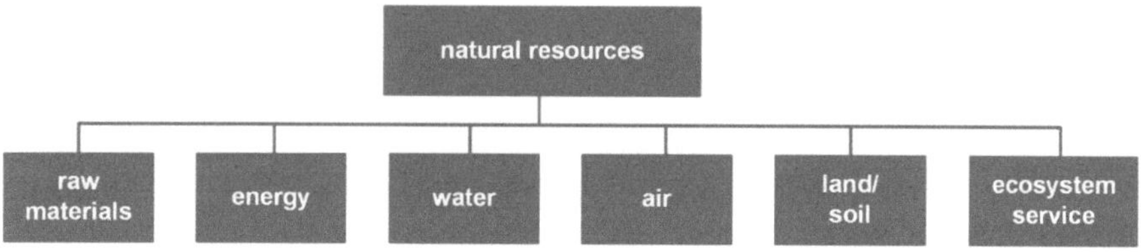

Fig. 29.3 Categories of natural resources (VDI 4800 Part 1)

Table 29.1 Strategies for increasing resource efficiency relating to process (taken from the VDI Standard 4800 Part 1)

	Related to		Influential parties in the company						Life phases with relevant effects					Life cycle analysis		
Strategy	product	production	product development	factory planning	operations planning	purchasing/procurement	production	sales	raw material processing	product manufacturing	use	recovery/disposal	transport	required	conditionally required	not required
Material choice / material substitution	●		●						●	●	●	●	●	●		
Lightweight design	●		●						●		●		●		●	
Mission match & safety	●		●						●		●					●
Miniaturisation	●		●						●	●	●	●	●	●		
Production-oriented product design	●		●		●		●			●					●	
Used-oriented product design	●		●								●		●	●		

phase(s) with relevant effects on resource efficiency and whether life cycle assessment is required (Table 29.1).

A product-related strategy for increasing resource efficiency is material substitution. This strategy plays an important role in, for example, the automotive sector, where new materials are used to make the car lighter and thus reducing petrol use. However, the production and recycling of the new material might also increase the use of primary resources and energy over the entire life cycle. Therefore, for each substitution, a life cycle assessment is required (see VDI ZRE 2016).

Process optimisation is a process-related strategy which leads to material and energy reduction in a production process. A best-practice example, presented in a VDI ZRE short film,[4] demonstrates how little improvements in the production process can achieve significant increases in material efficiency. The film shows a plastics-processing company that collects and recycles the incidental production wastage. Due to the changing quality of the plastic granulates an adaptive process

[4] http://www.resource-germany.tv/topics/plastics-processing/plastic-less-material-more-quality/

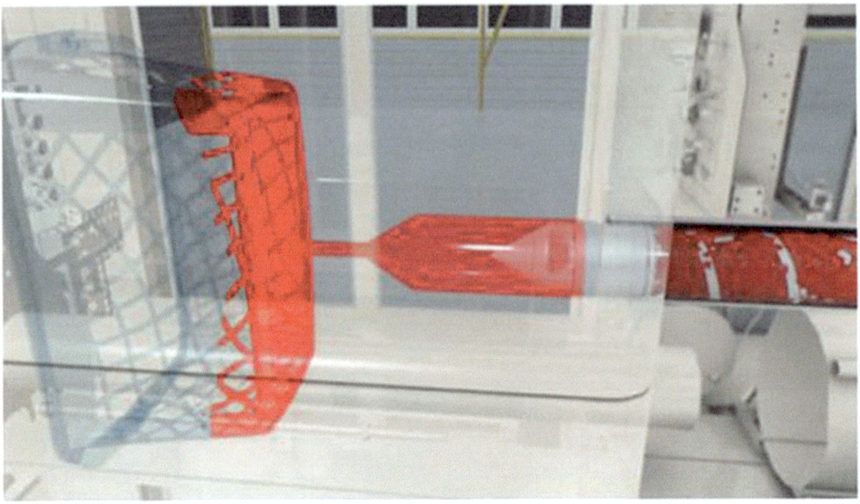

Fig. 29.4 Plastic recycling due to an adapted control system

control system was integrated. The pressure inside the extrusion machine is continuously adapted to the plastic granulate quality, thus always forcing the correct volume into the mould (Fig. 29.4).

29.3 Tools for Practice

The previous discussion shows just two examples of how the strategies and measures defined by the VDI Standard are translated into publications and tools published by VDI ZRE. The goal of this work is to develop tools that help enterprises implement measures for higher resource efficiency. They are provided free of charge and can be used online[5] or offline. Some of the main tools are described below.

29.3.1 Resource Check

Resource Checks are modular, structured checklists intended to support companies in increasing the efficiency of manufacturing processes, or reducing building resource consumption. A questionnaire of up to seven questions related to a specific technological process is provided. Upon finishing the test, a detailed assessment is given and measures to improve the resource efficiency of this process are provided. Various Resource Checks are available for companies in the manufacturing industry (18 Checks) and the building sector (5 Checks). An example is presented in Fig. 29.5.

[5] See www.ressource-deutschland.de, some tools are also available in English at www.resource-germany.com

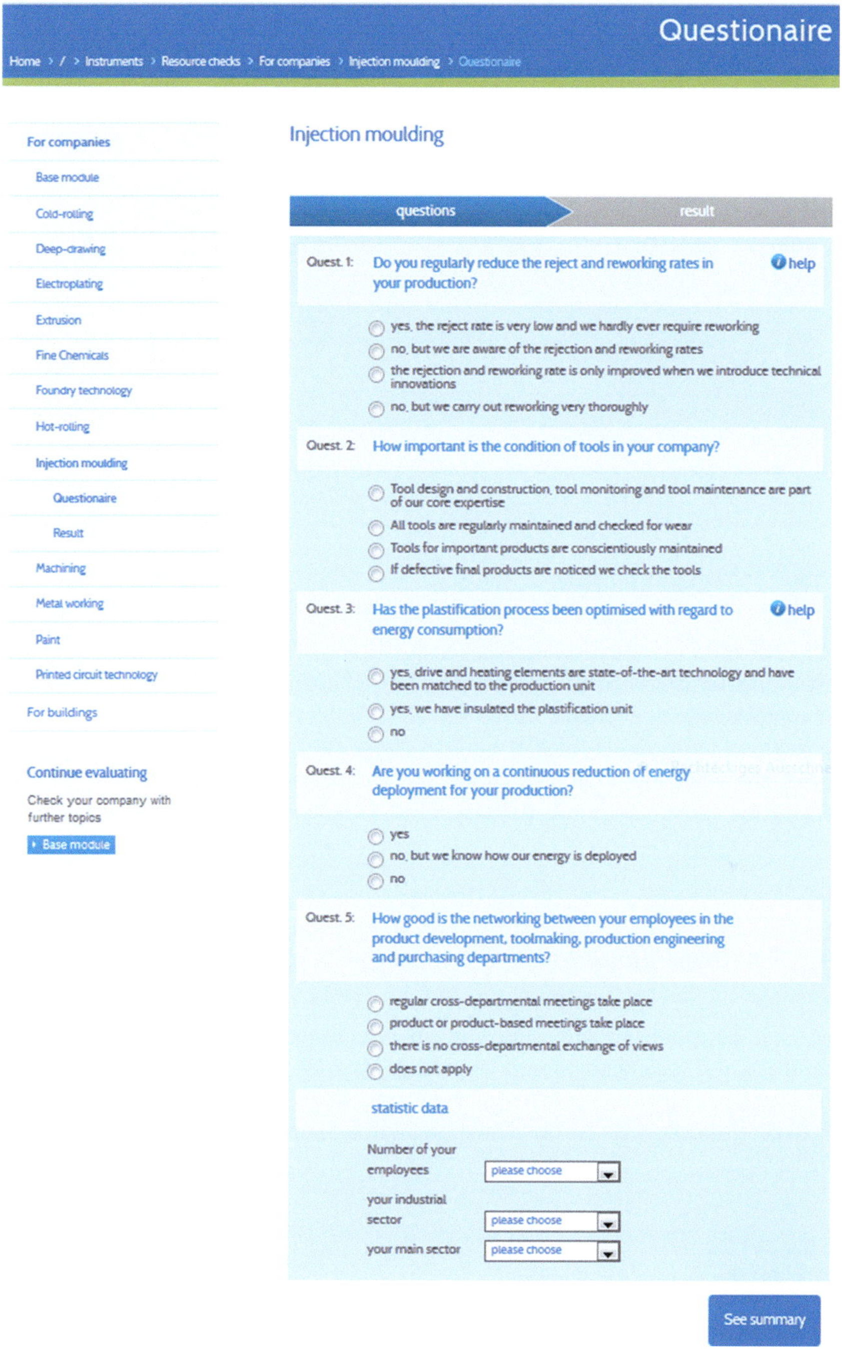

Fig. 29.5 Resource Check for injection moulding (see VDI ZRE a)

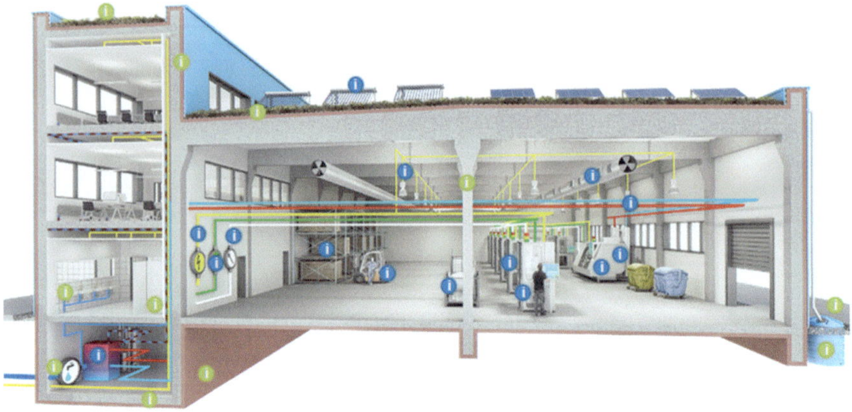

Fig. 29.6 Production infrastructure visualisation (see VDI ZRE b)

29.3.2 *Process Chain*

VDI ZRE offers visualised interfaces to its database which contains, for example, best available technologies, movies, and best practice examples. These interfaces are organised along Process Chains. They enable decision makers to quickly access the required information for every process step. At the moment, 15 Process Chain visualisations are available. In addition, the production infrastructure tool provides information about energy and material efficiency potentials in production and buildings (Fig. 29.6).

29.3.3 *Cost Calculator*

The Cost Calculator provides users in small and medium-sized businesses with a practical way of calculating resource-related costs. The tool analyses the company's cost structure as well as the material and energy flows, comparing them with industry data. Furthermore, the tool supports users in making investment decisions (Fig. 29.7).

29.3.4 *Innovation Radar*

The Innovation Radar provides information on the latest technology developments and process optimisations with the potential to sustainably reduce material and energy consumption.

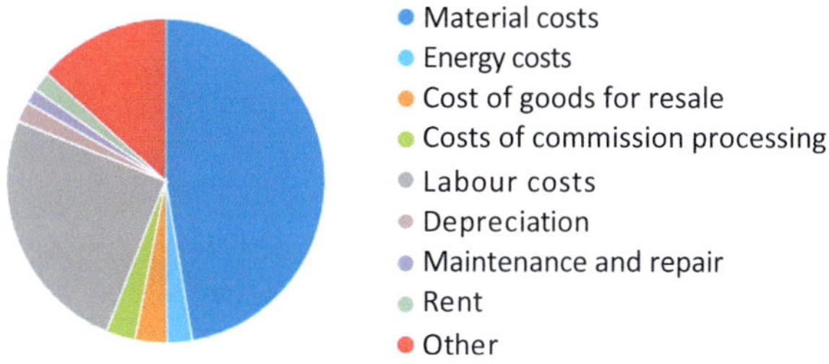

Fig. 29.7 Example of a cost structure analysis provided by the VDI ZRE cost calculator tool

29.3.5 Publications

The publications of the VDI ZRE provide a comprehensive overview of contemporary knowledge about resource efficiency in selected fields of technology. Currently 20 short analyses, 11 technological monitoring reports and nine studies are available.

29.3.6 Web Videomagazine

VDI ZRE has so far published 38 movies showing best practice examples for the successful implementation of resource efficient technologies and processes in companies. These short films can be either found on a specific website[6] or on YouTube.[7]

29.3.7 Qualification Courses

VDI ZRE offers qualification courses for consultants and employees of the manufacturing industry. These courses are carried out by VDI ZRE in cooperation with experienced resource efficiency consultants and local partners, such as associations or chambers.

29.3.8 Other Activities

VDI ZRE coordinates the nationwide Network Resource Efficiency (NeRess). Partners of the network are major stakeholders in the area of resource efficiency, such as, for example, regional administrations and networks, industrial associations,

[6] www.ressource-deutschland.tv

[7] YouTube channel: "Ressource Deutschland".

unions and environmental organisations.[8] VDI ZRE organises the biannual national network conference as well as up to 12 regional events per year. The competence centre also facilitates the Industry Club Resource Efficiency, a nationwide network of top-runner companies.[9]

29.4 Outlook

VDI ZRE is currently developing a comprehensive online guide to resource efficiency in SME based on the new VDI Standard 4801. It will connect the existing VDI ZRE tools with the content of this standard. After publication, this guide can be used for a step-by step analysis and the development of an in-house strategy to improve resource efficiency in companies.

29.5 Conclusion

VDI ZRE provides a broad portfolio of resource efficiency tools for companies. The VDI Standards on resource efficiency provide a methodical and systematic framework. These tools and standards contribute to the dissemination of resource efficiency in companies, thus increasing the competitiveness of the German manufacturing industry and improving environmental and climate protection.

Disclaimer This paper does not necessarily reflect the opinion or the policies of the German Federal Environment Agency.

References

German Ministry for the Environment, Nature Conservation, Building and Nuclear Safety BMUB (2012) Deutsches Ressourceneffizienzprogramm (ProgRess) - Programm zur nachhaltigen Nutzung und zum Schutz der natürlichen Ressourcen http://www.bmub.bund.de/themen/wirtschaft-produkte-ressourcen-tourismus/ressourceneffizienz/deutsches-ressourceneffizienz-programm/progress/. Accessed 27 Jan 2017

German Ministry for the Environment, Nature Conservation, Building and Nuclear Safety BMUB (2016a) Klimaschutzplan 2050. http://www.bmub.bund.de/fileadmin/Daten_BMU/Download_PDF/Klimaschutz/klimaschutzplan_2050_bf.pdf. Accessed 3 Jan 2017

German Ministry for the Environment, Nature Conservation, Building and Nuclear Safety BMUB (2016b) Deutsches Ressourceneffizienzprogramm II. Programm zur nachhaltigen Nutzung und zum Schutz der natürlichen Ressourcen. http://www.bmub.bund.de/fileadmin/Daten_BMU/Pools/Broschueren/progress_ii_broschuere_bf.pdf. Accessed 3 Jan 2017

[8] www.netzwerk-ressourceneffizienz.de (Only available in German).

[9] www.industrieclub-ressourceneffizienz.de (Only available in German).

Statistisches Bundesamt (2015) Industrial sector in Germany remains stronger than in many other EU Member States https://www.destatis.de/Europa/EN/Topic/IndustryTradeServices/IndustryTradeServices.html. Accessed 3 Jan 2017

Statistisches Bundesamt (2016) Kostenstruktur der Unternehmen des Verarbeitenden Gewerbes - Ältere Ausgaben https://www.destatis.de/DE/Publikationen/Thematisch/IndustrieVerarbeitendesGewerbe/AlteAusgaben/KostenstrukturAlt.html. Accessed 3 Jan 2017

VDI 4800 Part 1 (2016) Ressourceneffizienz – Methodische Grundlagen, Prinzipien und Strategien. Beuth Verlag GmbH, Berlin

VDI 4800 Part 2 [draft] (2016) Ressourceneffizienz – Bewertung des Rohstoffaufwands. Beuth Verlag GmbH, Berlin

VDI 4801 [draft] (2016) Ressourceneffizienz in kleinen und mittleren Unternehmen (KMU) – Strategien und Vorgehensweisen zum effizienten Einsatz natürlicher Ressourcen. Beuth Verlag GmbH, Berlin

VDI ZRE (2013) Analyse von Potenzialen der Material- und Energieeffizienz in ausgewählten Branchen der Metall verarbeitenden Industrie http://www.ressource-deutschland.de/fileadmin/user_upload/downloads/studien/Studie_ee_web.pdf. Accessed 3 Jan 2017

VDI ZRE (2016) Ressourceneffizienz im Leichtbau http://www.ressource-deutschland.de/fileadmin/user_upload/downloads/kurzanalysen/Kurzanalyse_17_Ressourceneffizienz_im_Leichtbau.pdf. Accessed 3 Jan 2017

VDI ZRE (a) Injection moulding. http://www.resource-germany.com/instruments/resource-checks/for-companies/injection-moulding/questionaire. Accessed 03 Jan 2017a

VDI ZRE (b) Produktionsinfrastrukur. http://www.ressource-deutschland.de/instrumente/prozessketten/produktionsinfrastruktur. Accessed 3 Jan 2017b

Woidasky J, Ostertag K, Stier C (2013) Innovative Technologien für Ressourceneffizienz in rohstoffintensiven Produktionsprozessen. Ergebnisse der Fördermaßnahme r2. Fraunhofer Verlag, Stuttgart

Chapter 30
Towards a Resource Efficient and Greenhouse Gas Neutral Germany 2050

Jens Günther, Harry Lehmann, Ullrich Lorenz, David Pfeiffer, and Katja Purr

Abstract As a study by the German Environment Agency (UBA) already demonstrated, a greenhouse-gas neutral Germany in 2050 is technical possible, which is an important and necessary first step. However, possible synergies and trade-offs must be examined in order to put these technical options into practice to facilitate a greenhouse-gas neutral economy and society. Especially the resources needed for the transition and for the maintenance of the new system must and will be considered as well as the society that will have to run and maintain such a transformed system. Bearing this in mind, a comprehensive policy mix needs to be designed. In this respect the new study "Greenhouse Gas Neutral and Resource Efficient Germany" will provide the scientific basis for the development of an integrated policy mix that systematically makes use of synergies and addresses conflicting targets and trade-offs. The fundamental setting of this new study is described below.

Keywords Greenhouse Gas Neutrality • Transformation pathways • Decarbonisation strategy • Resource efficiency • Policy mix • Synergies and trade-offs

30.1 Greenhouse Gas Neutrality – What Is Possible?

Germany aims for total greenhouse gas (GHG) emissions reductions of 80–95% by 2050, compared to 1990 levels. This comes rather close to a neutral Greenhouse Gas Balance. As one of the world's leading industrialised countries, Germany must play a significant role and take responsibility with regard to the global challenges concerning climate protection and lower its emissions until 2050 by 95%.

J. Günther (✉) • U. Lorenz • D. Pfeiffer • K. Purr
German Environment Agency (UBA), Wörlitzer Platz 1, DE-06844 Dessau-Roßlau, Germany
e-mail: jens.guenther@uba.de

H. Lehmann
Factor X/10 Club, German Environment Agency, Dessau-Roßlau, Germany

H. Lehmann (ed.), *Factor X*, Eco-Efficiency in Industry and Science 32,
https://doi.org/10.1007/978-3-319-50079-9_30

In order to stay below the internationally agreed two-degree limit on global warming, and to pursue efforts to achieve a level of 1.5 °C of temperature increase compared with pre-industrial values, and as well as to adapt to the unavoidable impacts of climate change, a profound fundamental transformation towards an economic management and lifestyle is required that largely does without any greenhouse gas emissions.

This change affects many sub-systems of our society simultaneously: the energy sector, transport, building, agriculture, industry and so forth. The transformation is therefore characterised by the combination and interaction of innovations, actors and political instruments under changing framework conditions. It must be embedded in a sustainable development and include originator-attributable solution pathways. Fossil fuels, nuclear energy, geo-engineering, and CO_2 capture and storage (CCS) are not components of a sustainable future. Rather the key measures in all fields of action should aim to promote or enable:

- full utilisation of the potential for avoiding emissions and increasing efficiency;
- the substitution of fossil fuels in all fields of application with energy from renewable sources;
- an efficient use of resources;
- sufficient lifestyles and business operations and
- climate-resilient (infra-)structures (Schmeja et al. 2016).

The greatest reduction contributions must come from the energy sector including industry and transport. This requires the utilisation of efficiency potentials, the increased use of renewable sources of energy for power generation, and withdrawal from fossil-fuel power generation, as well as the regenerative supply with electricity-based energy sources and resources for all fields of application.

The study by the German Environment Agency (UBA) entitled "Germany in 2050 – a greenhouse gas-neutral country" demonstrated that a reduction of GHG emissions of approximately 95% compared to 1990 or annual per capita emissions of around one tonne of CO_{2eq} is technologically feasible (Fig. 30.1) (Purr et al. 2014). It is based on a series of studies analysing various archetypes of a 100% electricity supply of Germany with renewable sources. It includes in its research all relevant emission sources. Alongside complete energy supply, including heating and transport, the study also looks at emissions from industry, waste disposal, agriculture and forestry as well as changes in land use. The study presents a target scenario: along with the electricity sector, the heat generation and transport sectors must become completely CO_2 neutral. Sectors like agriculture and certain industrial processes cannot eliminate all their emissions. The complete energy demand can be met by a system based fully on renewable energy sources. The resulting energy system is based largely on direct use of renewable electricity, hydrogen produced by electrolysis of water, and synthesised hydrocarbon compounds (i.e. methane and fuels) (Fig. 30.2).

Demonstrating the technical feasibility is an important and necessary first step towards a greenhouse-gas neutral country. However, to put these technical options into practice to facilitate a greenhouse-gas neutral economy and society, an examination on possible synergies and trade-offs must be considered. This should finally lead to an integrated decarbonisation strategy including a transformation roadmap.

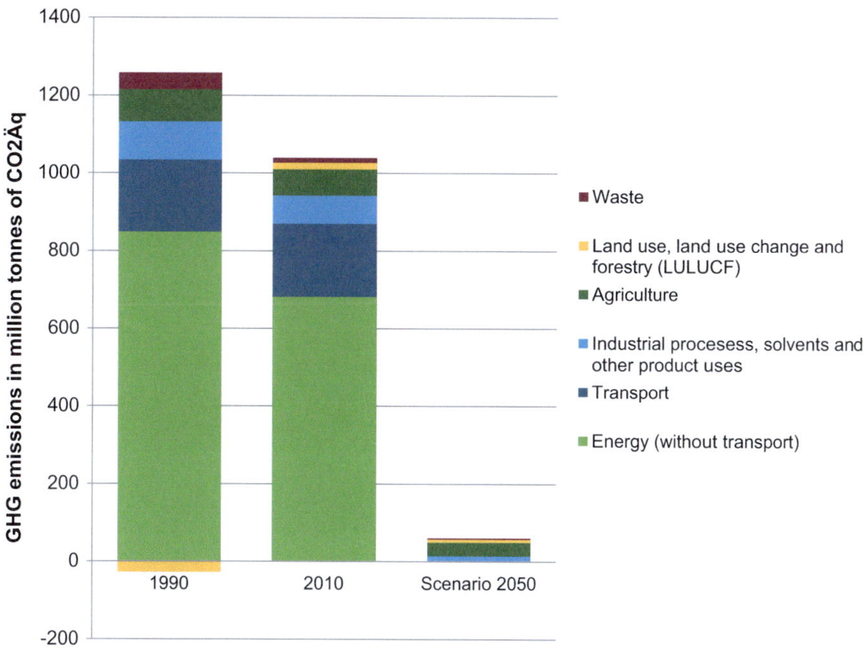

Fig. 30.1 Greenhouse gas emissions across all source categories according the UBA scenario "Germany in 2050 – a greenhouse gas-neutral country" (Purr et al. 2014)

30.2 Broaden the Picture – Linkages Between Resource Efficiency and GHG Neutrality

From a mass flow perspective, material use and greenhouse gases are closely interlinked (e.g. Giljum et al. 2011; Behrens 2016; UNEP 2016a). Most apparently, any burning of fossil fuels leads directly to more carbon dioxide but also to climate active precursor substances like e.g. NOx. Furthermore, there is also a series of industrial processes where either fossil substances are needed and transformed to carbon dioxide (e.g. carbon as reduction substance in steel cooking) or directly set free due to the chemistry of the process (e.g. calcination) (Worell et al. 2001). Especially the energy intensive extraction, processing and use of raw materials is directly linked to climate change under the current fossil fuel based energy system. For instance, the mining and refining of metals currently amount to around 7–8% of the total global primary energy consumption. On the other hand, metals recycling could reduce the energy consumption in a range between 55 and 90% compared to primary production (UNEP 2013). Beside these synergies between resource efficiency and energy savings respectively GHG emissions reduction, possible trade-offs also need to be taken into account. Most obvious, building up renewable energies systems are material intensive, especially critical raw materials such as

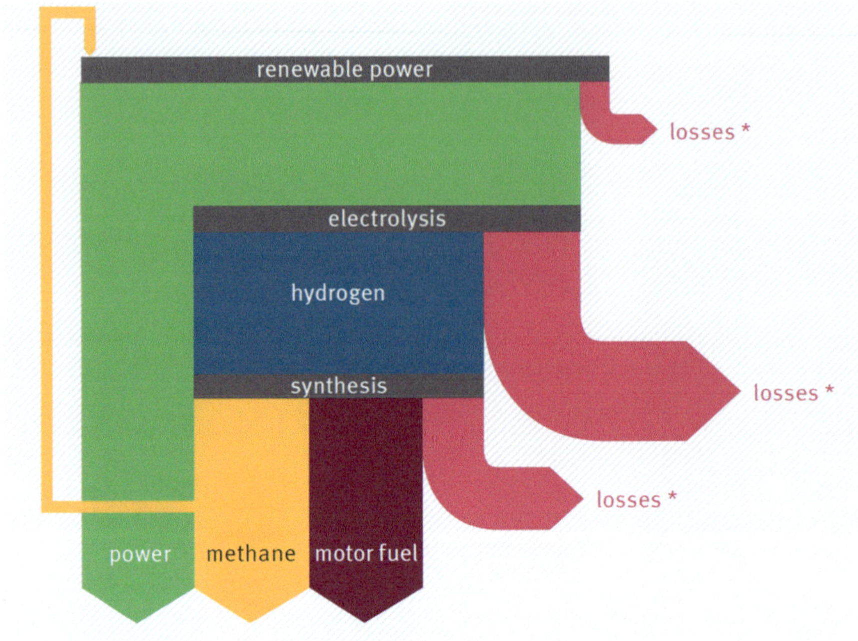

Fig. 30.2 Qualitative representation of the energy flow in the UBA scenario for 2050 including demand for renewable inputs for the chemical industry. Representation of energy flows are proportional to the energy flows required (Purr et al. 2014). * Including line losses, losses from reconverting methane into power and losses from converting biomass into power

copper, gold or neodymium are needed (e.g. UNEP 2016b). Also further technological measures to reduce energy consumption and GHG emissions like lightweight constructions of vehicles or building envelope insulation may lead to a higher material input or need innovative resource efficient recycling solutions (e.g. Werland et al. 2014).

But it is not only the energy intensive industry or the renewable energy sector that is part of the picture. The whole production and consumption, transportation and housing system currently relies to a large extent on energy and materials. As long as the energy system is based on fossil fuels, nearly any human activity emits greenhouse gases. In such a sense, the challenge is manifold: it is not only the way in which energy for production and consumption is provided, it is also the method of production of the actual goods, and the build-up of the necessary infrastructure and finally the respective life and consumption styles (Behrens 2016; UNEP 2010, 2016a).

While we have been able to demonstrate that greenhouse gas neutral Germany is technically possible, the resources needed for the transition and for the maintenance of the new system must and will also be considered as well as the society that will

Table 30.1 Impact categories of environmental assessments (Berger and Finkbeiner 2015)

Impact group	Impact category
Chemical impacts	Greenhouse gases
	Indoor emission
	Wastewater
	Hazardous substances emitted to air
	Diffuse emission of hazardous substances and nutrients
Physical impacts	Nuisance
	Radiation
	Mechanical killing of animals
Biological impacts	Health risks from pathogens
	Biological invasion
Impacts on resources	Raw materials/energy carriers
	Biotic resources
	Water consumption

have to run and maintain such a transformed system. Additionally, we know that the use of resources has impacts on the environment besides the greenhouse effect. Berger and Finkbeiner (2015) differentiate 13 impact categories clustered in four categories that include the nexus of materials and fossil fuels (Table 30.1).

A series of environmental effects can occur throughout the value chain, from the extraction of raw materials to consumption, usage and final deposition or recycling, partly indirectly by the accompanying infrastructure (e.g. in the case of streets: land use change/loss of soil, cut off of habitats, emission of hazardous substances, etc.) (e.g. UNEP 2013, 2016a, b).

The German National Programme for Resource Efficiency (ProgRess) (BMUB 2012, 2016) addresses these issues politically in Germany. The comprehensive aim of the programme is to reduce the overall consumption of natural resources to reduce the environmental impacts and to decouple economic growth from resource use. For this, the programme defines various measures and instruments along the whole value chain. ProgRess II (BMUB 2016) explicitly follows an integrated perspective on material and energy flows and efficiency. While from a general perspective the close relation of resource policy and climate change is clear, the concrete synergies and possible trade-offs or analysis of conflicting targets is still pending. This is the reason why the mere analysis of the technological feasibility of a greenhouse gas neutral Germany was just the first step.

Now, we need to find out what resources are needed to realise such a transformation. What are possible pathways towards the overarching goal of GHG-neutrality? Are there more resource efficient alternative technologies towards GHG-neutrality? How could society contribute, both by saving energy and materials and fostering repairing, sharing and recycling?

30.3 Building Blocks Towards a Greenhouse Gas Neutral and Resource Efficient Germany

The underlying challenges are manifold and the objectives for a transition towards a sustainable society are clear. It leaves no doubt that in the long run, there are no alternatives to a transition to a greenhouse gas neutral and resource efficient Germany. We also know that climate change cannot and must not be seen separately from resource efficiency policy. While the problem descriptions are comprehensive, a shift of the focus is necessary from the problem focus to potential solutions.

While we are already able to show some potential synergies and trade-offs of the two policy domains climate protection and sustainable resource use, a comprehensive policy mix needs to consider societal questions and cannot ignore the economic system. In this respect, UBA's intention is to provide the scientific basis for the development of an integrated policy mix that systematically makes use of synergies and addresses conflicting targets and trade-offs without ignoring the laws of thermodynamics:

- **While the reduction of GHG-emissions must be in total 95% compared to the level of 1990, resource use must also be reduced with the same level of ambition**. Since there are currently no absolute thresholds for raw material consumption in the sense of ecological viability that can be clearly described, a sustainable level of raw material consumption based on the precautionary principle needs to be applied. At least, this mean reducing raw material consumption where possible, making use of all efficiency potentials, avoiding rebound effects and preventing environmental impacts to end up in a currently discussed corridor of global per capita consumption between three and eight tonnes primary raw material use (e.g. UNEP 2011; Bringezu 2015).
- **All sectors in the German economy must be fossil-free by 2050**. This also includes the material base for petrochemical and plastic/polymer production. The carbon sources for polymers must either be biotic or derived by catalytic processes.
- The fundamental bases to achieve fully fossil free sectors are an **integrated energy system based on 100% renewable energy and Power-to-X technologies, e.g. electric mobility, power to heat with heat pumps, power to liquid** (Purr et al. 2016; Schmeja et al. 2016).
- **The energy sector is currently – next to mobility – the biggest emitter of GHG**. It is technically possible with today's technology to achieve 100% renewable energy. There exist numerous options to design the renewable energy system. After the 2050 target scenario, UBA will look at viable transformation paths towards a greenhouse-gas neutral and resource efficient future for Germany to deduct possible transformation steps for 2030 and 2040 to reach the 2050 targets. UBA will analyse and assess the different material needs including "rucksacks" for different technology mixes.

- **A GHG-neutral mobility sector needs an ambitious combination of technology changes and measures on traffic prevention, traffic shift and an efficient handling of traffic** (Bergk et al. 2016; Schmeja et al. 2016). Besides a considerable increase in energy efficiency of vehicles, a shift towards power based fuels and power units is needed. To achieve this, new infrastructures may be needed in addition to a traffic shift and efficient traffic handling. The construction of this infrastructure must be resource efficient right from the start. Moreover, the implementation of technical measures, like lightweight construction, and technological innovations (e.g. e-mobility) must consider raw material constraints, resource efficiency and recycling options as far as possible.
- **Process-related greenhouse gas emissions in industry must be reduced as far as possible**. This may require innovative processes and new production infrastructure. The evaluation of the raw material demand, also to avoid resource constraints, and possible environmental impacts of these new processes are essential to avoid negative lock-in effects.
- **A circular economy is a key element in a resource efficient and greenhouse gas neutral society**. In order to lower the use of materials, to close the material flows and avoid diffuse losses, a resource-efficient based product design must be adopted. UBA will follow – where possible – all options of a resource efficient economy. That includes: rethink, redesign, remanufacture, refuse, recycle, recover, reduce, reuse, remodel and increase the lifetime of the products.
- **A climate-neutral and resource efficient building stock is a key element both for sustainable resource use and GHG-neutrality**. A fast and comprehensive building renovation is needed to achieve a climate-neutral building stock (Bürger et al. 2016; Schmeja et al. 2016). To avoid an unnecessary high amount of raw materials use, measures on resource efficiency and sustainable construction must be implemented in the process of refurbishment as well as construction of buildings. The use of the anthropogenic stock and its path dependencies have to be considered (UBA 2013, S. 20ff).
- **Efficiency both in resource and energy use cannot be the only strategy**. From the consumption perspective, absolute reductions in energy and material use must be achieved.
- This requires also considering **potentials in behavioural changes** in society. This must not be seen uncoupled with societal acceptance. Attention must also be drawn towards societal dynamics and acceptance processes of the population. Special attention will be given to the areas of needs: mobility, construction and housing and diet. These areas are most relevant in terms of GHG emissions as well as resource use (UNEP 2010).
- **The agricultural sector will have to reduce the GHG emissions by at least 50% compared to 2010 and the use of nitrogen, potassium and phosphorus also have to decrease**. To achieve this, animal production needs to be reduced, the use of energy crops has to be stopped and organic farming has to be extended. In addition, an effective fertilizer management have to be in place.
- **The sector land use, land use change and forestry (LULUCF) has to be developed towards a carbon sink**. This will require the determination of an

optimal balance between the carbon stock in forests and carbon storage in wooden products. In addition, the restoration of high-carbon ecosystems is needed.

- **Technical solutions must not be achieved at the cost of environmental pollution**. To avoid inadvertent lock-in effects and negative environmental impacts, UBA will evaluate innovative technological solutions concerning possible synergies and trade-offs with special focus on resource use and further ecological side effects (UNEP 2016b).
- **While single technical solutions appear effective, we will systematically analyse the feedbacks and interferences with other solutions**. For example: Bauxit/Aluminium recycling will lead to the reduction of energy demand, but may translocate the energy demand.

The main directions, overall and in different relevant sectors, are clearly defined. And numerous measures, approaches and innovative technologies to move towards these objectives are already under discussion. But the major task will be to draw a consistent and coherent picture, using the aforementioned building blocks. In doing so, it will be necessary to identify and close relevant knowledge gaps, especially in terms of synergies and trade-offs between climate policy and sustainable resource use.

Disclaimer This paper does not necessarily reflect the opinion or the policies of the German Federal Environment Agency.

References

Behrens A (2016) The climate change impact of material use. Intereconomics 51(4):209–212

Berger M, Finkbeiner M (2015) Streamlined environmental assessments: picking the greenest option despite limited time and data at the Federal Environment Agency in Germany. J Environ Assess Policy Manag 17:1550024

Bergk F, Biemann K, Heidt Ch, Knörr W, Lambrecht U, Schmidt T, Ickert L, Schmied M, Schmidt P, Weindorf W (2016) Klimaschutzbeitrag des Verkehrs bis 2050. On behalf of the German Environment Agency. Texte 56/2016. Dessau-Roßlau. Download https://www.umweltbundesamt.de/sites/default/files/medien/376/publikationen/texte_56_2016_klimaschutzbeitrag_des_verkehrs_2050_getagged.pdf

BMUB – Federal Ministry for the Environment, Nature Conservation, Building and Nuclear Safety (2012) German Resource Efficiency Programme (ProgRess) – programme for the sustainable use and conservation of natural resources. Berlin

BMUB – Federal Ministry for the Environment, Nature Conservation, Building and Nuclear Safety (2016) German Resource Efficiency Programme II – programme for the sustainable use and conservation of natural resources. Berlin

Bringezu S (2015) Possible target corridor for sustainable use of global material resources. Resources 4(1):25–54

Bürger V, Hesse T, Quack D, Palzer A, Köhler B, Herkel S, Engelmann P (2016) Klimaneutraler Gebäudebestand 2050. On behalf of the German Environment Agency. Climate Change 06/2016. Dessau-Roßlau. Download https://www.umweltbundesamt.de/sites/default/files/medien/378/publikationen/climate_change_06_2016_klimaneutraler_gebaeudebestand_2050.pdf

Giljum S, Lutter S, Polzin Ch, Dittrich M, Bringezu S (2011) Resource use and resource efficiency in Emerging Countries. A pilot study on trends over the past 25 years. Commissioned by UNIDO under the "Green Industry" programme. Download https//www.unido.org/fileadmin/user_media/Services/Green_Industry/1256_11_Onscreen_Green Industry.pdf

Purr K, Strenge U, Will M, Knoche G, Volkens A (eds) (2014) Germany in 2050– a greenhouse gas-neutral country. Climate Change 07/2014. Dessau-Roßlau. Download https://www.umweltbundesamt.de/publikationen/germany-in-2050-a-greenhouse-gas-neutral-country

Purr K, Osiek D, Lange M, Adlunger K, Burger A, Hain B, Kuhnhenn K, Klaus T, Lehmann H, Mönch L, Müschen K, Proske Ch, Schmied M, Vollmer C (2016) Integration of Power to Gas/Power to Liquids into the ongoing transformation process. Position. Dessau-Roßlau. Download https://www.umweltbundesamt.de/sites/default/files/medien/377/publikationen/uba_position_powertoliquid_engl.pdf

Schmeja T, Purr K, Seel O, op de Hipt K (eds) (2016) Klimaschutzplan 2050 der Bundesregierung. Diskussionsbeitrag des Umweltbundesamt. Position. Dessau-Roßlau. Download https://www.umweltbundesamt.de/publikationen/klimaschutzplan-2050-der-bundesregierung

UBA – German Environment Agency (2013) What matters 2013 – annual report of the Federal Environment Agency. Dessau Roßlau. Download https://www.umweltbundesamt.de/sites/default/files/medien/376/publikationen/schwerpunkte_2013_eng_bf.pdf

UNEP (2010) Assessing the environmental impacts of consumption and production: priority products and materials, a report of the working group on the Environmental Impacts of Products and Materials to the International Panel for Sustainable Resource Management. Hertwich E, van der Voet E, Suh S, Tukker A, Huijbregts M, Kazmierczyk P, Lenzen M, McNeely J, Moriguchi Y

UNEP (2011) Decoupling natural resource use and environmental impacts from economic growth, a report of the working group on Decoupling to the International Resource Panel. Fischer-Kowalski M, Swilling M, von Weizsäcker EU, Ren Y, Moriguchi Y, Crane W, Krausmann F, Eisenmenger N, Giljum S, Hennicke P, Romero Lankao P, Siriban Manalang A, Sewerin S

UNEP (2013) Environmental risks and challenges of anthropogenic metals flows and cycles, a report of the working group on the Global Metal Flows to the International Resource Panel. van der Voet E, Salminen R, Eckelman M, Mudd G, Norgate T, Hischier R

UNEP (2016a) Resource efficiency: potential and economic implications. a report of the International Resource Panel. Summary for policy-makers. Ekins P, Hughes N. et al.

UNEP (2016b) Green energy choices: the benefits, risks and trade-offs of low-carbon technologies for electricity production. Report of the International Resource Panel. Hertwich EG, Aloisi de Larderel J, Arvesen A, Bayer P, Bergesen J, Bouman E, Gibon T, Heath G, Peña C, Purohit P, Ramirez A, Suh S (eds)

Werland S, Graaf L, Jacob K, Bringezu S, Bahn-Walkowiak B, Hirschnitz-Garbers M, Schulze F, Meyer M (2014) The Nexus between Resource Efficiency Policy and the Energiewende in Germany Synergies and conflicts. Policy Paper. Download http://www.ressourcenpolitik.de/2015/10/werland-stefan-graaf-lisa-jacob-klaus-bringezu-stefan-bahn-walkowiak-bettina-hirschnitz-garbers-martin-schulze-falk-meyer-mark-2014-the-nexus-between-resource-efficiency-policy-and-th/

Worell E, Price L, Martin N, Hendriks C, Meida LO (2001) Carbon dioxide emissions from global cement industry. Annu Rev Energy Environ 26:303

Chapter 31
Pope Francis' Encyclical Laudato Si' as a Catalyst for Societal Transformation? Critical Remarks and Presentation of an Inspired Exemplary Project as a Driver for Sustainability

Ulrich Bartosch, Christian Meier, and Till Weyers

Abstract There is a broad consensus among scientists and society that a "Great Transformation" towards more sustainability is of existential urgency for saving the living conditions of our planet. Nonetheless, the global political and economic reactions and efforts have been insufficient. In this context, the encyclical letter "Laudato Si' – on care for our common home" of Pope Francis (2015a)—in which he calls for a more sustainable lifestyle and economy and more social justice—received worldwide attention from all sectors of society with many welcoming the encyclical almost enthusiastically. Therefore, it seems worthy to explore the relevance, transformative potential and impact of the statements of a religious leader. In this chapter, we will attempt to answer the question: can the encyclical Laudato Si' act as a catalyst for a global societal transformation towards more sustainability? First, we give some critical remarks and observations about this question, set it in the context of sustainability research and invite a broader discussion.

Drawing from sustainability science we begin by giving some remarks on transformations, their drivers and characteristics. For this we use the flagship report "World in transition" by the German Advisory Council on Global Change (WBGU 2011b). Second, we present and reflect on the encyclical. We start with some general remarks on the document, continue with a presentation of statements on sustainable resource use and then comment on the question of "Laudato Si'" as a potential catalyst for transformation. Third, we present the project "Laudato Si' – The papal encyclical in discourse for a Great Transformation" of the Catholic

U. Bartosch (✉) • C. Meier
Katholische Universität Eichstätt-Ingolstadt, Fakultät für Soziale Arbeit,
Kapuzinergasse 2, 85072, Eichstätt, Germany
e-mail: ulrich.bartosch@ku.de; christian.meier@ku.de

T. Weyers
Vereinigung Deutscher Wissenschaftler e.V., Marienstraße 19/20, 10117, Berlin, Germany
e-mail: till.weyers@vdw-ev.de

H. Lehmann (ed.), *Factor X*, Eco-Efficiency in Industry and Science 32,
https://doi.org/10.1007/978-3-319-50079-9_31

University of Eichstätt-Ingolstadt (KU) and the Federation of German Scientists (VDW). The project takes the encyclical and its statements on socio-ecological sustainability as a starting point. Through transdisciplinary, communicative and scientific-educative activities the project re-contextualizes the statements of the encyclical in the scientific field and thereby acts as a sustainability transformative driver, building a bridge between religion and science in the field of transformation. It is shown that the encyclical has considerable transformative relevance. Some impacts can already be noticed, even if long-term and global impacts are difficult to determine, yet.

Keywords Laudato Si' • Pope Francis • Great transformation • Transformation drivers • Change agents • Sustainability • Sustainable resource use • Transformative education

31.1 The Urgency of a "Great Transformation", Its Key Elements and Transformative Drivers

Before analysing Laudato Si' and its relevance for transformation processes it seems reasonable to first establish a brief technical basis regarding great transformations. First, the need for such a transformation is explained. Second, a definition of the "Great Transformation" is given. Third, transformative drivers are specified. Fourth, characteristics of these drivers are identified.

31.1.1 Why Is a "Great Transformation" Necessary?

Many scientists and NGOs have called for more sustainable development for decades.[1] With the *United Nations Framework Convention on Climate Change* (UNFCCC), the *Millennium Development Goals* and the subsequent *Sustainable Development Goals* (SDGs), the world has tried to find a global consensus on climate protection, sustainable development and transformations since the early 1990s (The United Nations 2015).

The motivation behind these global efforts is the conviction that we need a transformation. But why is a "Great Transformation" necessary? To highlight the situation we follow the flagship report "World in Transition. A Social Contract for Sustainability" by the *German Advisory Council on Global Change* (WBGU 2011b). It simply and drastically answers that a Great Transformation is necessary, because otherwise humankind will soon lose its possibilities to live on this planet.

[1] To name just one example: *The St. James's Palace Memorandum* from 2009, published by 60 Nobel Laureates.

It is necessary to avoid dangerous climate and environmental change and a collapse of the essential guardrails of the earth system. It is mandatory—already for normative, ethical reasons—to save the planet, at least in its current state, and to pass it along in such a state that future generations still have a fair chance for a good life. For this reason, more equality between the global north and south must be reached, respecting sustainable development. Value changes are already noticeable, cf. Chap. 2 of the WBGU report "Changing Values: The Global Transformation of Values has already begun" (WBGU 2011b). Technical knowledge about climate change and sustainability urgencies is widespread. However, we are far from the desired and necessary state. The shift from thinking to acting is long and often difficult, on the individual and especially on the political and economic level, as there are complex path dependencies (WBGU 2011b).

Therefore, transformation efforts must ask and answer the central question: how do you get from *knowing* to *transformative acting*? It is to a large part underlying attitudes and values that cause people to act. Consequently, it seems reasonable to also focus on the ethical, normative and moral level addressed by the encyclical Laudato Si.

31.1.2 The Great Transformation: Key Elements and Transformative Drivers

The WBGU report represents one example out of many important scientific sources both for the investigation and explanation of great transformations generally and concretely for the foundation of the transformative KU/VDW Project, "Laudato Si' – the papal encyclical in discourse for a Great Transformation", presented in part 3 of this chapter.[2] The WBGU report seems appropriate for this as "[t]he pathway towards this successful transformation is mapped out in detail" (WBGU 2011c, p. 4) in it and "[t]he transformation into a low-carbon society is examined in its entirety" (WBGU 2011b, p. 28).

The WBGU defines the Great Transformation as "imminent change in politics, economy and society that is required to master the global challenges" (WBGU 2011b, p. 81)—principle among these being global warming and climate change. It concerns the "global remodelling of economy and society towards sustainability" (WBGU 2011b, p. 322). Its key element is a global energy turnaround and decarbonisation of the energy systems by 2030 (cf. WBGU 2011b, pp. 171–172, citing WBGU 2004). The general key measures include "the cooperation of the international community and the establishment of global governance structures" (WBGU 2011b, p. 6) and the mobilization of *Change Agents* at all levels (WBGU 2011b, p. 272).

---

[2]The project relies on a long list of important publications such as the *Sustainable Development Goals* ("SDGs"; The United Nations 2015), *The Limits to Growth* (Meadows et al. 1972), *Silent Spring* (Carson 1962) and the *Brundtland Report* (The United Nations/World Commission on Environment and Development 1987).

The transformation can be supported and facilitated by *transformative drivers*, which are accelerating *factors and actors*. Among others, "[s]ustainability-minded actors [...] have an important role to play as drivers of transformation, for they set the process in motion and endow it with the requisite legitimacy" (WBGU 2011a, p. 2). Change agents "support certain changes, actively driving them ahead" (WBGU 2011b, p. 243, referring to Kristof 2010).[3] Amongst other fields they can be active in science, research and education (cf. WBGU 2012, p. 1).

In order to illustrate how change agents work and how their messages can be put into practice, we must first describe the *characteristics* of transformative drivers. These characteristics represent evaluation standards, which are used and applied to the KU/VDW[4] Project (see Sect. 31.3) with the aim of illustrating that actors and initiatives in the field of research and education can be effective as change agents.[5] "'*Change agents have a convincing idea for change*'" (WBGU 2011b, p. 243, citing Kristof 2010). "They *network* and gain important fellow campaigners, [...] they gradually develop the idea further together" (WBGU 2011b, p. 243, citing Kristof 2010), and *join forces with other sustainability minded actors* (cf. WBGU 2011b, p. 245, referring to Grin et al. 2010). Transformative drivers and project actors should have the *self-perception of a catalyst for societal change and the aim to activate, convince and carry people along* (cf. WBGU 2011b, p. 272). *Dialog, discourse, cooperation with politics, civil society and economy for exploring and developing visions and solutions* is another characteristic of transformative drivers (cf. WBGU 2012, p. 1).

The above-mentioned characteristics are valid for all change agents; however, because the KU/VDW Project is active especially in the field of science, that is research and education, the focus will be put on this field. *Involvement of civil society in the discourse; sharing of "[t]he current status quo of scientific knowledge [...] actively, and especially participatively, [...] with society"* (WBGU 2011b, p. 352); and *making society participate in the education and transformation process* are characteristics of drivers in the field of research and education.[6]

Transformative research and its actors are attributed the characteristics *active advancement of the Great Transformation* (cf. WBGU 2011b, p. 22) through the *"development of solutions"* (WBGU 2011a, b, p. 351) and *use of "inter- as well as transdisciplinary procedure methods, including stakeholder participation"* (WBGU 2011b, pp. 351–352).[7]

According to the WBGU, *convincing* people of the meaningfulness of transformations seems to be one of the crucial criteria related to the efficacy of transformative

[3] Chapter 6 of the WBGU's flagship report of 2011 (pp. 241–264) is dedicated to these "Agents of Transformation: How Innovations Can Spread (Faster)" (p. 241).

[4] VDW is the acronym for the German name of the organisation "Vereinigung Deutscher Wissenschaftler e. V.".

[5] In this part of the essay (and in Sect. 31.3) the drivers' characteristics are highlighted by putting them in italic letters.

[6] The project, presented in Sect. 31.3, includes activities especially in the field of *education*.

[7] For affirmation see also WBGU 2011b, pp. 23, 323 and WBGU 2012, p. 2.

measures (cf. WBGU 2011b, p. 273). One important question therefore is, how can people be convinced and what takes them from knowing about the urgency of an action to acting in the "required" way. One possible way to achieve this is through *transformation education*. It involves four ascending steps for making people change their attitude and behaviour towards sustainability.[8] It begins with *(a) knowledge communication* (cf. WBGU 2011b, p. 23; see also pp. 241 and 352), followed by *(b) creating problem awareness* (cf. WBGU 2012, p. 1) and *transformation-relevant awareness* (cf. WBGU 2011b, p. 352).[9] In order to influence people to then change their behaviour, the *(c) attitudes and values have to be shaped and changed* (cf. WBGU 2011b, p. 7)[10] as they are the compass and guiding principles for people's behaviour and actions. From sustainability knowledge, awareness and values a corresponding *(d) responsibility* can result. *Transformative education* can contribute to this through *communication, promotion and creation of sustainable behaviour to take on collective and cross-generational responsibility of our "common home"* (WBGU 2011b).

The WBGU report and the illustrations above show that transformative actors can be active in the field of research and education. They play an important role in the transformation towards sustainability as they set it into motion and drive it forward through inter alia transformative knowledge communication, awareness rising and value building. Both Laudato Si' and the KU/VDW Project, which takes the encyclical as the starting point for its activities, reveal considerable potential for supporting transformation processes in the fields of education, research and communication.

31.2 Laudato Si' as a Catalyst for Transformation?

In this part of the chapter we want to examine the potential role of the encyclical Laudato Si' as a catalyst[11] for transformation, especially in the areas of education, values and communication.[12] First, we give some information about the encyclical letter, regarding its content and especially its messages on the sustainable use of resources. Second, we underline the relevance of Laudato Si' and its potential impact as a catalyst for societal transformation.

[8] It is assumed that actions and behaviour are determined by awareness and especially attitudes and values.

[9] The WBGU also makes suggestions on how to create the "right" awareness and to turn awareness into requisite action (see WBGU 2011b, pp. 7, 23, 241, 321, 355–357).

[10] For illustration and affirmation see also WBGU 2011b, pp. 67 and 352.

[11] In our examinations we use the term "catalyst" as something that provokes or accelerates a development.

[12] *Laudato Si'* means *Praise Be*! from the *Canticle of Brother Son, Sister Moon* of Francis of Assisi. An encyclical is a letter, formally addressed first to all the bishops, and is the highest level of teaching in the Catholic Church (Tucker 2015).

31.2.1 Laudato Si' and Its Statements on Sustainable Resource Use

The encyclical can be seen as another addition to the catholic social doctrine. The first social encyclical *Rerum Novarum* was published in 1891 by Pope Leo XIII (Pope Leo XIII 1891). Since then, a number of social encyclicals have been issued by subsequent popes. However, Laudato Si' marks the first time in the history of catholic social doctrine that a papal encyclical deals with the complex issue of the environmental challenges in a systematic and comprehensive manner and in connection with global development (cf. Bals 2016).

It must be underlined—in contrast to its coverage in the German media as an environmental encyclical—that Laudato Si' addresses much more than just environmental issues. The Pope argues that, "a true ecological approach always becomes a social approach; it must integrate questions of justice in debates on the environment, so as to hear both the *cry of the earth and the cry of the poor*" (49).[13] Thus, the text could be called the first socio-ecological encyclical.

The overarching theme of Laudato Si' is "care for our common home". It covers a wide variety of topics, ranging from resource efficiency to the care for the poor and a dialogue between religion and science. The encyclical is a theological text; however, it is addressed to "every person living on this planet" (3). It has been described as doubly coded, both theologically as well as scientifically (for an analysis see Bals 2016). This chapter seeks to highlight the ways in which the encyclical Laudato Si' talks about the use of resources. In this sense, the chapter addresses the message in the encyclical that is directed not only to the Christian world but to all people. Moreover, the theological dimension will not be the focus here (for a theological discussion see e.g., Marx 2015, Hengsbach 2016, Schorlemmer 2016).

The papal document is divided into six chapters and encompasses 184 pages, which is rather long for an encyclical. The central topics of the encyclical, as described by Pope Francis himself, evolve around

> the intimate *relationship between the poor and the fragility of the planet*, the conviction that *everything in the world is connected*, the critique of new paradigms and forms of power derived from technology, the call to seek *other ways of understanding the economy* and progress, the value proper to each creature, the human meaning of ecology, the need for forthright and honest debate, the serious responsibility of international and local policy, the *throwaway culture* and the *proposal of a new lifestyle* (emphasis added) (16).

Chapter one "What is happening to our common home?" describes the current state of the world, laying out the problems that our world faces today. In this chapter, the pope talks about various issues, such as climate change (cf. 23–26); pollution, environmental degradation, waste and throwaway culture (cf. 20–22); issues pertaining to water (cf. 27–31) and biodiversity loss (cf. 32–42). These are some of

---

[13] Explanation: All the citations from the encyclical have numbers in brackets. These do not refer to page numbers but the numbered sections in the text. This is especially useful, because the quoted section is always easy to find in each edition of the encyclical in different languages.

the areas that threaten the "planetary boundaries" (cf. WBGU 2011b). Furthermore, the pope addresses global inequality (48–52) and the problem of weak (political) responses to the issues we face today.

Regarding *resources*, he criticises the throwaway culture by citing the example of the lack of complete paper recycling (cf. 22). He highlights that "our industrial system, at the end of its cycle of production and consumption, has not developed the capacity to absorb and reuse waste and by-products" (22). He points out the need for a circular economy and what has been called the "cradle-to-cradle" approach—in contrast to the current cradle-to-grave mode of production and consumption. The circular economy would be "capable of preserving resources for present and future generations, while limiting as much as possible the use of non-renewable resources, moderating their consumption, maximizing their *efficient use*, reusing and *recycling* them" (22, emphasis added). In sum, the Pope demands a radical change of the current conventional mode of production.

Pope Francis highlights the huge impact of climate change and underlines its effects on the carbon cycle. Per the encyclical (which refers to state of the art scientific findings), climate change "creates a vicious circle which aggravates the situation even more, affecting the availability of essential *resources* like drinking water, energy and agricultural production in warmer regions, and leading to the extinction of part of the planet's biodiversity" (24). Therefore, there "is an urgent need to develop policies so that, in the next few years, the emission of carbon dioxide and other highly polluting gases can be drastically reduced" (26). The Pope points to the need for more energy and sustainable resource use, e.g. in the transport and building sectors. Unfortunately, though, good practices (some of them described in part three of this publication and in this chapter) are still far from widespread.

Pope Francis specifically addresses the issue of water in the context of natural resource depletion. He states that "we all know that it is not possible to sustain the present level of consumption in developed countries and wealthier sectors of society, where the habit of wasting and discarding has reached unprecedented levels" (27). "Fresh drinking water is an issue of primary importance, since it is indispensable for human life and for supporting terrestrial and aquatic ecosystems" (28). The arguments in the subsequent sections follow the same line of reasoning. We are overusing natural resources and on top of that not using them in an efficient way and disposing most of them instead of recycling them.

The Pope puts emphasis on the need and obligation of each government to carry out "its proper and inalienable responsibility to preserve its country's environment and natural resources, without capitulating to spurious local or international interests" (38). He criticises that "every intervention in nature can have consequences which are not immediately evident, and that certain ways of exploiting resources prove costly in terms of degradation" (41).

Of great importance is how the Pope draws a link to social inequality and global injustice. He highlights that it is often the poor who suffer the consequences of climate change, environmental degradation, etc. (cf. 48, see also Bals 2016, p. 13). In this light, it is important to understand the encyclical as a socio-environmental encyclical—as the Pope puts it: "a true ecological approach [...] must integrate questions of justice in debates on the environment, so as to hear *both the cry of the earth and the cry of the poor*" (49). The injustice does not only exist on the individual level but also the country level, especially between the global north and the global south. The disproportionate use of resources over long periods of time between countries has created an "ecological debt" (51). Citing the Bishops of the Patagonia-Comahue Region (Argentina), the Pope criticises the unsustainable operation of multinational companies:

> [A]fter ceasing their activity and withdrawing, they leave behind great human and environmental liabilities such as unemployment, abandoned towns, the depletion of natural reserves, deforestation, the impoverishment of agriculture and local stock breeding, open pits, riven hills, polluted rivers and a handful of social works which are no longer sustainable (51).

Chapter two is dedicated to a more theological analysis, which will not be covered here.[14] *Chapter three* describes the human roots of the ecological crises, criticising the dominant "technocratic paradigm" and its focus on growth to the detriment of the environment (cf. 101 and 106).

In *chapter four*, the Pope calls for a paradigm shift, away from the dominant technocratic paradigm to a new approach, so-called "*integral ecology* [...] which clearly respects its human and social dimensions" (137, for a discussion of the paradigm shift see Bals 2016, pp. 43–59). The Pope highlights that "we are faced not with two separate crises, one environmental and the other social, but rather with *one complex crisis* which is *both social and environmental*. Strategies for a solution demand an *integrated approach* to combating poverty, restoring dignity to the excluded and at the same time protecting nature" (emphasis added, 139).

Chapter five talks about some possible solutions and gives some concrete proposals, also regarding resource efficiency. The Pope gives some advice on how to use resources more efficiently, albeit staying for the most part quite general in his observations.

A very good example of a solution to resource use is the development of cooperatives "to exploit renewable sources of energy which ensure local self-sufficiency and even the sale of surplus energy" (179). He points out that there are "no uniform recipes, because each country or region has its own problems and limitations", but that much still needs to be done, such as promoting ways of conserving energy" (180). Some of these measures should be "favouring forms of industrial production with maximum energy efficiency and diminished use of raw materials [...], improving transport systems, and encouraging the construction and repair of buildings aimed at reducing their energy consumption and levels of pollution" (180).

The Pope underlines that there are many possible (political) activities, which can be carried out on the local level. These include "modifying consumption", creating an "economy of waste disposal and recycling", "protecting certain species" and "planning a diversified agriculture and the rotation of crops" (180). The Pope gives a hopeful outlook with a vast array of possible alternative ways of producing, consuming

[14] As stated before, the theological dimension is discussed by authors such as Marx (2015), Hengsbach (2016) and Schorlemmer (2016).

and living. He states that "[n]ew forms of cooperation and community organization can be encouraged in order to defend the interests of small producers and preserve local ecosystems from destruction. Truly, much can be done!" (180).

But he also makes clear that there needs to be a good framework to guarantee more sustainable development. Environmental impact assessment must be an integral part of planning (cf. 183), decisions "must be made based on a comparison of the risks and benefits foreseen for the various possible alternatives" (184) and the precautionary principle, enshrined in the *Rio Declaration* of 1992,[15] should be respected (cf. 186). Also, "negative externalities have to be accounted for and paid for by companies" (195).

The Pope highlights that another mode of thinking and production is necessary and possible. He claims that "efforts to promote a sustainable use of natural resources are not a waste of money, but rather an investment capable of providing other economic benefits in the medium term" (191). In the same way, "more diversified and innovative forms of production, which impact less on the environment, can prove very profitable" (191). So in the end, resource efficiency proves useful in many ways. The Pope himself describes this aptly: "It is a matter of openness to different possibilities which do not involve stifling human creativity and its ideals of progress, but rather directing that energy along new channels" (191).

31.2.2 Laudato Si′—Reception, Relevance and Potential Impacts

Pope Francis made a valuable contribution to socio-ecological sustainability transformation processes with his encyclical letter "Laudato Si' – on care for our common home", completed in May 2015 and presented to the world on 18 June 2015. It has been widely received in the press, by environmental and climate activists, e.g. Bill McKibben, founder of 350.org and Naomi Klein, author of the book *This Changes Everything*, and in the scientific world (McKibben 2015; Klein 2015). Two leading scientific journals, *Nature* and *Science*, even published a response to the encyclical in their editorials (Campwell 2015). Major newspapers, such as the *Guardian*, *Süddeutsche Zeitung* and *FAZ* discussed Laudato Si' (for a good overview of global media coverage see Heimbach-Steins and Stockmann 2015). It has even been called a "*Magna Carta* of integral ecology as a reaction to humanity's self-destructive course" (Bals 2016).

The papal encyclical represents a new opening for potential dialogue and cooperation. The question is: Does the encyclical make a difference? And can the statements of a global religious leader on eco-sustainability be a catalyst for societal transformation?

[15] The *Rio Declaration* of 1992 states in Principle 15 that "where there are threats of serious or irreversible damage, lack of full scientific certainty shall not be used as a pretext for postponing cost-effective measures" which prevent environmental degradation (The United Nations 1992).

It is certain that the importance of the messages in Laudato Si' can be underlined by the fact that Pope Francis as a moral authority does not only speak to his 1.2 billion catholic followers but his analysis is addressed to all of humankind (cf. O'Riordan et al. 2015; Jamieson 2015). What makes the encyclical all the more relevant is that it does not see the required transformation to a low-carbon society merely as a technical issue (even though technical solutions are absolutely necessary—refer to other chapters of this publication) but as a challenging *moral* issue. Thus, the Pope calls for a change of heart and perspective through the process of ecological "conversion" (cf. O'Riordan et al. 2015).

The impact of the encyclical letter Laudato Si' can be observed in different areas. As Pope Francis is the highest leader of the Catholic Church, Laudato Si' is evidently addressed at the Catholic Church and its 1.2 billion followers. It has a revolutionary character because he brings socio-ecological issues—which until now had been discussed primarily on the margins of the Catholic Church—into the centre of attention and for the first time treats issues pertaining to the environment, climate change, ecological and social justice in a document of highest importance. The encyclical can have catalysing effects as an inspiration on the individual level and for the catholic organisation, being "a challenge, mission, confirmation and encouragement at the same time" (Arbeitsgemeinschaft der Umweltbeauftragten der deutschen (Erz-)Diözesen 2017).

However, it is difficult to measure exactly what *political* effect the encyclical has had until now and could still have. It is no coincidence, though, that the encyclical was released before the United Nations COP21 climate change negotiations in Paris in November 2015 and the negotiations on the SDGs in September of the same year. Also, the Pope brought his message both to the UN (Pope Francis 2015b) and the US Congress. In the Paris negotiations, he acted as a player in the background, even calling some head of states and government to respect the consensus when the deal could have failed in the last minute (information provided by Prof. Dr. Hartmut Graßl, a German meteorologist and chairman of the VDW). In doing so the pope established himself as a political player (Göring-Eckardt 2015, p. 20). *Ecologically*, it seems that the encyclical has had and will continue to have an effect, especially since many scientists and civil society organisations have welcomed Pope Francis' call for action. In a third area, *ethics*, the encyclical will create an even stronger link between ecological ethics and social justice. Working for more ecojustice is especially important for younger scholars, activists and religious leaders. In this field, seminaries and religiously based schools will draw on the encyclical for years to come (cf. Tucker 2015).

The expert interviews conducted as part of the KU/VDW Project reveal a similar picture of the potential impact of the encyclical letter (see Sect. 31.3). All of the scientists interviewed welcomed the publication of the encyclical and underlined the great reception it received in the scientific and civil society community as well as its importance. Some even called it "the first really important environmental text of the twenty-first century" (Jamieson 2015).

The encyclical Laudato Si' has given important impulses for transformation processes, inter alia because it explicitly invites conversation both with believers and non-believers. The implementation of these impulses must be carried out by already existing and new projects all around the world (see O'Riordan et al. 2015). Laudato Si' can be used to inspire the world to change for better but in the end—and as the Pope said himself—*it is people* who will make the necessary changes. The KU/VDW Project "Laudato Si' – The papal encyclical in discourse for a Great Transformation" attempts to add to this process. Like many projects around the world that are using the encyclical as an impulse and starting point for their work, such as the *Global Catholic Climate Movement* (*GCCM*) with their "Laudato Si' Week" in 2016[16] or a conference on investing in clean energy in Rome in January 2017[17] as well as catholic universities around the world (e. g. Santa Clara University in California,[18] Pontifical Catholic University of Chile[19]), the KU/VDW Project uses the impulse and puts the encyclical into context, debating its core messages in an attempt to bring them to a broader audience.

The changes catalysed by the encyclical and related projects will not be sudden but the impact of Laudato Si' over the next decade and beyond will be significant. "It is the long-term impact for change that we can envision" (Tucker 2015).

31.3 The KU/VDW Project as a Driver for Transformation

In this section we want to examine whether project groups active in the field of research and education can contribute to sustainability. The questions that shall be answered are: Can projects like the KU/VDW Project actually be transformative drivers? Are they able to act as change agents in the fields of research and education? And, are they suitable to support and drive the Great Transformation, by spreading sustainability impulses, statements and concepts and trying to make them effective? To answer these questions, the specific conceptual structure of the KU/VDW Project and its activities will be described in more detail. In a second step, the WBGU's characteristics of transformative drivers[20] will be assessed for the KU/VDW Project to show that the project acts as a change agent.

Our project benefits from being composed of *actors* from diverse backgrounds and disciplines: on the one hand a public university—the *Catholic University of Eichstätt-Ingolstadt* (KU)—and on the other, a scientific-academic organisation, the *Federation of German Scientists* (VDW), which works to promote responsible research in the fields of climate change, biodiversity, peace and security policy. The *Institute for Advanced Sustainability Studies* (IASS) in Potsdam and the *Potsdam*

---

[16] See the websites http://catholicclimatemovement.global and http://laudatosiweek.org for details.

[17] See the website http://www.laudato-si-investing.com for details.

[18] See information on the university website https://www.scu.edu/ and the thematic section on Laudato Si' at https://www.scu.edu/ourcommonhome/ for details.

[19] See the dedicated section on the website of the Pontifical Catholic University of Chile at http://www.uc.cl/es/enciclica-laudato-si

[20] See Sect. 31.1 for details.

Institute for Climate Impact Research (PIK) are the other main project partners. The project runs for 2 years (May 2016–April 2018).

The particularity of the KU/VDW Project is its *specific communicative structure*, which can be denominated as *contextualisation* and *re-contextualisation*. It takes impulses from two different sources (religious and scientific), brings them together in an "open forum", develops them further in a synergetic way and injects them enriched back into their areas of origin. This is expressed by the project name: "Laudato Si' – The papal encyclical in discourse for a Great Transformation".

In its self-perception, the KU/VDW Project acts as a transmission belt (In German: "Transmissionsriemen") and bridge. It takes the impulses of the encyclical from a "religious communication room" into another communication room.[21] In this second room, the project presents, discusses and develops the *issues* touched upon in the encyclical. Through an academic lecture series and other formats, the main *scientific* sources and expertise, which the Pope implicitly uses in his encyclical, are made more visible.[22] The scientific sources, together with the main topics of Laudato Si', are explained and contextualized by experts from economics, politics, religion, science and civil society organisations. All these actors analyse, critically reflect and constructively discuss the issues of sustainability and all activities are coordinated by the project team. The participating experts *enrich* the concepts and discourse with the current state of knowledge, their expertise and additional sustainability concepts. This enrichment process represents the second source of impulses, which are drawn on and catalysed by the project activities. The topics are advanced and the project tries to develop approaches for concrete solutions. These findings, enriched concepts and strategies for action are then injected back into the fields of science and other fields of origin. Through this *re-contextualisation*, the KU/VDW Project drives the dialogue and constructive cooperation between the religious discourse and the scientific discourse and research for sustainability.[23] Through this process, inspiring, synergetic forces and dynamics towards sustainability can be created. The project contributes to sustainability research and transition processes by presenting and using an innovative procedure and a source of input—the religious one—maybe underestimated until now.

---

[21] One important question in the project's research activities is: which role can religion and statements of religious leaders play for supporting and driving societal transformation processes? Which impacts can statements of religious leaders have on people's attitudes, values and resulting actions regarding sustainability, climate protection and also resource efficiency? This was discussed, e.g., at a scientific workshop at the IASS in Potsdam in September 2016; more information at http://www.laudato-si-transformation.de/veranstaltungen/internationaler-fachworkshop-am-iass/; see also Bergmann and Gerten (2010); Gerten and Bergmann (2013).

[22] E.g. the scientific expertise on the current state of knowledge on climate impact research of Prof. Dr. John Schellnhuber (PIK), one of the presenters of the encyclical to the public on 18 Jun. 2015, are presented by our project.

[23] See e.g. the joint research project of KU, VDW, IASS and PIK on "Religion and Climate/ Environmental Development" or the IASS/PIK/KU/VDW workshop on the encyclical and its transformative potential, 12–14 Sept. 2016 in Potsdam.

The main *formats* of the project activities are: (1) academic lectures and panel discussions (2) research and publications (3) documentation (and multiplication) of the findings via different media.

1. From May 2016 to Jan. 2017 five academic events have been hosted in Eichstätt, Berlin and Potsdam,[24] and more are planned for the year 2017.
2. KU, VDW, PIK and IASS have started a joint research project on "Religion and Environmental/Climate Change" with a first research workshop in May 2016. Two PhD projects are being pursued by the two project managers[25] under the auspices of the KU/VDW Project, with topics related to the project. Furthermore, about 15 interviews with people from different areas of expertise have been carried out, focusing on the encyclical, its perception, its role in transformation processes and its potential impacts.[26]
3. For the aim of documentation, all lectures, conferences and interviews are filmed and transcribed in written form. All results and findings are made accessible on the project's website, www.laudato-si-transformation.de (KU/VDW Project 2016a). In this way, the project team tries to spread and multiply transformative knowledge and impulses.

These structural elements of the KU/VDW Project reveal its potential as a catalyst and sustainability driver. In the following we will use the WBGU's characteristics of transformative drivers (cf. Sect. 31.1) and assess whether they apply to the KU/VDW Project. The change agent's characteristic *Having a convincing idea and vision for change* is apparent in the KU/VDW Project without any doubt. The common vision of KU, VDW, PIK and IASS is the transformation of the economy and society towards sustainability.[27] This is also expressed in the project's title and its guiding principle: "The aim is to face the global challenges of our time and develop solutions for a Great Transformation towards (more) sustainability" (KU/VDW Project 2016b).

For the realisation of this vision, the KU/VDW Project team joins forces with other sustainability-minded actors and institutions, first and foremost with PIK and IASS. Moreover, the project team networks on the regional, national and international level.[28] Within the Catholic University of Eichstätt-Ingolstadt and its faculties, the project tries to cooperate with actors already working for sustainability (cf. KU "Nachhaltigkeitsbericht" 2016). Through this, and via the project website and additional work on media relations and other communication measures, the project team

---

[24] For details see http://www.laudato-si-transformation.de/veranstaltungen/

[25] Namely Christian Meier and Till Weyers.

[26] For example Prof. Dr. Klaus Töpfer, Prof. Dr. Hartmut Graßl, Christoph Bals, Rev. John Patrick Ngoyi, Prof. Dr. Mary Evelyn Tucker, Card. Jaime Lucas Ortega, Prof. Dr. Hubert Weiger have been interviewed. Almost all interview partners were participants of the academic events.

[27] This is also the vision of the WBGU and is inherent in Laudato Si'.

[28] Partners on regional level include the "Department for the Creation and Climate Protection" and the "Department for the World-Wide Church ('Referat Weltkirche')" of the Diocese of Eichstätt, or also single actors such as the KU-Prof. Dr. Ingrid Hemmer and the KU-Prof. emer. Dr. Engelbert Groß. On the national level the project engages with *Wuppertal Institute for Climate, Environment and Energy*, *Germanwatch*, *Pax Christi* and the *BUND – Friends of the Earth Germany*.

tries to activate and gain fellows. These *network*-building activities are further characteristic of change agents.[29]

The next change agent's characteristic is closely linked: *Dialogue, discourse, cooperation with politics, civil society and economy for exploring and developing visions and solutions.* The above-mentioned activities of networking include the mutual support of the partners and the common goals and cooperation. Generating and conducting formats, platforms and events for dialogue comprises the core activity and communicative structure of the project, as is already expressed in the project title. Following the project's guiding principle "Key messages of the encyclical [...] are analysed and critically discussed with experts from the areas of science, economics, politics, churches and civil society organisations" (KU/VDW Project 2016b). Indeed, all the above-mentioned actors participate as speakers at the KU/VDW events.[30]

With this the KU/VDW Project team's activities also meet the characteristic, *Use of inter- as well as transdisciplinary procedure methods and approaches, including stakeholder participation.*[31] All the project activities mentioned have, in addition to other aims, the goal of *developing* (theoretical und practical) *solutions*, cf. the project's guiding principle: "The aim is to face the global challenges of our time and develop solutions for a Great Transformation towards more sustainability" (KU/VDW Project 2016b).

The *involvement of the civil society in the discourse, share the current status quo of scientific knowledge and make it participate in the education and transformation process* is also characteristic for the KU/VDW Project activities. There are almost always actors from civil society involved in the academic events organised. Furthermore, all the KU/VDW events are open to the public; the participation of the civil society is explicitly sought after.[32] All participants, among them many from the regional civil society, are invited to discuss and contribute at the end of every event.[33]

With all its activities, the KU/VDW Project team aims at the dissemination and realisation of eco-social sustainable thinking and acting. The project involves activities in the four fields or steps that bring people from "knowing to acting" (cf. also Sect. 31.1). First, in terms of (a) *knowledge communication*, in the project's discourse process and its events, main statements of the encyclical initially are "explained" and

[29] *They network and gain important fellow campaigners, they gradually develop the idea further together, and join forces with other sustainability minded actors.*

[30] From the areas of politics e.g. the Hessian undersecretary of state Dr. Beatrix Tappeser took part in a KU/VDW event, from the area of civil society e. g. Wiltrud Rösch-Metzler (federal chairwoman of *Pax Christi*). Prof. Dr. Ernst Ulrich von Weizsäcker will be key note speaker at a KU/VDW event on eco-social sustainable economy on 12 May 2017 as well as Prof. Dr. Hubert Weiger (federal chairman of the *BUND*) on 16 and 17 May 2017.

[31] This is also readily apparent in the project team setup, which includes organisations from various different backgrounds, and their joint research projects. It is also illustrated by the networking activities within the KU faculties.

[32] These results have been confirmed by a questionnaire with socio-geographic data filled out by the participants of the events.

[33] The involvement of civil society is also a key component of the networking activities and the information measures with and by the project partners.

contextualized by experts using the current state of knowledge.[34] All the presentations are accessible to the public via the project's homepage. Secondly, the project-team tries to (b) *create problem awareness and transformation-relevant awareness* with all its events.[35] Besides the events, the project team also gives weekly lectures at the KU on the encyclical and the Great Transformation in every summer semester as part of "education for the transformation".[36] This is intensified by excursions and background talks that help add depth to the topics.[37] With all this the KU/VDW team aims at (c) *shaping and changing attitudes and values towards more sustainability* and (d) *creating* a corresponding *collective and cross generational responsibility for our 'common home'*. In the best case goals, values and visions are generated to help guide the actions of individuals in a "necessary" sustainable direction. The project also meets the related characteristic, *self-perception as a catalyst for societal change and the aim to activate, convince and carry people along*. Some indicators for the impact of the project are: high attendance, at every event at least 50 people took part; and extensive media coverage in the form of 23 reports about the events of the KU/VDW Project as of December 2016.[38]

All the activities mentioned above show that the KU/VDW team can give impulses, create dynamics and support changes towards more sustainability and, in other words, *actively advance the Great Transformation.*

31.4 Final Remarks and Critical Reflection

In this chapter, the question, "Can Laudato Si' act as a catalyst for transformation?", has been examined in different ways. Without any doubt the attention of the media, believers, non-believers, science and great parts of society has been enormous.

---

[34] For example, Dr. Angelika Hilbeck (researcher at the ETH Zürich) started the KU/VDW event on "The Encyclical Laudato Si' and World Nutrition, Peace and Social Justice" with a presentation of the current state of knowledge on world nutrition and its unsustainable practices, cycles and relations.

[35] In particular the event „The Encyclical Laudato Si' and concrete transformation by Bamboo Reforestation on the Philippines" on 17 Oct. 2016 could contribute to this; at the beginning the concept of the "Great Transformation" had been explained by an expert of the IASS; all participants and especially civil society had actively been involved by fundraising possibilities; an awareness for the importance and power of seemingly insignificant actions of single individuals for transformation was created. For more details see http://www.laudato-si-transformation.de/veranstaltungen/die-welt-retten-transformation-konkret/

[36] See for example the lectures for Social Works Master Students on „Social Works and transformative science in the horizon of the Encyclical Laudato Si'" at the KU in the summer semester 2016.

[37] Background dialogues between the keynote speakers of the KU/VDW events and students, offered by the project team, can contribute to awareness building. To the same end, two student groups were taken on excursion to the "DBU Week of the Environment" at Berlin and took part in the KU/VDW panel "The Encyclical Laudato Si' and Science" at 8 Jun. 2016, see http://www.laudato-si-transformation.de/veranstaltungen/woche-der-umwelt/

[38] Some of the press reports can be reviewed at http://www.laudato-si-transformation.de/presse/

It has been illustrated that the statements of the Pope have some revolutionary character within the Catholic Church, as their highest leader puts the issue of the environment into the centre of attention for the first time and treats it in the form of an encyclical. At its core is the concept of "integral ecology" which underlines the connection of environmental and social issues. This along with the Pope's moral authority invites all people, regardless of their background, to dialogue and cooperation in mastering the environmental and social challenges of our time. For this reason, the encyclical is of great relevance—within the church and beyond.

Furthermore, the encyclical shows considerable potential to support synergetic socio-ecological transformation processes through bridge-building. In addition to the individual level, Laudato Si' has potential to impact transformations at the global political and economic level. However, these potential impacts have long-term character and can only be revealed in the future. Nevertheless, the Pope gave a very strong impulse with this document. It is up to us to realise a transformation towards more sustainability. Thus, the central issue is *how* the messages of Laudato Si' and scientific recommendations can be implemented. There are many scientific and practical examples, which offer answers to this question.

One example has been presented in this chapter. It has been shown that, in its actions and goals, the KU/VDW Project meets all the characteristics of a driver for transformation towards more sustainability in the field of transformative education, knowledge communication and values change. Active as a change agent in these fields, the KU/VDW Project is suited and able to support and drive the Great Transformation. Indeed, projects like the one presented here can contribute to awareness building and a global transformation towards more sustainability.

The second main finding involves the specific scientific method that the project uses to further discourse, namely contextualisation and re-contextualisation of the papal encyclical within the scientific discourse. Using these methods, the KU/VDW Project drives the dialogue and constructive cooperation between the religious discourse and the scientific discourse on sustainability. This supports the Great Transformation towards more sustainability in an innovative manner by opening, creating and driving new procedures of dialogue and cooperation.

References

Arbeitsgemeinschaft der Umweltbeauftragten der deutschen (Erz-)Diözesen (2017). http://www.kath-umweltbeauftragte.de. Accessed 16 Jan 2017

Bals C (2016) A successful provocation for a pluralistic global society. The encyclical Laudato Si' – A Magna Carta of integral ecology as a reaction to humanity's self-destructive course. Germanwatch Briefing Paper, Bonn

Bergmann S, Gerten D (eds) (2010) Religion and dangerous environmental change. Transdisciplinary perspectives on the ethics of climate and sustainability. LIT Verlag, Münster

Campwell P (2015) Hope from the Pope. Nature 522(391)

Carson R (1962) Silent Spring. Houghton Mifflin, Boston/New York

Gerten D, Bergmann S (eds) (2013) Religion in environmental and climate change. Suffering, values, lifestyles. Bloomsbury Academic, London

Global Catholic Climate Movement (2016) http://catholicclimatemovement.global/. Accessed 16 Jan 2017

Global Catholic Climate Movement (2017) Conference on Laudato Si' & Catholic Investing. Clean energy for our common home. http://www.laudato-si-investing.com/. Accessed 16 Jan 2017

Göring-Eckardt K (2015) Mehr als ein Weckruf, sondern Kairos. In: Laudato Si'. Die Umweltenzyklika des Papstes. Verlag Herder, Freiburg im Breisgau, p 19–34

Grin J, Rotmans J, Schot, J (2010) Transitions to Sustainable Development. New Directions in the Study of Long term transformative change. Routledge, New York/London

Heimbach-Steins M, Stockmann N (2015) Pope for Planet? Laudato Si' als „dringliche Einladung zum Dialog" (LS 14) und das weltweite Echo auf die Enzyklika. Sozialethische Arbeitspapiere des Instituts für Christliche Sozialwissenschaften (ICS AP Nr. 3). https://www.uni-muenster.de/imperia/md/content/fb2/c-systematischetheologie/christlichesozialwissenschaften/heimbach-steins/ics-arbeitspapiere/ics-ap_3_laudato_si_.pdf. Accessed 16 Jan 2017

Hengsbach F (2016) Laudato Si'. Ein ökosoziales Rundschreiben in euphorischem Überschwang? Frankfurter Arbeitspapiere zur gesellschaftsethischen und sozialwissenschaftlichen Forschung (FAgsF) Nr. 65. http://www.sanktgeorgen.de/nbi/fileadmin/redakteure/Dokumente/FAgsFs/FAgsF65_Internetversion.pdf. Accessed 16 Jan 2017

Jamieson D (2015) Why laudato si' matters, environment. Sci Policy Sustain Dev 57(6):19–20. doi:10.1080/00139157.2015.1089140

Katholische Universität Eichstätt-Ingolstadt (2016) Nachhaltigkeitsbericht. http://www.ku.de/unsere-ku/nachhaltigehochschule/nachhaltigkeitsberichte/. Accessed 16 Jan 2017

Klein N (2015) A Radical Vatican? New Yorker, 10 Jul. 2015

Kristof K (2010) Wege zum Wandel. Wie wir gesellschaftliche Veränderungen erfolgreich gestalten können. oekom, Munich

KU/VDW Project (2016a) http://www.laudato-si-transformation.de. Accessed 16 Jan 2017

KU/VDW Project (2016b) Leitbild. http://www.laudato-si-transformation.de/Leitbild. Accessed 16 Jan 2017

KU/VDW Project (2016c) Presse. http://www.laudato-si-transformation.de/presse. Accessed 16 Jan 2017

KU/VDW Project (2016d) Veranstaltungen. http://www.laudato-si-transformation.de/veranstaltungen. Accessed 16 Jan 2017

KU/VDW Project (2016e) Veranstaltungen. Die Welt retten? Transformation konkret! http://www.laudato-si-transformation.de/veranstaltungen/die-welt-retten-transformation-konkret. Accessed 16 Jan 2017

KU/VDW Project (2016f) Veranstaltungen. Internationaler Fachworkshop am IASS. http://www.laudato-si-transformation.de/veranstaltungen/internationaler-fachworkshop-am-iass. Accessed 16 Jan 2017

KU/VDW Project (2016g) Veranstaltungen. Woche der Umwelt. http://www.laudato-si-transformation.de/veranstaltungen/woche-der-umwelt. Accessed 16 Jan 2017

Laudato Si Week (2016). http://laudatosiweek.org. Accessed 16 Jan 2017

Marx Cardinal R (2015) Den Fortschritt neu denken. In: Laudato Si'. Die Umweltenzyklika des Papstes. Verlag Herder, Freiburg im Breisgau, p 9–18

McKibben B (2015) The Pope and the Planet. http://www.nybooks.com/articles/archives/2015/aug/13/pope-and-planet. Accessed 16 Jan 2017

Meadows DH, Meadows DL, Randers J, Behrens WW III (1972) The Limits to Growth. Universe Books, New York

O'Riordan T, McGowan A, Hamann R, Myanna L, Zhang L (2015) The legacy of the papal encyclical, environment. Sci Policy Sustain Dev 7(76):2–5. doi:10.1080/00139157.2015.1089135

Pontifical Catholic University of Chile (2016) Laudato Si'. http://www.uc.cl/es/enciclica-laudato-si. Accessed 16 Jan 2017

Pope Francis (2015a) Encyclical Letter Laudato Si' of the holy father Francis on care for our common home. Libreria Editrice Vaticana, Vatican City

Pope Francis (2015b) Ansprache bei den Büros der Vereinten Nationen in Nairobi am 26. November 2015. http://de.radiovaticana.va/news/2015/11/26/papst_franziskus_bei_der_uno/1189744. Accessed 16 Jan 2017

Pope Leo XIII (1891) Rerum Novarum. Encyclical of Pope Leo XIII on capital and labor. Libreria Editrice Vaticana, Vatican City

Santa Clara University (2016a) https://www.scu.edu Accessed 16 Jan 2017

Santa Clara University (2016b) Our common home. SCU responds to Laudato Si'. https://www.scu.edu/ourcommonhome. Accessed 16 Jan 2017

Schorlemmer F (2016) Unsere Erde ist zu retten. Haltungen, die wir jetzt brauchen. Verlag Herder, Freiburg im Breisgau

The Guardian (2015) The Guardian view on Laudato Si': Pope Francis calls for a cultural revolution. The pope links the destruction of the environment with the exploitation of the poor. The world should pay attention. https://www.theguardian.com/commentisfree/2015/jun/18/guardian-view-on-laudato-si-pope-francis-cultural-revolution. Accessed 16 Jan 2017

The St. James's Palace Nobel Laureate Symposium (2009) The St James's Palace Memorandum. "Action for a Low Carbon and Equitable Future" London, UK, 26–28 May 2009. University of Cambridge, Cambridge, UK/Potsdam Institute for Climate Impact Research

The United Nations (1992) The Rio Declaration on Environment and Development. The United Nations, New York

The United Nations (2015) Sustainable Development Goals. The United Nations, New York

The United Nations/World Commission on Environment and Development (1987) Our Common Future (also known as "Brundtland Report"). The United Nations, New York

Tucker ME (2015) Ecological challenges evoke ethical response. Environment. Sci Policy Sustain Dev 57(6):25–27. doi:10.1080/00139157.2015.1089144

WBGU – German Advisory Council on Global Change (2004) Towards Sustainable Energy Systems. Flagship Report 2003. Springer, Berlin/Heidelberg/New York

WBGU – German Advisory Council on Global Change (2011a) Factsheet No. 1/2011. A Social Contract for Sustainability. WBGU, Berlin

WBGU – German Advisory Council on Global Change (2011b) World in Transition. A Social Contract for Sustainability. Flagship Report. WBGU, Berlin

WBGU – German Advisory Council on Global Change (2011c) Factsheet No. 3/2011. Global Megatrends. WBGU, Berlin

WBGU – German Advisory Council on Global Change (2012) Factsheet 5. Research and Education. Drivers of Transformation. WBGU, Berlin

Index

H. Lehmann (ed.), *Factor X*, Eco-Efficiency in Industry and Science 32,
https://doi.org/10.1007/978-3-319-50079-9

MIX
Papier aus verantwortungsvollen Quellen
Paper from responsible sources
FSC® C105338

If you have any concerns about our products,
you can contact us on
ProductSafety@springernature.com

In case Publisher is established outside the EU,
the EU authorized representative is:
Springer Nature Customer Service Center GmbH
Europaplatz 3, 69115 Heidelberg, Germany

Printed by Libri Plureos GmbH
in Hamburg, Germany